Assessing Revolutionary and Insurgent Strategies

RESISTANCE AND THE CYBER DOMAIN

Kristen Ryan, Study Lead

Johns Hopkins University Applied Physics Laboratory (JHU/APL)

JHU/APL Contributing Authors

Bruce Milligan	Chapter 1
Summer Agan and Bruce Milligan	Chapter 2
Matthew Dinmore and Chris Scheidler	Chapter 3
Patrick Allen and Anthony Guess-Johnson	Chapter 4
Kimberly Glasgow, Ian McCulloh, and Patrick Allen	Chapter 5
Christopher Scheidler	Chapter 6
Jason Spitaletta	Chapter 7
Patrick Allen and Anthony Guess-Johnson	Chapter 8
Erin Hahn and W. Sam Lauber	Chapter 9
Anthony Guess-Johnson	Chapter 10

United States Army Special Operations Command

Resistance and the Cyber Domain is a work of the United States Government in accordance with Title 17, United States Code, sections 101 and 105.

Cite me as:

Ryan, Kristen, et al. *Resistance and the Cyber Domain*. Fort Bragg: US Army Special Operations Command, 2019.

Published by Conflict Research Group.

First published by USASOC in 2019

CONFLICT
RESEARCH
GROUP

ASSESSING REVOLUTIONARY AND INSURGENT STRATEGIES

The Assessing Revolutionary and Insurgent Strategies (ARIS) series consists of a set of case studies and research conducted for the US Army Special Operations Command by the National Security Analysis Department of the Johns Hopkins University Applied Physics Laboratory.

The purpose of the ARIS series is to produce a collection of academically rigorous yet operationally relevant research materials to develop and illustrate a common understanding of insurgency and revolution. This research, intended to form a bedrock body of knowledge for members of the Special Forces, will allow users to distill vast amounts of material from a wide array of campaigns and extract relevant lessons, thereby enabling the development of future doctrine, professional education, and training.

From its inception, ARIS has been focused on exploring historical and current revolutions and insurgencies for the purpose of identifying emerging trends in operational designs and patterns. ARIS encompasses research and studies on the general characteristics of revolutionary movements and insurgencies and examines unique adaptations by specific organizations or groups to overcome various environmental and contextual challenges.

The ARIS series follows in the tradition of research conducted by the Special Operations Research Office (SORO) of American University in the 1950s and 1960s, by adding new research to that body of work and in several instances releasing updated editions of original SORO studies.

RECENT VOLUMES IN THE ARIS SERIES

Casebook on Insurgency and Revolutionary Warfare: Volume I: 1927–1962 (2013)
Casebook on Insurgency and Revolutionary Warfare: Volume II: 1962–2009 (2012)
Human Factors Considerations of Undergrounds in Insurgencies (2013)
Undergrounds in Insurgent, Revolutionary, and Resistance Warfare (2013)
Understanding States of Resistance (2019)
Legal Implications of the Status of Persons in Resistance (2015)
Threshold of Violence (2019)
"Little Green Men": A Primer on Modern Russian Unconventional Warfare, Ukraine 2013–2014 (2015)
Science of Resistance (forthcoming)

TABLE OF CONTENTS

CHAPTER 7. HUMAN FACTORS CONSIDERATIONS IN UNDERGROUNDS IN CYBER RESISTANCE 201

LIST OF ILLUSTRATIONS

EXECUTIVE SUMMARY

Important weapons of the next major war will be the acquisition, denial, and employment of information. The explosive growth of the cyber domain, with its abilities to vector large quantities of information to billions of Internet users worldwide at trivial cost, exacerbated the importance of information operations across the spectrum of communications and conflict. In a world that witnesses the rise of insurgencies across the globe, the Internet is a weapon used not only by hackers seeking to empty the bank accounts of unwitting victims but also by governments dedicated to the defeat, or overthrow, of their enemies. This work explores the methods, successes, and failures of some recent resistance movements, as well as the efforts of their adversaries.

The authors identified the pertinent research in the cyber domain with subject matter experts to determine the areas of focus for this book. Within these focus areas, we considered the mission sets relevant to the US Army Special Operations Forces (ARSOF) soldier and assessed the operational relevance of resistance in the cyber domain. Upon analysis of the required knowledge for the ARSOF soldier, the chapters develop key takeaways. These key takeaways are for the ARSOF soldier to digest and integrate into their planning and operational actions in the field. A more applied perspective of the key takeaways can be found in the concluding chapter in case studies and fictional scenarios.

Resistance movements leverage the cyber domain since information technology began to intertwine the globe with unprecedented scope, speed, and accessibility. Several important characteristics distinguish today's resistance movement. One of these characteristics is the use of online social media platforms to frame the messaging of the resistance. Additionally, the new technology diminishes the role of formal organizations. While some argue that new media is simply a faster and more resource efficient means of communicating and organizing, others maintain that new media changes how resistance movements mobilize participants, as well as the role of formal organizations, and the resulting political outcomes. This new form of resistance relies heavily upon new media to mobilize and organize, while diminishing the role of formal organizations in favor of leaderless, networked structures, as discussed in the social network and human factors chapters. Ultimately, the cyber domain forever changed the resistance landscape in how they emerge, diffuse, and operate, yielding new advantages and creating new vulnerabilities.

In the development of this book, we articulated the role of the cyber domain in today's resistance movements while focusing on those lessons that influence a soldier's decisions or actions in this domain. We started by historically placing cyber operations within ARSOF information operations. Chapter 2 analyzes the science of resistance in two broad categories: organizations or individual resistances that use new media and those that leverage the domain for offensive and defensive operations in pursuit of political or security objectives. After laying the historical and scientific groundwork in the first two chapters, chapter 3 provides a concise overview of the key terms and concepts in cyber operations drawn from hacker culture and emerging military doctrine.

Chapters 4 through 6 focus on specific areas within the three dimensions of the information environment: cognitive, informational, and technical. Chapter 4 considers how cyberspace provides many new communication and information avenues for resistance movements. However, these new cyber mechanisms involve inherent security risks. Chapter 5 depicts various aspects of narratives, social media, and social networks within the cognitive dimension. This chapter introduces readers to concepts at the intersection of social media, social psychology, and network science, all of which contribute to understanding military operations in the online environment. Chapter 6 focuses on cyber systems within the physical dimension of cyberspace. The interconnected network of information technology infrastructures operating across cyberspace are a part of the US national critical infrastructure, and their security vulnerabilities are considered.

Chapters 7 through 9 focus on important considerations that shape the planning and execution of cyber operations: actors, attribution, and legal environments. Chapter 7 outlines how cyber affects the traditional underground functions of leadership and organization, recruiting, intelligence, financing, logistics, training, communications, security, subversion and sabotage, and psychological operations. Chapter 8 provides the reader with a foundational knowledge of the technologies and tactics involved in managing attribution, both from the standpoint of users seeking to hide their identities and others seeking to uncover hidden activities. In addition, chapter 9 considers the thresholds for the application of certain bodies of law, which can often be particularly difficult to determine in cyber activities. The chapters develop key takeaways throughout to provide the reader with guiding principles to inform their actions in the cyber realm. Case studies and fictional scenarios demonstrate the employment of these key takeaways, which are crucial for the development of successful cyber operations and appropriate security protocols.

Chapter 10 is an applied lessons chapter with case studies and fictional scenarios. The scenarios provide the reader with some scene setting information. Follow-on questions then hint at pertinent key takeaways. In the some examples, the answers reveal historically accurate life events. While the fictional scenario has a hypothetical answer, this is by no means the only correct answer. The chapter provides an opportunity for the ARSOF solider to critically think through the takeaway messages from across the various chapters and think holistically about an operation.

This book offers readers an overview of a new realm of warfare, particularly as it applies to resistance movements and information operations. Few suspected that one day cyber would become a major weapon of war, with its own tactics and principles. Now we find ourselves with the need to study and understand these tactics to employ the full breadth of possible impacts. Information has always been powerful, but that power is amplified by cyberspace. We recognize that information is far more accessible now, and knowledge of how to employ cyber operations increases the potential to disrupt an adversary.

CHAPTER 1.
HISTORY OF US ARMY SPECIAL OPERATIONS FORCES (ARSOF) AND INFORMATION OPERATIONS

INTRODUCTION

"Airplanes are interesting toys, but have no military value." Ferdinand Foch, one of the most famous generals of World War One, uttered these seemingly insignificant words in 1911 when the nations had little understanding of the potential power of air supremacy in combat operations.[1] Less than seven years later, Foch was the supreme commander of the French, British, Belgian, and American armies on the Western Front, and the French alone produced almost 68,000 combat aircraft.[2] By 1918, few major military operations could be undertaken and expected to succeed without the attacker first securing control of the airspace over the battlefield.

When the Internet was first commercialized, little thought was given to the implications of a worldwide system that could connect to every nation, organization, and for that matter, individual computer, on the globe. As the Internet became a global phenomenon, in the late 1980s and early 1990s, although its commercial and social advantages were quickly recognized, few suspected that one day it would serve as a weapon of war, with its own tactics and basic principles. Much like the airplane in 1911, we have been slow to understand the significant power of the cyber domain in the future operating environment. This study seeks to acknowledge that power and harness fundamental knowledge of the cyber domain as it applies to resistance studies.

Important weapons of the next major war will be the acquisition, denial, and employment of information. The growth of the cyber domain, with its abilities to vector large quantities of information to billions of Internet users worldwide at trivial cost, exacerbated the importance of information operations across the spectrum of communications and conflict.

The 2018 National Defense Strategy calls for the dedication of funds toward "cyber defense, resilience, and continued integration of cyber capability into the full spectrum of military operations."[3] In a world that witnesses the rise of insurgencies across the globe, the Internet is a weapon used not only by hackers seeking to empty the bank accounts of unwitting victims but also by governments dedicated to the defeat, or overthrow, of their enemies. More recently, resistance movements, from the Arab Spring in the Middle East, to the Zapatistas in Mexico, to Falun Gong in China, to the Occupy Movement in the United States, leverage the many benefits offered to resistance groups by the cyber domain in their pursuit of political, religious, and military victory against governments they consider oppressive or evil. This work explores some of these movements and highlights their methods, successes, and failures, as well as the efforts of their adversaries.

This book offers readers an overview of cyber operations, a new realm of warfare, particularly as it applies to resistance movements and information operations. This first, introductory chapter offers a brief history of information operations, as well as an overview of the historical use of information operations—with a focus on the cyber domain—by US Army Special Operations Forces.

A SHORT HISTORY OF INFORMATION OPERATIONS

"IO is not everything, but everything we do has an IO effect."[4]

This lesson from the Tactical Commander's Handbook on "Information Operations: Operation Iraqi Freedom (OIF)" demonstrates that information operations should be a constant consideration in operations. However, this is not a new phenomenon. Information warfare is a concept as old as warfare itself. One needs only to recall the story of the Trojan Horse, from Homer's *Iliad*, or the words of the legendary Chinese military sage, Sun Tzu, who once wrote, "All warfare is based on deception."[5]

Dr. Tom Rona coined the term 'information warfare.' He first used the term in a report to the Boeing Company in 1976, entitled, "Weapon Systems and Information War." In this article, Dr. Rona noted that information itself, as it became a more critical component of the national economy and military infrastructure of the United States, was as a consequence an important, vulnerable target.[6]

According to the handbook[7] issued to US Army company commanders during Operation Iraqi Freedom, information operations:

- "Is a tool to influence the use of information to meet your intent in every operation
- Is a horizontal staff synchronization process through all elements of combat power and conducted within the construct of the military decision-making process (MDMP)
- Aligns the use of information by all the unit's existent

operations and focuses on the commander's intent

Figure 1-1. Chinese military philosopher, Sun Tzu.

- Is both lethal and non-lethal, directed under the S3/G3/C3, and vertically integrated with higher and lower plans and with coalition and host nation (HN) information operations (IO) efforts
- Synchronizes effects"

Note that despite these characteristics, the implementation of information operations, at least at the outset, clashes with traditional American values, notably as they relate to the means by which information is obtained (e.g., honesty or falsehood).

A successful use of information warfare can be traced back to World War One. In 1917, the Germans loaded Vladimir Lenin and his comrades on a train headed to the Russian city of Petrograd. One commentator called this an example of "human malware."[8] By facilitating the arrival of Lenin, the Germans brought about the destruction of the Russian monarchy. Lenin was the human malware, an agent carrying information intentionally designed to cause damage to a network. In this case, the network was the Russian monarchical government.

Thus, the arrival of Lenin and his associates in Russia brought about the eventual collapse of the Russian army, the fall of the monarch, and the Russian Revolution that same year.

The closing of the State Department's Cipher Bureau in 1929 by Secretary of State Henry Stimson, accompanied by his legendary statement, "Gentlemen do not read each other's mail."[9] This kind of separating American military operations from political matters resulted in retarding the widespread acceptance of information operations as a form of warfare.

Over time, however, information operations, and the employment of its active component, information warfare, became a vital part of modern war. During the Second World War, major combatants employed information operations. The British, in particular, excelled at its use. One example was Operation Mincemeat, prior to the allied invasion of Sicily in 1943. At the time, although the German and Italian high commands expected an allied invasion of continental Europe, after being driven out of North Africa, they were unable to decide where such an invasion might come. Possible targets ranged from the coast of southern France to the Balkans and Greece. Operation Mincemeat involved a plan to convince the Germans that an important courier with documents of great importance died when his plane crashed off the coast of Spain. Using the corpse of a dead vagrant, whose papers identified him as "Captain William Martin," the British planted the body, along with a briefcase packed with false planning documents, close to Spain, where it was retrieved by Spanish fishermen. As hoped, the body and its bogus materials eventually found its way into the hands of the Germans. According to the documents they read, the Germans were led to expect an invasion in either Greece, Sardinia, or both, which resulted in mostly undefended beaches of the actual target of the invasion, Sicily.[10]

Another example of the utility of information operations in World War Two was the creation of General George S. Patton's "ghost army" created before the D-Day invasion. Patton, whom the Germans considered one of the better American field generals, was appointed to "command" the fictional First US Army Group (FUSAG), which allegedly included over a million men and thousands of tanks and other vehicles (see Figure 1-2).[11]

Figure 1-2. Picture of inflatable tank during World War I

The FUSAG's deception operations were extremely effective. A critical analysis of the FUSAG conducted by Major Donald J. Bacon reveals operational success by implementing deception in the information operations that the FUSAG conducted. Brown's study further identifies that the allies were able to control the key channels of information and collect enemy intelligence on the deception operations. Brown concludes that by strategically aligning deception planning to strategic and operational objectives, the FUSAG centralized high-level planning, maintained the required secrecy, and executed sound techniques in the allotted timeframe."[12]

Numerous other examples of the effective employment of information operations occurred both during World War Two and in succeeding wars. The Cold War tactics included the use of radio, in the forms of such networks as Radio Free Europe, Radio Liberty, and the Voice of America, to convey information and political opinions from the US government to the peoples of regions such as Eastern Europe and the Soviet Union.

Estimates claim that Radio Free Europe reached over thirty-five million listeners[13] in the Soviet-occupied countries of Eastern Europe, and during the Cold War, many citizens considered it their primary source of accurate, objective news. The network was seen as a beacon by the citizens of their respective nations during the Hungarian uprising against the Soviets in 1956, as well as the "Prague Spring" in Czechoslovakia in 1968.[14]

The Morale Operations (MO) branch of the Office of Strategic Services (OSS) implemented information operations tactics. The MO Branch generated undercover and misinformation propaganda against the axis powers. One of its first significant operations was a rumor campaign, during which the group spread memorable rumors that appealed to the emotions with the intent to cause fear, confusion, and doubt.[15] For example, one rumor touted the capture of a high-level Nazi leader. During the war, the OSS and the Political Warfare Executive (British MO equivalent) churned out about twenty rumors per week and measured the effectiveness through publicity in the press.[16]

Today, information operations encompasses a range of communication media. According to Joint Publication 3-13, information operations incorporates the following: electronic warfare (EW), computer network operations (CNO), psychological operations (PSYOP), military deception (MILDEC), and operations security (OPSEC). The aforementioned IO activities are employed with the intent to influence, disrupt, corrupt, or interfere with adversaries' capabilities.[17]

During the first Gulf War, information operations impacted not only the planning of US and coalition forces but also the results of the campaign itself.[18] One of the most memorable deceptions of these endeavors was the public embarkation of US Marine forces off the Kuwaiti coast, reinforced by frequent news reports of Marines practicing for an amphibious invasion. This publicity resulted in the positioning of many Iraqi units along the coast in anticipation of such an attack.[19]

Information operations employment in a military context can control the flow of information to and from an adversary. This flow can deceive enemy commanders about the intentions of one's own forces or inspire friendly forces, neutral nations, civilians, or those yet unengaged in a current conflict. This book focuses on resistance in the cyber domain, but cyberspace itself is a new, albeit far more impactful, tool among many to be used in the information operations arena.

THE US MILITARY AND THE CYBER DOMAIN

The cyber domain's strengths and weaknesses increasingly hold relevance to commerce, national defense, and warfare.[20] Therefore, the US Department of Defense established a command to devote its resources solely to the military aspects of cyberspace. In May 2018, US Cyber Command (USCYBERCOM) became the nation's 10th Unified Combatant Command. Its mission is to direct, synchronize, and coordinate cyberspace planning dedicated to the defense of the United States and to the furtherance of US policies and goals worldwide.

Figure 1-3. Seal of USCYBERCOM.

Accessing and controlling information has always been a valuable tool on the battlefield. However, the Information Age resulted in a requirement that such information be available, controllable (if need be), and immediate. Near-peer countries, such as Russia and China, stood up similar organizations, recognizing the threats and inherent value in cyberspace.

Underscoring the importance of the cyber domain, the 2017 National Security Strategy (NSS) emphasizes the fundamental responsibility of the federal government to the American people is to defend the critical infrastructure from "malicious cyber actors." The NSS continues:

> A democracy is only as resilient as its people. An informed and engaged citizenry is the fundamental requirement for a free and resilient nation. For generations, our society has protected free press, free speech, and free thought. Today, actors such as Russia are using information tools in an attempt to undermine the legitimacy of democracies. Adversaries target media,

political processes, financial networks, and personal data. The American public and private sectors must recognize this and work together to defend our way of life. No external threat can be allowed to shake our shared commitment to our values, undermine our system of government, or divide our Nation.[21]

Cyberspace and its vulnerabilities are referenced a number of times in the NSS. Referencing cyber attacks, the NSS demonstrates the challenges to information, technology, military dominance, and economic prosperity, concluding that that, "America's response to the challenges and opportunities of the cyber ear will determine our future prosperity and security.[22]

A modern combatant can access an adversary's computer networks to gather information. However, having secure access to reliable information—without manipulated data, cyber attacks, or even disinformation campaigns—is increasingly important. In short, one's enemies can now be anywhere within one's own network, and therefore one's own defenses.

Because information is so much more accessible today, information operations, and thus information warfare, offers an enterprising combatant far more potential to disrupt enemy operations, or enhance one's own, than ever before.

ARSOF AND INFORMATION OPERATIONS

While the concept of information operations itself is longstanding, the US Army, and the American military in general, first used the term "information superiority" in 2010. The Army's specific definition of information superiority, as defined in ATP 3-13.1, "The Conduct of Information Operations," states:

Information operations is the integrated employment, during military operations, of information-related capabilities in concert with other lines of operation to influence, disrupt, corrupt, or usurp the decision-making of adversaries and potential adversaries while protecting our own.[23]

The Special Operations Forces Reference Manual[24] lists military information support operations (MISO) as one of the core activities of special operations. The manual further describes MISO in this manner:

> MISO convey selected information and indicators to foreign audiences to influence their emotions, motives, objective reasoning, and ultimately the behavior of foreign governments, organizations, groups, and individuals. The purpose of MISO is to induce or reinforce foreign attitudes and behaviors favorable to the joint force commander's objectives. Dramatic changes in information technology and social networking have added a new, rapidly evolving dimension to operations, and the ability to influence relevant audiences is integral to how SOF address local, regional, and transnational challenges.

In practical terms, US Special Forces should begin perceiving information operations—particularly as they involve cyberspace and social media—as a new, albeit non-kinetic, weapon, one that offers enormous potential to those who can exploit it properly. This study highlights some historic uses, and successes, of cyber warfare, and new examples of its use as a tool in modern diplomacy and conflict appear frequently.

Social media can be employed inexpensively and directed at a far-flung audience to promote or belittle a cause. People can use social media to spoof or dilute similar attempts at messaging from one's foes or organize demonstrations for or against a standing government. Cyber attacks can be used to paralyze or disrupt an enemy's communications network or to access databases or records previously considered secure.

In the modern social and political environments, a unit commander—even at the lowest level—is concerned with the perceptions, and therefore the perceived effects, of the actions of his troops, and of his enemy, upon public opinion across the world. It is increasingly easier for members of the public to access information on any subject; therefore, it is far easier to influence that public and, consequently, its perceptions.

In the growing digital age, "information superiority," a concept that did not exist two decades ago, is now considered a vital aspect of modern warfare. An example would be the Russian war in Chechnya in the 1990s. This example exemplifies information operations and/

or superiority because the information campaign had the power to change the outcome of the conflict.

When the Russians first invaded Chechnya, they encountered strong armed resistance from the populace and suffered unexpectedly heavy casualties, despite targeting specific groups and individuals, not the population at large. When they returned to the area three years later, in a second invasion, their intrusion into Chechnyan territory was preceded by a massive propaganda campaign to make it clear that they were not targeting the population at large, but rather those Russians perceived as terrorists. As a result, they encountered far less resistance and secured the assistance of many members of the local population who previously might have been hostile to them.[25]

AN OVERVIEW OF THE CHAPTERS IN THIS BOOK

Each chapter of this book focuses on a specific area related to cyber operations, and the phenomenon of cyberspace itself, and their effect on resistance movements around the world. We begin (in Chapter 2) with a discussion of the science of resistance in the cyber domain.

Chapter 2 focuses on the use of cyber operations along the spectrum of resistance. First, there is an analysis of organizations or disaffected individuals that use cyberspace, particularly via media disseminated via the Internet, to further communication between their members and like-minded individuals, as well as with the rest of the world, regarding their political and religious objectives. The discussion proceeds to more dynamic uses of the Internet to leverage digital information tools in offensive or defensive operations in support of political, military, or religious objectives. This chapter also describes the features in the cyber domain that impact and motivate a successful resistance, including a sense of commitment to the cause, affiliation with those of similar sympathies, and membership in social networks.

Chapter 3 provides a foundation of the key terms and concepts of cyber operations. Although some of the information provided is necessarily technical, the bulk of this chapter focuses on new military doctrine, blended in some cases with computer hacker culture, and the creation of powerful weapons within the non-kinetic cyber domain.

Chapter 4 discusses the positive and negative implications of cyber operations for information security. This chapter compares and contrasts cybersecurity principles between the resistance movement and state security forces. The discussion addresses opportunities of rapid, inexpensive, and widespread information sharing and dissemination. In turn, the risks of using the cyber domain to transmit valuable information are also highlighted, as this new realm of technology can often dispatch its user as easily as an intended target.

Chapter 5 introduces readers to the cognitive dimension of cyberspace. "Cognition" is a scientific, or academic, word that implies the process of thought itself. When we think, we collect information, use our existing knowledge to analyze it, and then make decisions based upon the suggestions yielded to us from the combination of that experience and the new information. This chapter discusses the intersection of social media, social psychology, and network science. Without a clear understanding of this process, the success of developing the proper courses of action in the realm of military operations becomes unlikely. Social media, once a form of entertainment, now represents persuasive political tools, as well as a real weapon on the battlefield. This tool is used to keep the loyalty of one's own people and military, attract the loyalty and support of the undecided, and weaken the loyalty of one's foes. "Hearts and minds" are still the target that offers the greatest rewards to a successful strategy, and as such, a very strong understanding of its primary battlefield—cyberspace—is critical to victory.

Chapter 6 focuses on the physical, non-cyber, aspects of cyberspace. Supervisory control and data acquisition (SCADA) systems are used across the telecommunications and cyberspace industries to monitor computer systems by collecting data in real time (notably, telecommunications and cyberspace) from a group of dispersed assets. This chapter reviews past attacks against SCADA systems and discusses protective measures to reduce such attacks in the future.

Chapter 7 dissects the theme of human factors as they relate to cyber resistance movements. This chapter discusses the effects of cyber warfare on the conduct of irregular warfare, without necessarily revoking or changing its fundamental principles. Chapter 7 intends to review the fundamental building blocks of traditional resistance movements and then assess the effect of access to, and a strong understanding of, the cyber domain on key resistance tenets in the realms of organization, recruiting, intelligence, financing, training, and many others. Drawing upon previous ARIS works, this chapter provides a cyber "lens" to the basics of irregular warfare and underground resistance movements.

In Chapter 8, the authors discuss that, unlike in earlier wars, it is often very difficult, if not impossible, to discover the identity of one's enemies, at least when it comes to specific attacks. Nonattribution, misattribution, and tools offering anonymity make conducting "false flag" and similar attacks much easier than in the past. This chapter investigates attackers who might leverage various cyber tools to mask their attacks, but also the use of these same tools to defend against and unmask the attackers.

Chapter 9 explores the legal implications of offensive and defensive operations in cyberspace. It includes discussion of laws that might apply to a cyberattack and which laws (national, international, or a mix) lend themselves to a cyber defense. This chapter also describes different interpretations of certain laws, depending upon who is attacking whom and the level, or lack thereof, of general proclivity for violence in the resistance movements conducting cyberattacks.

Finally, Chapter 10 details historical and hypothetical scenarios, with suggestions as to which principles of the new realm of cyber war, drawn from the previous chapters, might apply. Readers are encouraged to consider their own solutions and are provided with the actual response to the historical event or a suggested response to those fictional scenarios.

ENDNOTES

1 Wikiquote.org, "Ferdinand Foch," https://en.wikiquote.org/wiki/Ferdinand_Foch.

2 The Aerodrome.com, "The Aircraft of World War 1," http://www.theaerodrome.com/aircraft/statistics.php.

3 Jim Mattis "Summary of the 2018 National Defense Strategy of the United States of America: Sharpening the American Military's Competitive Edge," US Department of Defense, 2018.

4 Lawrence H. Saul, "Tactical Commander's Handbook Information Operations: Operation Iraqi Freedom (OIF)," Center for Army Lessons Learned (CALL) Combined Arms Center, Fort Leavenworth, Kansas, May 2015, FOR OFFICIAL USE ONLY.

5 Sun Tzu, "The Art of War," translated by Lionel Giles, from http://classics.mit.edu/Tzu/artwar.html.

6 General Abe C. Lin, "Comparison of the Information Warfare Capabilities of the ROC and PRC," *Infowar* (2000), https://cryptome.org/cn2-infowar.htm.

7 Saul, "Tactical Commander's Handbook."

8 Nick Brunetti-Lihach, "Information Warfare Past, Present and Future," *RealClear Defense* (November 14, 2008), https://www.realcleardefense.com/articles/2018/11/14/information_warfare_past_present_and_future_113955.html.

9 Olga Khazan, "Gentlemen Reading Each Others' Mail: A Brief History of Diplomatic Spying," *Atlantic* (June 17, 2013).

10 Mark D. Vertuli and Bradley S. London, eds., *Perceptions are Reality: Historical Case Studies of Information Operations in Large-Scale Combat Operations*, (Fort Leavenworth, KS: Army University Press, 2018). An image of the ID card of this bogus identity can be found in Dwight Jon Zimmerman, "Operation Mincemeat: The Story Behind 'The Man Who Never Was' in Operation Husky," *DefenseMediaNetwork* (September 9, 2013), https://www.defensemedianetwork.com/stories/operation-mincemeat-the-story-behind-the-man-who-never-war-in-operation-husky/.

11 Elinor Florence, "D-Day Dummies and Decoys," *Blog* (May 28, 2014), http://elinorflorence.com/blog/d-day-decoys.

12 Donald J. Bacon, "Second World War Deception – Lessons Learned for Today's Joint Planners," *Air Command and Staff College* (December 1998), https://apps.dtic.mil/dtic/tr/fulltext/u2/a405884.pdf. For more information, consult Headquarters, Department of the Army, "The Conduct of Information Operations," ATP 3 13.1 (October 2018).

13 Encyclopedia Brittanica, eds., "Radio Free Europe: United States Radio Network," https://www.britannica.com/topic/Radio-Free-Europe.

14 Emily Thompson, "RFE During the Occupation of Czechoslovakia," RadioFreeEurope RadioLiberty Pressroom (June 28, 2018), https://pressroom.rferl.org/a/rfe-in-1968/29325749.html.

15 Central Intelligence Agency, "The Office of Strategic Services: Morale Operations Branch," (July 29, 2010), https://www.cia.gov/news-information/featured-story-archive/2010-featured-story-archive/oss-morale-operations.html.

16 Ibid.

17 US Joint Chiefs of Staff, "Chapter III: Authorities, Responsibilities, and Legal Considerations," III-2, "Chapter IV: Integrating Information-Related Capabilities into the Joint

Operations Planning Process," IV-1–IV-5, "Information Operations," Joint Publication 3-13 (November 20, 2014).

[18] Nick Brunetti-Lihach, "Information Warfare Past, Present and Future," *RealClear Defense* (November 14, 2008), https://www.realcleardefense.com/2018/11/14/ information_warfare_past_present_and_future_305202.html.

[19] Daniel L. Breitenbach, "Operation Desert Deception: Operational Deception in the Ground Campaign," Naval War College (June 19, 1991).

[20] Department of Defense, "Summary: Department of Defense Cyber Strategy 2018," (September 2018).

[21] Donald J. Trump, President of the United States, "The National Security Strategy of the United States of America," (December 2017): 14.

[22] Ibid., 12.

[23] Army, "The Conduct of Information Operations."

[24] John Alvarez, Robert Nalepa, Anna-Marie Wyant, and Fred Zimmerman, eds., *Special Operations Forces Reference Manual*, Fourth Edition (MacDill AFB, FL: The JSOU Press, June 2015).

[25] Frederick C. Gottschalk, "The Role of Special Forces in Information Operations," (Master of Military Art and Science thesis, U.S. Army Command and General Staff College, June 2000).

CHAPTER 2.
THE SCIENCE OF RESISTANCE IN THE CYBER DOMAIN

INTRODUCTION

"Resistance is defined in this work as a form of contention or asymmetric conflict involving participants' limited or collective mobilization of subversive and/or disruptive efforts against an authority or structure."

— Conceptual Typology of Resistance

This chapter presents an overview of the science of resistance and its relation to resistance in the cyber domain. The cyber domain is defined as "a global domain within the information environment consisting of the interdependent networks of information technology infrastructures and resident data, including the Internet, telecommunications networks, computer systems, and embedded processors and controllers."[1] Resistance in the cyber domain is analyzed in two broad categories. The first is organizational or individual resistance, particularly new media and social media, as a means of communication and coordination to further political objectives. The second category is devoted to organizations or individuals that leverage the cyber domain for offensive and defensive operations in pursuit of political or security objectives. Finally, this chapter explores particular features of the cyber domain that impact the push and/or pull factors motivating resistance, including commitment, affiliation, and social networks.

Resistance movements have leveraged the cyber domain since information technology began to intertwine the globe with unprecedented scope, speed, and accessibility. One of the first resistance movements to successfully pursue its political objectives with an information warfare strategy grounded in these new technologies emerged in an unlikely place in the early 1990s—the hills of the Chiapas state of Mexico.

> 2.1 *Resistance movements have leveraged the cyber domain since information technology began to intertwine the globe with unprecedented scope, speed, and accessibility.*

The Movimiento Civil Zapatista, known as the Zapatistas, formed in the Mexican state of Chiapas in 1994. The Zapatistas, represented by the iconic figure Subcomandante Marcos, adopted a repertoire of violent tactics. Primarily, this meant guerrilla warfare, but they also subordinated violence to their informational strategy. Led by Marcos, the Zapatistas combined an anemic strategy of armed resistance with a robust information strategy. This strategy helped them build coalitions with non-government and government organizations across the world, using nascent cyber technologies, and subsequently, engendered strong support.

It is unlikely that the Zapatistas would have been as successful as they were, as quickly as they were, without the aid of these new technologies. As of 2018, the Zapatistas still thrived and controlled much of the Mexican state of Chiapas with the tacit, if not enthusiastic, acceptance of the Mexican federal government.[2]

More recently, the widespread adoption of social media platforms by billions of users across the world ushered in a new era of resistance in the cyber domain. The first impact of these new technologies began after the 2008 global recession, when protesters in Iceland took to the streets after the country's economy collapsed. These protests resulted in a constitutional reform process that used Internet and social media technologies to crowdsource the reforms. Several years later in 2011, massive protests began in Tunisia, sparked by the self-immolation of a street vendor who was harassed by Tunisian security forces. The civil resistance that followed toppled the entrenched authoritarian regime in that nation.

Resistance movements in Iceland, Tunisia, Iran, Egypt, and even the United States all share several important characteristics that separate them from their predecessors. The most striking of these were those that employed previously unknown technologies, and the means which those technologies provided, to encourage mass mobilization. Corollary characteristics entailed the new technologies to frame these struggles and, at the same time, the diminished role of formal organizations.

There are three primary arenas of communication in the information environment pertinent to analyzing resistance in the cyber domain. These arenas include public communications, regime communications, and resistance communications.

> *2.2 The most striking characteristics of new forms of resistance is that they have employed previously unknown technologies, and the means which those technologies provided, to encourage mass mobilization.*

Public communications include those generated by non-state actors, particularly private corporations or media outlets. While most observers analyzed how non-violent resistance movements, such as the Arab Spring that began in Egypt and Tunisia in late 2010, help to shape resistance against the governments in target countries, violent resistance movements also supplement their information strategies with new media platforms. Figure 2-1 depicts the spread of the Arab Spring across the North Africa and the Middle East.

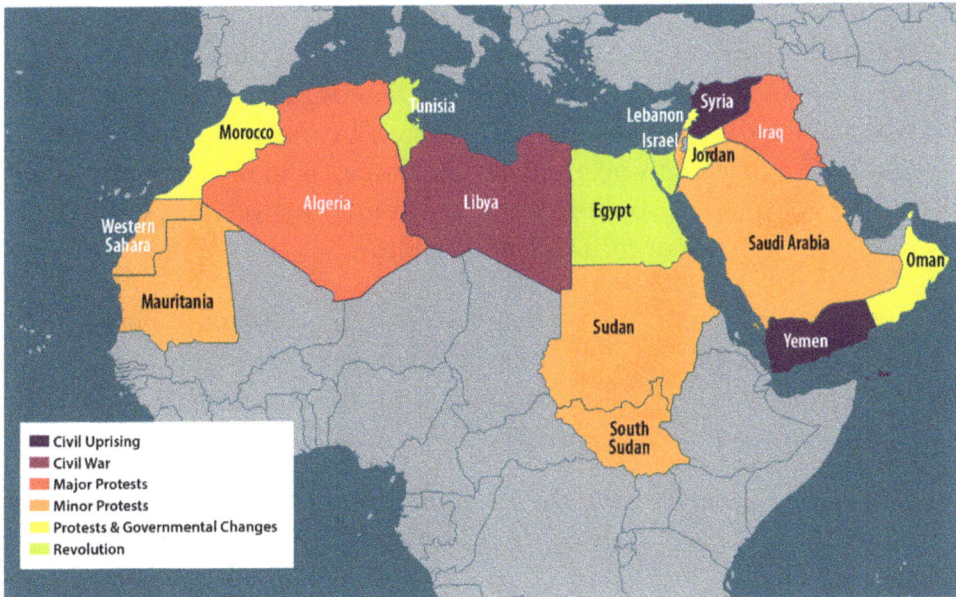

Figure 2-1. The Arab Spring began in Tunisia and Egypt in 2010, and quickly spread across North Africa and the Middle East.

Social scientists conducted initial research of new media shaping resistance activities. While some argue that new media is simply a faster and resource-efficient means of communicating and organizing, others maintain that new media changed the measures with which resistance movements mobilize participants, as well as the role of formal organizations, and the resulting political outcomes. This new form of resistance is called connective action, and it relies heavily upon new media to mobilize and organize, while diminishing the role of formal organizations in favor of leaderless, networked structures.

The three primary forms of resistance described in this chapter incorporate these features to varying extents. Organizationally brokered networks include strong, formal organizations that direct action, communication, and collective identities while using new media to manage participation and coordination. Organizationally enabled networks feature looser organizational structures with less central direction of action and communication. Their communication strategies rely on new media platforms that center on generating personalized identities to mobilize high numbers of participants.

The last of the three crowd-enabled networks exhibit little to no formal organizational structures; some describe these as "leaderless" networks, and they employ layers of new media technologies to support personal expression and identity associated with a wide variety of contentious issues. Adopting loose or leaderless organizational structures may be strategic decisions, but the decision by movements as to which to adopt also appears to be an ideological one, as many movements signaled their distrust of political and economic institutions, even those including any sort of organizational structure.

> *2.3 "Leaderless" networks exhibit little to no formal organizational structures but do employ layers of new media technologies to support personal expression and identity associated with a wide variety of contentious issues.*

Beginning in the late 2000s, resistance movements leveraging organizationally brokered networks, organizationally enabled networks, and crowd-enabled networks achieved various unconventional warfare (UW) objectives, whether it was to disrupt, coerce, or overthrow. Organizationally enabled action, by combining features of tighter coordination and personalized identity through new media technologies, was successful in mobilizing hundreds of thousands to participate in public protests during the Arab Spring protests, which helped to disrupt and overthrow several governments. The governments that were overthrown included those of Egypt, Libya, Tunisia, and Yemen.

The capacity for rapid mass mobilization, however, also underscores the fragility of this new form of resistance. The emphasis on loose or leaderless structures prevents many resistance movements from transitioning out of the revolutionary phase to roles in main-stream politics, as the movements lack the organizational resources to set agendas, develop common platforms, and act decisively.

While the loosely networked resistance movements formed during this period demonstrate the emergence of a new sort of resistance, it is important to emphasize that the older, more conventional forms of resistance continue to thrive all over the world. To perceive the contemporary science of resistance, it is vital to turn our gaze to this powerful tool. Resistance grounded in the cyber domain helped to topple several entrenched authoritarian regimes, sparked a brutal civil war, influenced policy-making decisions in powerful governments, and changed global conversations about issues surrounding environmental and economic injustice.

PUBLIC, REGIME, AND RESISTANCE COMMUNICATIONS IN THE INFORMATION ENVIRONMENT

Although this volume focuses upon resistance occurring in the cyber domain, it is important to note that such resistance activities occur in the larger domain of the information environment (IE). The cyber domain, a nebulous concept, is just one platform, or medium, through which information and action pass. This concept more closely maps to Russian military perspectives on the IE than those of the US military. While the United States relegates actions related to cyber warfare to a separate, discrete domain, the Russian military views confrontations within the IE as occurring at near-constant frequency. As a result, Russian cyber operations carefully align with traditional military operations at the strategic, not the operational or tactical level. Moreover, the targets of such actions include not only foreign militaries or adversaries, but also the societies in which those actors operate, to "prepare potential battlespace."[3]

While observers placed a great deal of attention on the role that social media played in resistance efforts, activity on social media rarely occurs in isolation from other media platforms. In the Egyptian Arab Spring, for instance, while protesters initially leveraged Facebook and Twitter, after the government restricted access to these platforms, the protesters transitioned to older forms of media, including leaflets and simple word of mouth through offline social networks.

Traditional media were also important prior to media restrictions because not all segments of Egyptian society had ready access to the Internet. One scholar observed that taxi drivers are a common mode of transportation in Egypt's crowded cities and held the same level of importance as Facebook as actors engaged in spreading the word about planned demonstrations.

Word about protests also spread in soccer fields, mosques, and coffee houses.[4] A survey of early Tahrir Square protesters found that about 25 percent of them first heard of the protests on Facebook.[5] This chapter later introduces a categorization of different forms of media—new and analog media—that should help to identify the characteristics of recent information technology developments that distinguish it from its predecessors. The novel characteristics of new media, in some cases, changed the emergence, diffusion, and operation of organized resistances.

Public Communications

Three primary arenas of communication are pertinent to understanding resistance in the cyber domain.[6] The first is public communication, which encompasses new and analog media generated by non-state actors, including private corporations or other non-state entities. Some examples of public communication include the common social media platforms YouTube and Twitter, as well as media outlets such as *The New York Times.*

Because the most prevalent of the platforms are based in the United States or other Western countries, access restrictions are rare unless supporting organizations or actions are illegal in the country. The algorithms used by social media platforms such as Twitter and Facebook present some challenges for resistance in the cyber domain. While billions of people use the Internet, only a few corporations such as Google and Facebook act as gatekeepers for Internet activities.[7]

For advertising purposes, Internet providers require access to immense amounts of user data. As such, they also have enormous power in deciding which content or products receive attention by manipulating their policies and algorithms. In this regard, the policies and algorithms act as a control mechanism that selects, ranks, and personalizes content according to the preferences of user accounts revealed by the user's prior behavior. Although corporations clearly state their policies, the algorithms are proprietary so resistance movements have little insight into the factors that can help push their content to wider audiences.[8]

During the Arab Spring, the algorithms and policies of Twitter and Facebook presented unique challenges for the protest organizers. While some social media platforms, such as Reddit or 4chan, allow their users to remain anonymous, Facebook follows a "real name" policy. This policy resulted in the deactivation of the "We are All Khaled Said" Facebook page that helped galvanize thousands of Egyptians to become involved in political action for the first time. Khaled Said was a young Egyptian man who was beaten to death during his arrest by Egyptian police outside a cyber café. Activists obtained pictures taken by family members of Said's corpse, which were posted on the Facebook page several days after Said's death. A pseudonym was used to set up the account to protect its creator from repercussions by the Egyptian government. Facebook deactivated the account after discovering that a pseudonym had been used to create it.

Although most people use their real names on Facebook, an estimated 20 percent do not. Facebook relies on community policing to identify breaches of its real name policies. For most users, this is not a significant issue. Adversaries of resistance movements, however, such as those that reported the false Facebook account, have incentive to report the behavior. The community policing model presents real challenges to activists trying to skirt unfavorable platform policies.

Algorithms are complex software programs that sift through massive troves of content to make decisions about which is prioritized. An example is Facebook's user newsfeed, populated according to the company's algorithms. The algorithm favors videos, mentions of people, and comments. Moreover, when newsfeeds or posts bury a story or post, a feedback loop occurs when the story is not shared or commented on by users that further pushes it to the margins.

This process occurred at the height of the protests against the shooting of Michael Brown by police in Ferguson, Missouri. While stories, videos, and comments on Brown's death trended heavily in Twitter, the more emotionally uplifting story of the ALS ice bucket challenge on Facebook overshadowed media coverage of the shooting. The lack of transparency and the high complexity of social media algorithms make it difficult to know which content will be widely distributed. When the policies and algorithms are unfavorable to a resistance movement's efforts, there are few mechanisms available to appeal decisions by private corporations.

The algorithmically driven flow of information can also drive to ideological isolation so that users are exposed only to information that conforms to their beliefs or preferences. The effect of the isolation contributes to the development of online echo chambers, which contributes to social or political extremism.[9] Chapter 5 includes an in-depth discussion on the impact of online platforms on social psychology, including the theories of majority illusion, pluralistic ignorance, social conformity, and network conformity.

Regime Communications

The second arena of communications within the IE, regime communications, encompasses those communications generated by the state, as well as the communication strategies that rely upon monitoring activity in resistance and public communications. State-generated media are more common in regimes relying upon authoritarian mechanisms of control, such as North Korea, which maintain their own intranet to avoid penetration of foreign or non-state-approved information. A 2016 glimpse at the North Korean intranet revealed it contains just twenty-eight websites.[10]

While relatively weaker non-state actors first leveraged social media platforms, many authoritarian regimes now recognize the utility of the platforms as another mechanism of societal control.[11]

China is among the most adept at blending technologies developed in the private sector, with state censorship that co-opts, disrupts, or subverts social media rather than relying upon strict suppression of online discourse. The popular Chinese app WeChat, for example, allows its 850 million users to seamlessly document details about their everyday activities, from making doctor's appointments, to receiving medical results, and lunch dates with friends.

Although convenient to its users, the Chinese government regularly monitors activity on the app, providing it with unprecedented amounts of data on its citizens. WeChat monitors private and group chats, banning sensitive topics such as Tiananmen Square or the Falun Gong, and meticulously logs suspicious conversations. Some users were incarcerated for conversations or pictures shared on the app.[12] A large-scale, multiple-source analysis of Chinese social media found that the Chinese government was more likely to censor material believed to spur collective mobilization against authorities.[13]

Social Media and Authoritarian Stability

Authoritarian regimes rely upon four mechanisms to use social media to further its political stability: counter-mobilization, discourse framing, preference divulgence, and elite coordination. The first, counter-mobilization, is a mechanism that halts any resistance against the regime by mobilizing the regime's own domestic support base. This base may include those groups that benefit from government patronage, such as the military or business elites, as well as citizens motivated by feelings of patriotism or trust in the government. During protests in Bahrain associated with the Arab Spring, Bahraini authorities mobilized regime supporters to help them identify and arrest protesters in pictures posted online in an effort dubbed "Together to Unmask the Shia Traitors." Social media are powerful organizing tools, but the advantages offered resistance movements also benefit the state in countering challenges to its authority.[14]

The mechanism of counter-mobilization closely links to discourse framing. This type of framing shapes public opinion by adapting online discourse to align with regime objectives. The Chinese government employs hundreds of thousands of online commentators to write posts that support the ruling party while discrediting its critics. Similarly, Russian authorities, operating through the youth group Nashi, reportedly established a facility for Internet trolls that produce around a hundred posts each day on various forums discrediting the West and Russian opposition leaders.

Social media also aids authoritarian regime stability through preference divulgence and elite coordination. Most authoritarian regimes operate under conditions of information scarcity. These regimes repress public dissent, punish free speech, and lack transparency in government institutions, hindering the regime from making effective decisions in an environment of information asymmetry. Therefore, authoritarian leaders lack information on the private preferences of its citizens and do not have effective ways of gauging when individual resentment against its policies likely transform to organized resistance. Moreover, under authoritarian conditions, political authorities have few mechanisms to provide feedback on the effectiveness and performance of local elites outside the central government.[15]

In authoritarian governments such as China's, social media helped the central ruling party overcome the problems associated with information scarcity. Rebecca MacKinnon calls the contemporary Chinese government "networked authoritarianism" because it allows a measure of citizen expression on blogs, websites, and other social media platforms to gain a clearer picture of public concern, sometimes using the information to address ineffective or unpopular policies.[16]

Resistance Communications

The third important arena of communication in the IE is resistance communications. The communication flows occur on platforms established in public communications but also in the regime communications arena. Some public communications are not monitored by the state but administered primarily through the private sector. Although less likely to occur in new media platforms, resistance movements may also generate their own analog media platforms.

This study defines "new media" as the various relatively modern means of mass communication that make use of the Internet, including emails, websites, and social media, such as Facebook, Twitter, and similar sites. "Analog" media are more traditional means of communication, notably including print publications. Resistance movements are more likely to produce analog platforms, such as, for example, al Qaeda's newsletter *Inspire* or Islamic State in Iraq and Syria's (ISIS's) similar newsletter *Dabiq*.

Some resistance movements also produce social media platforms. ISIS's Dawn of Glad Tidings, first launched in April 2014, was a Twitter-based app that distributed tweets with links, hashtags, and images.[17] In June 2014, when ISIS marched into Mosul, app users generated sufficient tweet volumes so that searches for Baghdad on Twitter first returned images of an ISIS fighter gazing at the Baghdad skyline, where an ISIS flag flew. The image read, "We are Coming," an information strategy intended to intimidate Baghdadis. The group also used this strategy successfully before its initial assault on Mosul in its #AllEyesOnISIS hashtag campaign.[18] The group's communication strategies allowed the group to overstate its online grassroots support through the impact of the majority illusion and pluralistic ignorance effect, as will be discussed in Chapter 5, allowing the group to "punch above their weight" or employ asymmetric threats to achieve maximum effects against a more powerful adversary.[19]

Some communications leverage existing new and analog platforms but without a direct threat of interdiction by the state or other adversaries. Moreover, this strategy is also highly resource efficient because existing platforms are leveraged. The resources and capabilities of resistance movements, in addition to the movement's preferred operational strategies and the legal environment in which it performs, shaped these ideal conditions.

Resistance movements adopting nonviolent operational strategies in states that protect political and civil rights are the most likely to manage communications from this ideal position. However, movements that adopt violent strategies and operate in restrictive environments are not well positioned to manage their organizational communications, particularly if the state possesses a robust intelligence or counterinsurgent capacity. More often, the state, or state apparatus, monitors or fully controls the communication of these resistances.

2.4 *Movements that adopt violent strategies and operate in restrictive environments are not well positioned to do so, particularly if the state possesses a robust intelligence or counterinsurgent capacity.*

RESISTANCE MOVEMENT EARLY INTERNET COMMUNICATION STRATEGIES

The first resistance movements that consciously integrated advanced information technology into their operational strategy emerged in the 1990s. One, the Zapatistas, adopted some violent tactics. The other, the transnational, anti-globalization resistance movement that coalesced around opposition to the World Trade Organization (WTO) summit in Seattle, culminated in several days of largely nonviolent protest, known as the "Battle of Seattle," that successfully disrupted the WTO's meetings.

A core component of the Zapatista information strategy was communicating its message transnationally through rapid communication platforms available on the Internet at the time. The Zapatistas were pioneers in the use of information technology as the basis for non-violent resistance. The group relied on telecommunications, videos, and other computer-mediated communication to spread its message.

The Zapatistas received abundant support among the Mexican society, which also helped to catapult a weak resistance movement to international attention. While the Zapatistas were armed and used guerrilla tactics against the Mexican security forces, its armed tactics contributed little to the group's success. Instead, the widespread support and dense transnational networks hindered the Mexican government from employing repressive countermeasures against the group.[20]

The relentless media attention on the relations between the Mexican government and the Zapatistas encouraged the government to enter into negotiations with the group. After a series of negotiations and public consultations in the mid-1990s, the Zapatistas successfully coerced the Partido Revolucionario Institucional (PRI), Mexico's ruling party, to address grievances in Indian communities, including implementing constitutional reforms, and helped to break the PRI's dominance in Mexican politics by igniting political debate on the party's long history of corruption and support for economic policies that favored elites.[21]

30

Figure 2-2. Comandanta Ramon is perhaps the most famous female Zapatista actor.

The "Battle of Seattle" occurred during protests against the WTO Ministerial Conference of 1999. This meeting intended to begin the next round of trade negotiations between the signatory countries. On the first day of the conference, around forty thousand to sixty thousand protesters converged in downtown Seattle, where the meeting occurred. During this historical event:

> protesters not only attacked targets beyond the nation-state but began to experiment with a new and imaginative repertoire of contention. They combined peaceful and violent performances, face-to-face and electronic mobilization, and domestic and transnational actions.[22]

The tens of thousands of protesters were part of a broad coalition of organizations, nearly seven hundred in all, that spanned the globe.[23] Local organizations also played a significant role, including the region's powerful trade unions, through a long history of strikes, protests, and labor militancy.[24] Trade unions joined the protest movement on the platform of global workers' rights and welfare, and later, the movement gained a variety of other organizations representing global environmental issues, students, faith-based groups, and academics.[25]

Contingents of protesters used direct-action measures, such as blockading entrances to hotels and convention centers, to effectively shut down the meeting. Although the protests were largely peaceful, factions within the larger protest movement adopted violent tactics, including the destruction of property, which incurred a heavy response from the Seattle Police Department. Police used tear gas, pepper spray, rubber bullets, and percussion grenades to disperse crowds.[26] In keeping with the protesters' anti-globalization grievances, violence was especially directed against corporate businesses such as Nike, The Gap, and Starbucks retail locations in the central downtown area.

Mobilization in the cyber domain aided the organizational coalitions surrounding the WTO protests in Seattle. Similar to the leaderless resistance movements leveraging social media platforms today, a large portion of the protests' online organization initiated in isolation by different individuals and groups. Websites such as SeattleWTO.org and Seattle99.org linked activists and resources. Similarly, listservs enabled communication between transnational activists. Activists leveraged the listservs' low barrier of entry and the absence of restrictions on redistribution to facilitate many-to-many interaction.[27]

> *2.5 Violent resistance movements benefited from new media as much as nonviolent movements.*

While a significant amount of research on resistance in the cyber domain featured nonviolent resistance, violent resistance movements also benefited from new media. Violent movements have the same advantages as nonviolent ones, although the former are more likely the target of restrictions by Western private corporations that own the platforms. As one journalist noted, "Never before in history have terrorists had such easy access to the minds and eyeballs of millions."[28] Al Hayat, ISIS's media arm, publishes a newsletter, *Dabiq*, that is produced and distributed from a centralized location, an example of traditional media.

New media allowed for individual ISIS members, using Twitter or other social media accounts, to generate and distribute propaganda without directives from centralized leadership. In this regard, ISIS operates more like a conglomerate than a hierarchical military organization with centralized command and control (C2). In a 2015 report, ISIS members and its followers had over 70,000 Twitter accounts producing over 200,000 tweets, sometimes called "mujatweets," a day. At that time, around 90 percent of the group's social media activities occurred on the site.[29] Moreover, the group, using media teams dispersed from West Africa to Afghanistan, produces on average thirty-eight batches of propaganda a day, including videos, photo essays, articles, and audio programs.[30] The volume and sophistication of its social media activity, exemplified in Figure 2-3, meant that its near peers were well-known Western brands, marketing firms, and publishing outfits.[31]

Figure 2-3. ISIS made extensive use of Western-style advertising and public relations tools.

ISIS used Twitter for various purposes, including recruitment and messaging its enemies. Its messaging is gruesome, but ISIS also used the site to normalize its members and behavior, often demonstrating its native fluency in Western popular culture for greater appeal. The individual accounts were "more personal, emotional, and therefore appealing for example young potential recruits."[32]

ISIS online propaganda also leveraged the popularity of first-person shooter games and Hollywood action movies in the West to connect with sympathetic target audiences in Europe and North America.[33] ISIS's efforts to infiltrate Western culture in this regard were highly successful. In 2015, the group had a reported 20,000 foreign fighters, 2,000 of whom came from Western countries, making ISIS one of the largest foreign fighter armies in the world at the time of writing.[34]

NEW MEDIA AND ANALOG MEDIA

A key feature of research on resistance in the cyber domain is the extent to which new media platforms altered the emergence, diffusion, and operation of resistance movements. While the influence of these new media on resistance is debated, several characteristics of new media distinguish it from older, analog forms. This section introduces a categorization of new and analog media that contextualize media developments as they shape resistance.

Numerous novel characteristics of new media distinguish it as an analytically important category. Transmedia, content and intellectual property that migrate across communication platforms, requires collaboration among producers across the media spectrum.[35] New media also shifted or blurred boundaries associated with analog media, which demonstrated clear distinctions between producers and audiences.[36]

Other attributes include its digitization, interactivity, and networked audience. Digitization occurs when input data are coded into numbers and output data are reviewed through online sources, digital disks, or memory drives on display screens or disseminated again. As a result, data can be compressed into very small spaces and accessed at high speeds. It is easier to manipulate than analog media so that the:

> scale of this quantitative shift in data storage, access
> and manipulation is such that it has been experienced

as a qualitative change in production, form, reception, and use of media.[37]

Interactivity allows users, through new media to interact with the data to alter it. This quality differs significantly from analog media, such as books, in which consumers passively engage the text without any ability to directly intervene with the material to alter it for themselves and other consumers. The boundary between consumers and producers shifted, because users, rather than a centralized production company, often produce the content.

The only broadcasting producers in the 1970s through 1990s that reached a broad audience were the big broadcasting companies Columbia Broadcasting System (CBS), American Broadcasting Company (ABC), National Broadcasting Company (NBC), and to a lesser degree, the Public Broadcasting System (PBS). By contrast, the producers in new media are individual users, with limited budgets, with the capacity to reach millions through platforms such as YouTube. As of 2018, several YouTube videos yielded over a billion views. While many of the most-watched videos on YouTube were centrally produced music videos, individually produced videos reached astonishingly broad audiences.

> 2.6 *The producers in new media are individual users, with very limited budgets, who nevertheless have the capacity to reach millions.*

Finally, audiences of new media are networked, or dispersed, to mass audiences. Their content is not distributed to audiences as a mass, but instead to a "dispersed mediasphere."[38] A dispersed media sphere distribution means that while the audiences for new media are potentially quite large, they are nevertheless more segmented and individualized such that the messages they receive are no longer simultaneous or uniform.[39] While that centralized stream is still important, new media fosters a flatter, decentralized communication strategy that allows for flexibility and adaptability in its content.

New Media: New Communication or New Organization?

One of the key debates that emerged in studies that investigate the impact of new media on resistance is whether it is simply a newer, faster method of communication or if the differences in new media altered the organization and mobilization of people in resistance movements.

Research that emphasizes the impact of new media strictly in functional terms suggests that new media are a faster, more efficient way of communicating and organizing than occurred in the past. For resistance groups, this is advantageous.

The lowered costs and enhanced efficiency of communication in new media impact resistance movements in two crucial ways. First, organizations are able to do more with less. The advantages of new media are described in terms of "affordance" but explain the advantages in a manner similar to resources.[40] New media allows simultaneous communication with networked individuals all over the world enabling rapid information sharing. It also negates the need for individuals to meet face to face to coordinate or participate in activities. Therefore, participation most of the time requires less effort and poses fewer risks.

> 2.7 *The use of new media, while in some ways merely permitting resistance units to do what they have always done, but more quickly and at much lower costs, fundamentally changed the science of resistance, if only for those two reasons.*

On the other end of the spectrum, there are those who argue that much of the resistance occurring in the cyber domain has undergone several fundamental shifts. Some resistance movements and related organizations use new media in ways similar to those previously described, as a tool for communication that signals a shift in scale or degree, but not in ways that suggest we need new theories to understand resistance in this domain. Others are better positioned to leverage new media in innovative ways that makes the platforms game changers.

While the research on resistance in the cyber domain is relatively new, the evidence suggests that we need to rethink our usual explanations for why people opt to participate, the sorts of participatory actions they take once they are involved, and the role that older forms of organizing play in this type of resistance.

Personalized Communication

In analyzing resistance in the cyber domain, it is important to understand that it is not just technology that changed. Resistance movements leverage social and new media platforms not only because the tools are force multipliers, but because the platforms are the most effective means to reach publics that pursue resistance differently than their predecessors.

The most noticeable shift, as discussed later, is that many people seek unconventional ways to press their political claims. Previously, most people joined traditional organizations, but today many look for ways to mobilize outside those constraints. By "traditional organizations," we mean labor unions, special interest groups, non-governmental organizations (NGOs), churches, or even political parties. The organizations mobilized participation and pursued their political objectives by aligning the goals, interests, and actions of their members. As a result, joining those organizations meant adopting the collective positions cultivated by the organization's leadership. Membership was also carefully regulated; people who wished to join underwent an enrollment process, perhaps paying dues, and administrators kept records of membership rolls.

Conventional forms of membership are, however, on the decline. The decline began before the widespread adoption of the Internet and social media. Robert Putnam first described the trend in his book *Bowling Alone*, which charts how American civic participation plummeted in the latter decades of the twentieth century. Membership rolls in civic organizations across the spectrum spiraled downward, from political parties, labor unions, and even bowling leagues.[41] Putnam's research focused on the United States, but civic participation is also truncated in other regions of the world, albeit for different reasons. In authoritarian countries such as Mubarak's Egypt, the regime's violently repressive policies stunted participation in certain organizations.[42] The results in each example, however, are similar in that reaching these sorts of groups requires different strategies and technologies.

Under these conditions, resistance movements use personalized communication to mobilize groups with specific interests. Personalized communication differs from past communication strategies in that formal organizations, in prior times, sought to unify their message. Chapter 5 discusses how formal organizations used narratives to help them form identities that overcome barriers to participation in collective action. Personalized communication, however, allows opportunities for people to mobilize on their own terms, not terms dictated by organizational leadership. In practice, personalized communication is characterized by symbolic inclusiveness and technological openness. The former term refers to personalization that provides opportunities for participants to customize their engagement, advocating for issues and engaging in actions meaningful to them.

In 2017, the Women's March, scheduled to vocalize concern over women's issues and to advocate for women's rights, coincided with the inauguration of President Donald Trump, who had been involved in a public scandal involving speaking controversially about women on an audio recording. Women's March advocates personalized communication mobilized hundreds of thousands of protesters in the Washington, DC, around the country, and across the globe. Protesters mobilized on a wide diversity of issues, including climate change, voting rights, affordable healthcare, and various sexual rights, rather than on a single platform.[43]

COLLECTIVE AND CONNECTIVE ACTION

Connective action differs from traditional resistance in several ways. It is a form of protest for a highly individuated public that does not prefer to forge a collective or group identity to facilitate mobilization processes. Simplified master frames enable "large-scale, networked action" to rapidly spread, both offline and online, through "easy personal associations."[44]

Connective action has several important advantages over collective action in terms of mobilization. Rather than rely upon the formation of collective action identities through formal organizations, the interactive technologies that connective action leverages enables mobilization through highly personal, inclusive frames that generate more opportunities for mobilization because they appeal to a broader audience.[45]

The interactive technology also replaces formal organizations as the directors or leaders of the movement, distributing the burden of mobilization from one or several formal organizations to individual participants. In achieving these advantages, connective action relaxes the requirements for the development and maintenance of a unifying message and communication processes.

Collective action is typically defined as action taken under direction from an organization's leadership, whereas connective action results from decisions made at a more grass-roots level, by the members of an organization or by individuals or smaller groups that possess no formal leadership or organization at all. The desirability of the relaxed requirements and advantages of connective action is demonstrated in the large-scale mobilization of people in movements such as the Arab Spring. It is still unclear, however, as to whether connective action is ultimately as politically effective as resistance based more exclusively on collective action logic.

Formal organizations play a large role in resistance movements, in helping to overcome significant barriers to mobilization and affecting successful political outcomes. Successful organizations adeptly leverage available resources to bring people together, engage publics, set an agenda, and sustain networks across time and through periods of adversity.

Engagement strength captures an organization's capacity to align its interests, goals, and interpretations of current events with those of the public it seeks to mobilize. Similarly, agenda setting refers to a resistance movement's ability to communicate to its followers, sympathizers, and adversaries clearly articulated goals or objectives. Lastly, when mobilizing large numbers of people, resistance movements must often form coalitions among many different smaller organizations and networks. Solidifying these crucial relationships requires agreement on the content of messages and specific goals and communication strategies.

> 2.8 *When mobilizing large numbers of people, forming coalitions from many different smaller organizations and networks can be a necessity.*

Connective action, which does not rely heavily on formal organizations, approaches the core competencies of formal organizations in a different manner with varying levels of success. The interactivity of new media enables flexible communication, action, and identities, but it also erodes organizational control over a movement's messaging and action, which impacts a movement's ability to set a clear agenda.

The struggle between flexibility and control is often apparent in a central communication hub for many resistance movements operating—its website. In a movement's efforts to attract followers, its public website offers many different avenues for entry, linking to numerous other organizations and individuals who in turn contribute content and links. The result is a robust network, but one which often provides a diluted message and claims. Thus, while a movement's engagement strength is vigorous, it makes it more difficult for the movement to set a singular agenda or clearly communicate its cause. Lacking a singular clearly communicated cause also impacts a movement's ability to sustain cohesive coalitions with other organizations, especially those that are rich in resources crucial to effective collective action.

In a comparison between two similar resistance movements, researchers found that some of these concerns about resistance taking place primarily through new media channels was unfounded. The researchers compared two movements, relying on evidence gathered from the movements' presence in the cyber domain (public websites), that held protests against the 2009 G20 summit that occurred in the aftermath of the 2008 recession. The two groups, the Put People First (PPF) and the Meltdown, shared similar goals but agreed to arrange protests on different days of the summit.

The PPF relied more heavily on connective action than the Meltdown, which yielded more involvement by formal organizations and rigid collective identities. Relying more on personalized frames, the PPF's website evidenced strong tendencies toward interactivity and tolerance of differing messages.

Despite the flexibility of personal identity frames and new media interactivity, which theoretically erodes organizational control, the PPF managed to mobilize more supporters, communicate clear goals in a twelve-point policy platform, and form a strong network of resource-rich organizations representing a wide variety of issues.

The logic of collective action often explains the process of mobilization in more traditional resistance movements that are not digitally enabled or those that rely upon information technologies as resources to relieve burdens of communication and coordination. In the logic of collective action, the primary obstacles to persuading individuals to join a resistance movement are the high costs of participation versus the potential gains, particularly when others can "free ride" on the efforts of others.[46] The conundrum of the free rider occurs due to the public goods produced by resistance movements. When a resistance movement is successful, such as overthrowing a government or changing its policies, everyone benefits, even those that did not perform the hard work of resisting. As researchers W. Lance Bennett and Alexandra Segerberg describe it in *The Logic of Connective Action*,

> The familiar concern is that the gains of connective action such a rapid scalability and adaptability may be paid for by a loss of capacity to set agendas, achieve policy change, and continue to mobilize and coordinate action in the face of adversity over time.[47]

The ability of connective action to rapidly mobilize, arguably the greatest strength of this new form of resistance, also yields potential vulnerabilities. Turkish activist and scholar Zeynap Tufecki likened new media to the legendary Sherpas, the mountain guides that assist climbers in ascents on Mt. Everest. The Sherpas, through their resources and experience, "give a boost to people who might not have otherwise be fully equipped to face the challenges that routinely occur above eight thousand meters."[48] Similarly, digital media provides the resources to rapidly mobilize a movement, but it does not develop the prior organizational capacities that are crucial resources for assuming authority in political institutions.[49]

2.9 *The ability of connective action to rapidly mobilize, arguably the greatest strength of this new form of resistance, also yields potential vulnerabilities. While digital media provides the resources to rapidly mobilize, it also permits such action without developing crucial organizational capacities that must precede the accession of power in political institutions.*

This conundrum is aptly captured in the events of the Arab Spring in Egypt. After President Mubarak was ousted from office, the movement mobilized millions of participants in Tahrir Square and in other urban areas of the country. The movement emerged as the only viable formal organization in Egyptian civil society and gained important political posts, including the election of President Mohamad Morsi in July 2012. Kefaya, a prior pro-democracy movement in Egypt, provided activists with vital experience in organizing and coordinating protest events, but it demonstrated little serious internal dialogue about the details of assuming state power once the immediate goals of ousting President Mubarak were met. Kefaya also formed a horizontal, leaderless structure that replaced strong, central leadership with weak, coordinating positions such as a steering committee.

A profound lack of institutional trust is a feature of many of the resistance movements that emerged in the last decade, making the decision to adopt leaderless structures one of emotive appeal, not strategic consideration.[50] One former Kefaya member and Arab Spring revolutionary began to recognize the need for greater organizational strength as the movement dwindled:

> We celebrated Kefaya for its new "form"—horizontal, loose, and flexible—because it was everything that traditional political parties were not. The problem now, however, is that Kefaya does not exist beyond the event. In other words, Kefaya is very successful at organizing a rally or a demonstration; it attracts people, emotions rise high. However, once the event (demonstration) is over, there is nothing left. Those newly mobilized, especially the youth, who are so inspired, full of energy and desire to do something, really have nothing to do.[51]

After the mass protests in early 2011, activists formed the Revolution Youth Coalition (RYC), which included youth from across the ideological spectrum, including those from the Muslim Brotherhood. The RYC, however, failed to establish any collective identity or brokered coalition that could overcome the deep ideological rifts between the liberal and conservative contingents. In part, the difficulties stemmed from the emphasis of ideology divided along religious and secular lines, which presented significant challenges to developing a consensus.[52]

The participation costs of resistance movements are high for numerous reasons. In repressive environments, the costs might include arrest, imprisonment, torture, or even death. However, even in more democratic or open environments, participation still demands significant financial and emotional costs. In addition to the more obvious costs, there are obscure costs associated with identity or culture; for example, participation can draw disapproval from the larger society, social networks, or family.

In the logic of collective action, organizations help to overcome these barriers to mobilization. One theory of how organizations accomplish this work is through resource mobilization.[53] Organizations deploy resources critical to a resistance movement that are not available to individuals in isolation as organizations contain the aggregate, or combined, resources of many individuals. The aggregate resources help to reduce the costs of participation through the provision of selective incentives.

In this regard, organizations aid mobilization by making participation less costly because rates of participation increase when the costs associated with participation decrease. For instance, the stigma associated with resistance is diminished as more people participate. Moreover, as more people join, the risk of experiencing state repression first hand is spread over more individuals. Thus, as mobilization increases, the costs for others to join falls, creating opportunities for individuals with a lower threshold of tolerance for such costs to participate.[54]

In conventional protests, formal organizations also perform the central role of mobilization through the formation of collective identities by leveraging framing strategies and brokering coalitions among organizations active in the field of resistance. Resistance is sufficiently risky in many cases such that "people do not risk their skin or sacrifice their time to engage in contentious politics unless they have good reason to do so. It takes a common purpose to spur people to run the risks and pay the costs of contentious politics."[55] Investment in the cause further enables mobilization and participation.

In isolation, individuals may think about the complaints they harbor against authorities, but resistance as it is meant in the science of resistance occurs when people come together collectively, and publicly, to protest, or more, in opposition to existing authorities and power structures. Coming together collectively not only requires that people hold common grievances or interests but also that they are aware that they share those grievances or interests. A collective identity helps to motivate individuals to participate, align different goals and interests of the participants, and coordinate and sustain collective action.

> *2.10 A collective identity helps to motivate individuals to participate, align different goals and interests of the participants, and coordinate and sustain collective action.*

The *los indignados,* or 15M, protests began in Spain in 2011 following the G20 summit meetings in London the previous year. The 15M resistance movement made the shift in the relationship between organizations and participants increasingly apparent in digitally mediated resistance in the twenty-first century. Whereas previous resistance movements relied on formal organizations, whether churches, labor unions, political parties, or NGOs, the 15M participants advocated a leaderless movement that distanced itself from formal organizations with definitive memberships and agendas. As a result, 15M signaled that it viewed these conventional organizations as part of the problem, not part of the solution.

According to a survey of 15M protesters, the movement differed in three critical ways from more conventional resistance movements that relied formal organizations for mobilization:[56]

- In conventional protests, most participants acknowledged the role that formal organizations played in furthering the objectives of the movement. In 15M, only 38 percent of the participants did so.

- Moreover, only 13 percent of the formal organizations cited by 15M protesters offered any means for individuals to become members or affiliates in contrast to conventional protests where membership and affiliation were critical.

- The age of organizations associated with the 15M protests is also striking. In conventional protests, most organizations range from ten to forty years old, but organizations associated with the 15M protests were on average only three years old.

Although the long-term effects of the 15M movement are still open to discussion, the replacement of two-thirds of Spain's parliament, clearly indicates that 15M's new modes of resistance made a difference. In 2015, there was a widespread electoral success of new elected representatives who had never held office before—including members of the 15M movement—to positions in the Spanish parliament, city offices, and even mayorships. Some observers also credit 15M with being an inspiration for the Occupy movement that came soon after in the United States.[57]

Success of Connective Action

Although connective action has advantages in terms of greater organizational flexibility and enhanced mobilization, it is unclear if these advantages contribute to enhanced capacity for achieving political objectives for resistance movements that rely more on connective action.[58] Connective action efforts in the past decade have been notable due to their speed of mobilization, the broad scope of issues they have addressed, and their ability to rapidly spread their message to the general public.

During the Arab Spring, the world witnessed several successes of organizationally enabled and connective action movements. In Egypt, large-scale, organizationally enabled mobilizations helped to topple the authoritarian President Hosni Mubarak. Likewise, in Tunisia, civil resistance led to the ouster of the longstanding authoritarian ruler President Zine El Abidine Ben Ali. Despite these successes, however, some of the weaknesses of connective action are also evident in each case.

The resistance movements in Tunisia, especially Egypt, had difficulties translating their successes to stable democracies that protect the freedoms denied their respective societies during the period of authoritarian rule. In the case of Egypt, the lack of formal organization among the Arab Spring protesters handicapped the movement when compared with the resource-rich Muslim Brotherhood and Egyptian military. After Mubarak's fall from power, these two powerful organizations wrested control of the revolution, and the future of the country, from the Arab Spring protesters.

Most people think about change facilitated by resistance movements as occurring strictly through political institutions, but noted Spanish sociologist Manuel Castells demonstrated that this is not always the case, particularly for large-scale mobilizations driven by connective action. Sometimes, the effects of a resistance movement are on the public mind, not necessarily the political institutions represented by the move. Castells used the unrest surrounding the contested 2009 presidential elections in Iran as an example.

According to some perceptions, because President Mahmoud Ahmadinejad remained in power, the mobilization was ineffective. However, in the 2013 presidential elections, candidate Hassan Rouhani gained the support of the youth factions that mobilized against Ahmadinejad. Although the lack of transparent public opinion data in Iran makes it difficult to precisely pinpoint the causal mechanisms behind support for Rouhani, it is plausible that the 2009 unrest helped generate a "mental transformation" among the Iranian public that paved the way for greater support of moderate candidates such as Rouhani.[59]

The road from the hope for political to change to its implementation through connective action depends on several factors. The first relates to the government's tolerance for meeting a movement's demands and, in turn, the movement's willingness to engage in back-and-forth negotiations and compromise with political authorities. When these conditions are not met, the facilitation of consensus-based reform directed through political institutions is challenging.

Despite these difficulties, resistance movements mobilizing through the logic of connective action can still indirectly impact politics. The nuance is captured in the doctrinal definition of UW in Joint Publication 3-05 "Special Operations" (2014), which identifies its objectives as "coerce, disrupt, or overthrow." The objectives of coercing and overthrowing are largely directed toward political institutions. Disruption, however, recommends more ambiguous objectives that indirectly impact political authorities. Disruption generates political vulnerabilities, which in turn present further political opportunities for opposition groups to press their demands.

In practice, the power law distribution in organizationally brokered collective action appears in the concentration of resources with formal organizations. Organizational leadership retains control of the resources necessary for sustained collective action, including the unified message that informs their identity, strategies, and goals.[60]

Finally, resistance movements relying on connective action may yield dispersed power signatures. Similar to the moderate power signature, the dispersed power signature involves power sharing, but to a greater extent than is evident in other power signatures. Graphically, the distribution of power is flat. In practice, this translates to the so-called leaderless movements with no discernible head but are instead "all tails."[61]

The Occupy movement provides a good example of a movement characterized by a flat power distribution (see Figure 2-4). In this type of movement, digital media acts as a "stitching technology" that helps to facilitate flows of information and action across interconnected but dispersed networks. The movement's dispersed power signature was evident in the diffusion of the "we are the 99%" personalized action frame. The "we are the 99%" frame initially appeared on a Tumblr microblog, originated by a single Occupy sympathizer, that invited visitors to post their own experiences of economic injustice and related issues.[62]

Figure 2-4. The Occupy movement offers a good example of a movement characterized by a flat power distribution.

The frame then migrated to Occupy sympathizers on Twitter. Eventually, the frame became a fixture at the physical Occupy campsites in New York and around the globe. Unlike the Robin Hood tax campaign, the Occupy movement did not emerge around a coalition of established organizations. Instead, it formed around loosely connected online networks that overlapped with offline physical campsites.

The diffusion of personalized action frames, such as "we are the 99%," alongside digital media, helped the movement to sustain action over an extended period. Unlike the Robin Hood tax campaign, the Occupy movement struggled to influence the adoption of policies specifically addressing issues of inequality. However, the Occupy movement was highly successful in bringing inequality to the attention of the general public. In December 2011, a Pew Research poll measuring public opinion reported that the percentage of Americans that perceived conflicts between the rich and the poor doubled since the previous survey in 2011, placing concern over economic inequality over other social conflicts, such as immigration or race-related issues.[63]

KEY TAKEAWAYS

2.1 Resistance movements leveraged the cyber domain because information technology began to intertwine the globe with unprecedented scope, speed, and accessibility.

2.2 The most striking characteristics of new forms of resistance are the employment of previously unknown technologies and the means which those technologies provided to encourage mass mobilization.

2.3 "Leaderless" networks exhibit little to no formal organizational structures but employ layers of new media technologies to support personal expression and identity associated with a wide variety of contentious issues.

2.4 Movements that adopt violent strategies and operate in restrictive environments are not as well positioned to do so, particularly if the state possesses a robust intelligence or counterinsurgent capacity.

2.5 Violent resistance movements benefited from new media as much as nonviolent movements.

2.6 The producers in new media are individual users, with very limited budgets, who nevertheless have the capacity to reach millions.

2.7 The use of new media, while in some ways merely permitting resistance units to perform ongoing tasks, but more quickly and at much lower costs, fundamentally changed the science of resistance, if only for those two reasons.

2.8 When mobilizing large numbers of people, often forming coalitions from many different smaller organizations and networks can be a necessity.

2.9 The ability of connective action to rapidly mobilize, arguably the greatest strength of this new form of resistance, also yields potential vulnerabilities. While digital media provides the resources to rapidly mobilize, it does so without developing crucial organizational capacities that must precede the accession of power in political institutions.

2.10 A collective identity helps to motivate individuals to participate, align different goals and interests of the participants, and coordinate and sustain collective action.

ENDNOTES

1   US Department of Defense, "Dictionary of Military and Associated Terms," (September 2018).

2   John Vidal, "Mexico's Zapatista Rebels, 24 years on and Defiant in Mountain Strongholds," *Guardian*, February 17, 2018, https://www.theguardian.com/global-development/2018/feb/17/mexico-zapatistas-rebels-24-years-mountain-strongholds.

3   Michael Connor and Sarah Vogler, "Russia's Approach to Cyber Warfare," CNA Analysis and Solutions, March 2017, https://www.cna.org/CNA_files/PDF/DOP-2016-U-014231-1Rev.pdf, 4–5.

4   Merlyna Lim. "Clicks, Cabs, and Coffee Houses: Social Media and Oppositional Movements in Egypt, 2004–2011," *Journal of Communication* 62, no. 2 (2012): 231–248.

5   Zeynep Tufecki, *Twitter and Tear Gas: The Power and Fragility of Networked Protest* (New Haven, CT: Yale University Press, 2017), 133.

6   The figure is adapted from Kirk A. Duncan, "Assessing the Use of Social Media in a Revolutionary Environment," (Master's thesis, Naval Postgraduate School, 2013), 21.

7   Tufecki, *Twitter and Tear Gas*, 135.

8   Stefania Milan, "When Algorithms Shape Collective Action: Social Media and the Dynamics of Cloud Protesting," *Social Media+ Society* 1, no. 2 (2015): 2056305115622481.

9   Pablo Barberá et al., "Tweeting from Left to Right: Is Online Political Communication more than an Echo Chamber?" *Psychological Science* 26, no. 10 (2015): 1531–1542.

10 Cara McGoogan, "North Korea's Internet Revealed to Have Just 28 Websites," *Telegraph*, September 21, 2016, http://www.telegraph.co.uk/technology/2016/09/21/north-koreas-internet-revealed-to-have-just-28-websites/.

11 Tufecki, *Twitter and Tear Gas*, 225–226.

12 Kiyo Dörrer "Hello, Big Brother: How China Controls Its Citizens Through Social Media," *Deutsche Welle*, March 31, 2017, http://www.dw.com/en/hello-big-brother-how-china-controls-its-citizens-through-social-media/a-38243388.

13 Gary King, Jennifer Pan, and Margaret E. Roberts, "How Censorship in China Allows Government Criticism but Silences Collective Expression," *American Political Science Review* 107, no. 2 (2013): 326–343.

14 Seva Gunitsky, "Corrupting the Cyber-Commons: Social Media as a Tool of Autocratic Stability," *Perspectives on Politics* 13, no. 1 (2015): 42–54.

15 Ibid.

16 Rebecca MacKinnon, "China's 'Networked Authoritarianism,'" *Journal of Democracy* 22, no. 2 (2011): 32–46.

17 J. M. Berger, "How ISIS Games Twitter: The Militant Group that Conquered Northern Iraq is Deploying Sophisticated Social Media Strategy," *Atlantic*, June 16, 2014, https://www.theatlantic.com/international/archive/2014/06/isis-iraq-twitter-social-media-strategy/372856/.

18 Shiraz Maher and Joseph Carter, "Analyzing the ISIS 'Twitter Storm," *War on the Rocks*, June 24, 2014, https://warontherocks.com/2014/06/analyzing-the-isis-twitter-storm/.

19 Ibid.

20 Manuel Castells, *The Power of Identity*, 2nd edition (Malden, MA: Blackwell Publishing Ltd, 2004), 83–84.

21 Castells, *The Power of Identity*, 85–86.

22 Sidney Tarrow, *Power in Movement: Social Movements and Contentious Politics* (Cambridge: Cambridge University Press, 1994), xv.

23 Stephen Gill, "Toward a Postmodern Prince? The Battle in Seattle as a Moment in the New Politics of Globalisation," *Millennium* 29, no. 1 (2000): 131–140.

24 The maritime unions along the west coast of the United States are among the most militant labor unions in the world.

25 Margaret Levi and David Olson, "The Battles in Seattle," *Politics & Society* 28, no. 3 (2000): 309–329.

26 Ibid.

27 Matthew Eagleton-Pierce, "The internet and the Seattle WTO protests," *Peace Review* 13, no. 3 (2001): 331–337.

28 Brendan I. Koerner, "Why ISIS is Winning the Social Media War," *Wired*, April 2016, https://www.wired.com/2016/03/isis-winning-social-media-war-heres-beat/.

29 Khuram Zaman, "ISIS has a Twitter Strategy and It is Terrifying," *Medium*, November 20, 2015, https://medium.com/fifth-tribe-stories/isis-has-a-twitter-strategy-and-it-is-terrifying-7cc059ccf51b.

30 Charlie Winter, "Documenting the Virtual Caliphate," *Quilliam*, October 2015, http://www.quilliaminternational.com/wp-content/uploads/2015/10/FINAL-documenting-the-virtual-caliphate.pdf.

31 Koerner, "Why ISIS is Winning the Social Media War."

32 Thomas Elkjer Nissen, "Terror. com – IS's Social Media Warfare in Syria and Iraq," *Contemporary Conflicts: Military Studies Magazine* 2, no. 2 (2014).

33 Anne Speckhard, "The Hypnotic Power of ISIS Imagery in Recruiting Western Youth," *International Center for the Study of Violent Extremism*, October 20, 2015, http://www.icsve.org/research-reports/the-hypnotic-power-of-isis-imagery-in-recruiting-western-youth/.

34 Zaman, "ISIS has a Twitter Strategy and It is Terrifying."

35 Martin Lister et al., *New Media: A Critical Introduction, Second Edition* (New York, NY: Routledge, 2009), 9–10.

36 Ibid.; Scott W. Ruston and Jeffry R. Halverson, "'Counter' or 'Alternative': Contesting Video Narratives of Violent Islamist Extremism," in Carol K. Winkler and Cori E. Dauber, eds., *Visual Propaganda and Extremism in the Online Environment* (Carlisle Barracks, PA: United States Army War College Press, 2014), 116.

37 Lister et al., *New Media*, 18.

38 Ibid.

39 Ibid., 31.

40 Jennifer Earl and Katrina Kimport, *Digitally Enabled Social Change: Activism in the Internet Age* (Cambridge, MA: The MIT Press, 2011), 10.

41 Robert D. Putnam, *Bowling Alone: The Collapse and Revival of American Community* (New York, NY: Simon and Schuster, 2001), 31–48.

42 Jason Brownlee, "The Decline of Pluralism in Mubarak's Egypt," *Journal of Democracy* 13, no. 4 (2002): 6–14.

43 Joanna Walters, "Women's March on Washington Set to be One of America's Biggest Protests," *Guardian*, January 17, 2017, https://www.theguardian.com/us-news/2017/jan/14/womens-march-on-washington-protest-size-donald-trump.

44 W. Lance Bennett and Alexandra Segerberg, "Introduction," in *The Logic of Connective Action: Digital Media and the Personalization of Contentious Politics* (Cambridge: Cambridge University Press, 2013), 2.

45 W. Lance Bennett and Alexandra Segerberg, "Personalized Communication in Protest Networks," in *The Logic of Connective Action: Digital Media and the Personalization of Contentious Politics* (Cambridge: Cambridge University Press, 2013), 59.

46 Mancur Olson, *The Logic of Collective Action: Public Goods and the Theory of Groups* (Cambridge, MA: Harvard University Press, 1965).

47 W. Lance Bennett and Alexandra Segerburg, "Networks, Power, and Political Outcomes," in *The Logic of Connective Action: Digital Media and the Personalization of Contentious Politics* (Cambridge: Cambridge University Press, 2013), 149.

48 Tufekci, *Twitter and Tear Gas*, xii.

49 Ibid.

50 Jeff Jarvis, #OccupyWallStreet and the Failure of Institutions," *HuffPost*, October 3, 2011, http://www.huffingtonpost.com/jeff-jarvis/occupywallstreet-the-fail_b_991928.html; and Tufekci, *Twitter and Tear Gas*, xxiv.

51 As quoted in Maha Abdelrahman, "In Praise of Organization: Egypt between Activism and Revolution," *Development and Change* 44, no. 3 (2013): 569–585.

52 Ibid.

53 John D. McCarthy and Mayer N. Zald, "Resource Mobilization and Social Movements: A Partial Theory," in *Social Movements in an Organizational Society: Collected Essays* (New Brunswick, NJ: Transaction Publisher, 1987), 15–42.

54 Earl and Kimport, *Digitally Enabled Social Change*, 70.

55 Tarrow, *Power in Movement*, 11.

56 W. Lance Bennett and Alexandra Segerberg, "The Logic of Connective Action," in *The Logic of Connective Action: Digital Media and the Personalization of Contentious Politics* (Cambridge: Cambridge University Press, 2013), 21.

57 "Five Years on, the Indignados have changed Spain's Politics," *LocalES*, May 14, 2016, https://www.thelocal.es/20160514/five-years-on-spains-indignados-have-shaken-up-politics.

58 Bennett and Segerberg, *The Logic of Connective Action*, 58-59.

59 Castells, *Networks of Outrage and Hope*, 272-273.

60 Bennett and Segerberg, *The Logic of Connective Action*, 152–156.

61 Ibid., 160–164.

62 Ibid.

63 Richard Morin, "Rising Share of Americans See Conflict Between Rich and Poor," *Pew Research Center Social & Demographic Trends*, January 11, 2012, http://www.pewsocialtrends.org/2012/01/11/rising-share-of-americans-see-conflict-between-rich-and-poor/.

CHAPTER 3.
AN INTRODUCTION TO CYBER OPERATIONS

INTRODUCTION

There are a number of definitions of cyberspace. Science fiction author William Gibson coined the term itself in the book "Neuromancer," describing it as a "consensual hallucination." Noted political scientists P.W. Singer and Allan Friedman defined it simply as "the realm of computer networks (and the users behind them) in which information is stored, shared, and communicated online"[1]. The Department of Defense (DoD) defines cyberspace as "the global domain within the information environment consisting of the interdependent network of information technology (IT) infrastructures and resident data, including the Internet, telecommunications networks, computer systems, and embedded processors and controllers."[2] The Internet is an electronic communications network that connects computer networks and organizational computer facilities around the world.[3] However, the dry description understates the overall impact of the creation of the Internet: "The Internet is at once a worldwide broadcasting capability, a mechanism for information dissemination, and a medium for collaboration and interaction between individuals and their computers without regard for geographic location."[4] Cyberspace is inextricably linked with the concept of the information environment. The DoD defines the information environment as "the aggregate of individuals, organizations, and systems that collect, process, disseminate, or act on information"[5] and further categorizes the information environment into three dimensions:[6]

- *Physical Dimension*: The information environment aggregates individuals, organizations, and systems that collect, process, disseminate, or act on information. This dimension contains the physical platforms and the communications networks that connect them—such as computers and network servers—and includes human beings, media, computational devices, and network infrastructure.

- *Informational Dimension*: The informational dimension specifies the location and processes by which information is collected, processed, stored, disseminated, and protected. The dimension exercises the C2 of modern military forces and conveys the commander's intent. Actions in this dimension affect the content and flow of information. Examples include a specific message, such as an email, which could be edited by an adversary (affecting content) or forwarded through an adversary-controlled computer en route to its destination (affecting flow).

- *Cognitive Dimension:* The cognitive dimension encompasses the minds of those who transmit, receive, and respond to or act on information. In this dimension, people think, perceive, visualize, understand, and decide. The cognitive dimension considers and evaluates the mind of an adversary commander.

While the information environment splits between three dimensions, Joint Publication 3-12, "Cyberspace Operations," categorizes cyberspace into three layers:[7]

- *Physical Network Layer:* Data travels through the physical network layer, which is further subdivided into the geographic component and the physical network component. Geospatial intelligence activities primarily focus on this layer.

  ◊ *Geographic Component:* The geographic component entails the location in land, air, sea, or space where elements of the network reside. For example, radio-wave propagation between antennas dictate the extent of a wireless network geographic component. The latitude and longitude of a cable landing station for an undersea cable is another example of the geographic component.

  ◊ *Physical Network Component:* The physical network component is comprised of the hardware, systems software, and infrastructure (wired, wireless, cabled links, Electromagnetic Spectrum (EMS) links, satellite, and optical) that supports the network and the physical connectors (wires, cables, radio frequency, routers, switches, servers, and computers). Thus, the fiber optic line itself is the physical network component and can run underground or on the seabed in an undersea cable. Likewise, a satellite communications network extends through space in electromagnetic waves between ground stations and the satellites in orbit.

- *Logical Network Layer:* The logical network layer consists of those elements of the network related to one another in a way abstracted from the physical network, i.e., the form or relationships are not tied to an individual, specific path, or node. Joint Publication (JP) 3-12 further provides the example of a website, accessed through a single uniform resource locator (URL), hosted on multiple servers in separate physical locations. Requests to access a single website, a single entity in the logical network layer, are answered by different nodes, such as web servers.[8]

- *Cyber-Persona Layer:* The cyber-persona layer consists of the people interacting with each other on the network. Cyber personas emulate an actual person or entity, incorporating biographical or corporate data, email and Internet protocol (IP) address(es), web pages, phone numbers, etc. However, one individual may have multiple cyber personas, which vary in the degree to which they are factually accurate. A single cyber persona can characterize multiple users. Cyber personas include Facebook and Twitter accounts, email addresses, Internet group accounts such as Meetup, and online banking accounts. See Figure 3-1 for a graphical depiction of the layers of cyberspace.

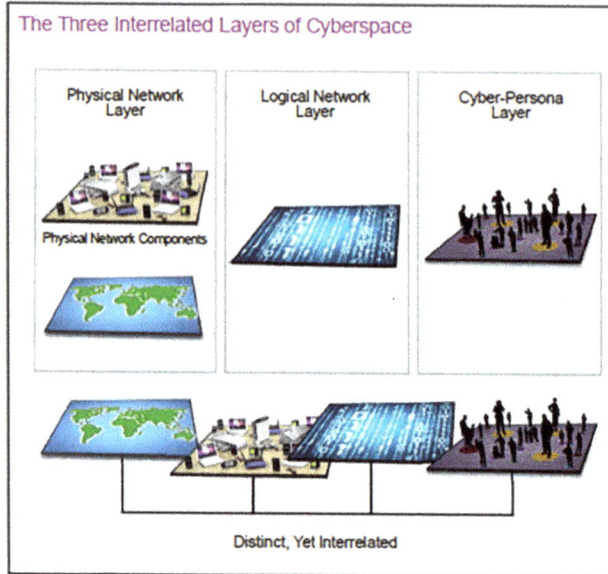

Figure 3-1. Three layers of cyberspace.

Based upon these DoD definitions, the layered information environment, with its dimensions, and cyberspace, intersects in several ways. Table 3-1 contains examples of entities that may occupy the intersection of an information environment dimension (physical, informational, and cognitive) and a cyberspace layer (physical network, logical network, cyber-persona).

These intersections also show the multiple effects of one entity using cyberspace operations or an information-related capability (IRC).[9] For instance, IRCs include key leader engagements, or ways to exert influence and ultimately affect the cognitive dimension, as well as the mind of the key leader. However, a similar effect could be generated through cyberspace using targeted messages disseminated to the key leader's social media accounts. Likewise, a website favoring an insurgent group might promote certain narratives detrimental to friendly messaging. These narratives might be countered through public affairs, or a cyberspace operation could target the website itself, preventing it from disseminating the message. Therefore, information operations, cyberspace operations, or cyberspace operations in support of information operations[10] all offer options to affect the information environment.

Table 3-1. Cyberspace and the information environment.

	Physical Layer	**Logical Layer**	**Cyber-Persona Layer**
Physical Dimension	C2 system hardware/ software Geospatial location of communications nodes	C2 system IP address or URL Wi-Fi service set identifier (SSID)	Network credential Social media account
Informational Dimension	Database hardware/ software Computer systems in combat operations center	Public affairs website An email	Database administrator credential
Cognitive Dimension			Any online identity

MILITARY OPERATIONS IN CYBERSPACE

Over the past decade, the DoD began the process of building its doctrine for operations in cyberspace. This section provides a brief overview of the key documents and concepts of this doctrine to establish a foundation and terminology for discussing military operations in cyberspace.

The primary joint reference for DoD cyberspace doctrine is JP 3-12.[11] The components correspond to cyber-focused doctrine: Army Field Manual FM 3-12,[12] Air Force Doctrine Document (AFDD) 3-12,[13] Navy Warfare Publication (NWP) 3-12,[14] and Marine Corps Interim Publication (MCIP) 3-40.02.[15] In general, the component-level views and terminology are largely the same as that found in JP 3-12, which will be the focus of this overview.

The DoD defines three primary types of cyberspace operations: offensive cyberspace operations (OCO), defensive cyberspace operations (DCO), and DoD information network (DODIN) operations. Offensive operations are those taken to apply force against an adversary, while defensive cyberspace operations defend friendly forces in cyberspace. DODIN operations are activities to build, secure, and operate the DoD networks. Additionally, all six basic warfighting functions—C2, intelligence, fires, maneuver, sustainment, and protection—have applicable parallels in cyberspace operations.

Operations in cyberspace generally follow a common life cycle from initiation to some conclusion. There are several models of this cycle; one of the most well recognized is the "cyber kill chain,"[16] which breaks a cyber operation into seven steps, as illustrated in Figure 3-2.

Phases of the Intrusion Kill Chain

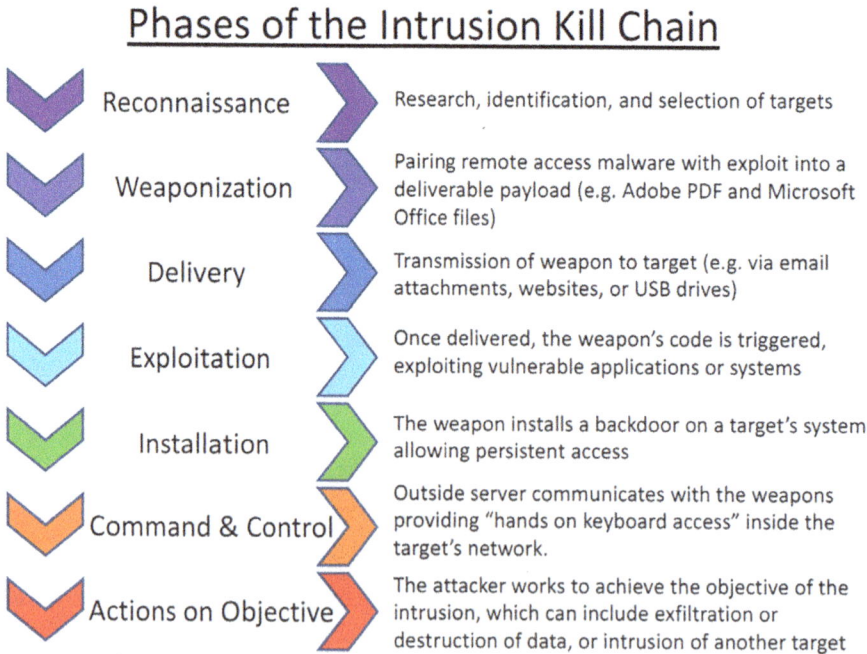

Reconnaissance	Research, identification, and selection of targets
Weaponization	Pairing remote access malware with exploit into a deliverable payload (e.g. Adobe PDF and Microsoft Office files)
Delivery	Transmission of weapon to target (e.g. via email attachments, websites, or USB drives)
Exploitation	Once delivered, the weapon's code is triggered, exploiting vulnerable applications or systems
Installation	The weapon installs a backdoor on a target's system allowing persistent access
Command & Control	Outside server communicates with the weapons providing "hands on keyboard access" inside the target's network.
Actions on Objective	The attacker works to achieve the objective of the intrusion, which can include exfiltration or destruction of data, or intrusion of another target

Figure 3-2. Phases of the cyber kill chain.

The kill chain serves as both a framework for planning and executing an OCO as well as for defending against one (DCO). Each point in the kill chain offers an opportunity to detect an operation and "kill" it, thus the name. Successful execution requires a combination of defensive design and practice (DODIN operations) and active attempts to find or "hunt" the adversary (DCO).

> *3.1 The kill chain is a common model for a cyber operation. The objective of a defender is to detect and disrupt the kill chain at the earliest phase of an operation.*

Cyberspace operations must be integrated, both across the spectrum of DODIN operations, DCO, and OCO, but more importantly, with operations in all domains (air, land, maritime, space, and cyberspace[17]). Because cyberspace intersects all of the physical domains, and the services all claim access to it for their operations, synchronization and deconfliction are essential to preventing operations from inadvertently "stepping on" each other, the equivalent of cyber fratricide. While it may not cost lives, this type of cyber fratricide would result in lost accesses to cyber targets, compromised cyber capabilities, and the loss of cyber intelligence sources. The requirement to synchronize and deconflict cyber operations also applies to the commander's broader operational plan: though it is sometimes possible to achieve the desired effects in or through cyberspace alone, more often cyber is best applied in conjunction with non-cyber actions to achieve operational objectives. The need for synchronization and deconfliction extends beyond the military domain, as other government and nongovernmental entities can easily access the same cyberspace. It is therefore essential that the modern soldier consider attack and defense in both physical and cyber space simultaneously and holistically and remain aware of the adversary's ability to exploit any friendly force failures.

A particularly complex aspect of warfighting in cyberspace is the evolving set of legal authorities, policies, and associated considerations, as will be discussed further in Chapter 9. Because of the close relationship between intelligence activities (Title 50 of the US Code) and warfighting (Title 10 of the US Code) in the cyber domain, it is important to clearly define missions and align them to the appropriate authorities. International cyber law also continues to evolve. The Tallinn Manual[18] is one of the first documents to thoroughly explore the applicability of international law to cyber.[19] As an illustration, the important question of whether a "cyber attack" by an adversary on a North Atlantic Treaty Organization (NATO) member state requires a response from the alliance under Article 5, the collective defense article of the Washington Treaty that established NATO, was debated for years, resulting in a declaration in 2014 that such aggression could be addressed under Article 5 on a "case by case basis" depending on the severity of the attack.[20] It has been noted, though, that there is significant ambiguity in determining whether an event is an attack or such an attack's severity.[21] Such policy limitations may hamper nation states operating within legal norms but may serve as an asymmetric ally to those nations and non-state actors who do not.

Finally, it is important to understand the relationship between cyberspace operations and IO. IO are activities undertaken to affect an adversary's decision-making through any number of information-related capabilities.[22] Cyberspace operations can provide a mechanism for realizing many of these and may be particularly effective due to speed, volume, and access provided by cyberspace. However, cyber is not the only means of delivering IO effects, nor is IO the only kind of operation undertaken in cyberspace, further emphasizing that cyberspace is another domain in which warfighting can and will occur, but that the fight is inherently multi-domain.

Categories of Operations

IT network operators and defenders often characterize their actions as protecting the confidentiality, integrity, and availability of their networks.[23] Whether from the point of view of an attacker or a defender, cyberspace operations can be categorized similarly using the "CIA triad":

- *Confidentiality*: Confidentiality is defined as "preserving authorized restrictions on information access and disclosure, including means for protecting personal privacy and proprietary information."[24] Protecting confidentiality means preventing unauthorized access to information, whether that information is in transit or at rest. For example, confidentiality seeks to ensure that an adversary cannot access a classified operation order or a cyber-criminal cannot access bank account credentials.

- *Integrity:* Integrity is "guarding against improper information modification or destruction, and includes ensuring information non-repudiation[25] and authenticity."[26] Protecting integrity means that unauthorized personnel will not tamper with such information. Such tampering, or an integrity attack, could include cyber-criminals using ransomware, which is a type of malware that infects a computer and restricts a user's access to the computer, to tamper with data by encrypting the data and holding it hostage until the affected user or users pay(s) the offender to release the encrypted data.[27]

- *Availability:* Availability is "ensuring timely and reliable access to and use of information."[28] Availability ensures that users can access required information resources. Common examples of attacks on availability are a denial of service (DOS) or a distributed DOS (DDoS) attack. These attacks occur when an attacker attempts to overload a network resource with requests, such as attempts to view a website.[29] Compromised devices, such as a hacked computer connected to the Internet, may be used to launch such attacks.[30]

In addition to methods categorized according to the "CIA triad," the layers of cyberspace provide another set of classifications:

- *Physical Layer:* Sophisticated technical defenses in the other layers of cyberspace (e.g., firewalls[31]) can often be bypassed simply through physical access to a network resource.[32] Physical layer attacks may include physically entering a facility and accessing the network wirelessly or from a computer terminal located inside a security perimeter.

- *Logical Layer:* Attacks at the logical layer are often thought of as "hacking." An attacker can misuse network functionality remotely, to include over the Internet, to access a network. Common techniques include phishing, such as sending emails to a target that include malicious software (malware) itself or a link to a URL hosting malware.[33,34] Direct attacks on systems seek to exploit vulnerabilities, or flaws, in the software of the system, either the operating system or applications running on it. Software developers and security researchers attempt to prevent or discover these vulnerabilities so they can be "patched," or fixed, by the software vendor before hackers can take advantage of them (or at least allow defenders to detect the signatures of attacks against the vulnerability before it is fixed). This is why it is critical to maintain software updates provided by manufacturers, as well as antivirus updates, to detect attack signatures. When hackers discover and exploit a vulnerability of which vendors are unaware, and therefore have not been able to fix, it is referred to as a "zero day" attack; the defenders had zero days to fix or even detect attacks against the vulnerability.

- *Cyber-Persona Layer*: Authenticated access, typically through a network account, to computing resources resides at this layer of cyberspace. If an attacker can compromise an account, such as learning of a username/password combination, then the attacker has access to the same resources as the authorized user. As the "cyber-persona layer consists of network or IT user accounts" and "may relate directly to an actual person or entity,"[35] it may be profitable for an attacker to focus on the people in addition to the technology. The concept of "social engineering" addresses the "targeting and manipulation of human beings rather than technology or other mechanisms."[36] An example of social engineering leading to compromise of a cyber-persona occurred in the case of a journalist from *Wired* magazine.[37] An attacker used social engineering on Apple technical support staff. In that case, the attacker claimed to the journalist and convinced the Apple technical support personnel (despite being unable to answer the security questions) to reset the journalist's password (the attacker used a billing address and the last four digits of an associated credit card as bona fides). Using the new temporary password, the attacker assumed control of one of the journalist's email accounts and expanded this control to additional cyber-personas using similar password reset methods.

While more dependent upon actions in the logical layer, credential theft provides similar functionality. Credential theft attacks use a technique in which an attacker obtains stored "credentials from a compromised computer and then uses those credentials to authenticate to other computers on the network."[38] In this way, an attacker expands from the point of original infection to additional locations within the network. Finally, it is also important to emphasize that users within the network may cause it to be compromised, whether intentionally or inadvertently as a result of social engineering or poor practice. Both of these situations are referred to as "insider threats," and protection against these threats is an important subset of cyberspace operations.[39] Indeed, "insider threats are a significant concern to the joint force.

Because insiders have a trusted relationship with access to the DODIN, the effects of their malicious or careless activity can be far more serious than those of external threat actors."[40]

It is also important to understand certain key technologies and emerging capabilities in the cyber domain:

- *Malware*: Malicious software (malware) is a term that encompasses a variety of software, to include computer viruses and spyware, enabling the theft of passwords and personal information, as well as damage to hardware.[41] One common capability of malware is that of a "keylogger," which provides the ability to capture user keystrokes as they are entered and sending those keystrokes to another Internet user, informing that malicious user of a person's usernames and passwords to various online sites and accounts.

- *Social Media:* While social media sites have become nearly ubiquitous,[42] there are certain nuances associated with military operations. Social networks in cyberspace can incite popular support and disseminate ideological information.[43] For example, the Islamic State uses Twitter to disseminate propaganda.[44] From a military professional's perspective, these services can be used for both official purposes, such as public affairs, and unofficial purposes, such as personnel morale; however, they also pose operational security risks. Social media also represents a valuable source of information about adversaries, such as identifying demographic information about Islamic State supporters.[45] A burgeoning field of social media mining[46] applies capabilities such as machine learning[47] to perform such actions as measuring the influence of individuals in a social network.[48]

- *Onion Routing:* Onion routing is a method to anonymize Internet communications by forwarding encrypted Internet traffic through numerous nodes. The Onion Router (Tor) is likely the most well-known form of onion routing.[49] See Figure 3-3 for a graphical depiction of Tor forwarding encrypted traffic through several intermediate nodes between a server and client. Tor enables users to protect human rights from oppressive regimes[50] but also provides communications for such cyber-criminal organizations as the Silk Road.[51]

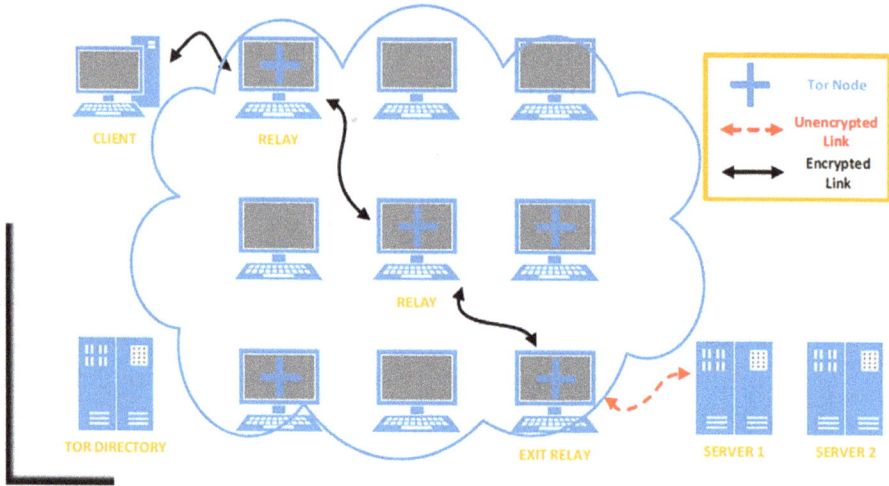

Figure 3-3. Tor operations.

- *Blockchains/Cryptocurrency:* A blockchain is a shared, distributed database built on cryptographically[52] authenticated transactions.[53] The cryptocurrency "Bitcoin" is built around a blockchain and, like other cryptocurrencies, offers the ability to conduct financial transactions with enhanced anonymity and security. Cryptocurrencies are commonly used on cyber-criminal marketplaces such as the Silk Road.[54]

PURPOSES

Actors undertake cyberspace operations for a variety of purposes, in many cases related to the organization in which the actor is a member. Such purposes include:

- *Hacktivism:* Combining "hacking" and "activism," hacktivists use cyberspace operations to support political goals, to include engaging in DDoS attacks, website defacements, and compromising and publishing confidential information.[55] Both violent and non-violent resistance movements like to use hacktivism, as the members of these movements typically view themselves as advancing or fighting for an ethical or even righteous social or political cause.

- *Crime:* Cyber criminals typically focus on monetary rewards. Groups of cyber criminals have become increasingly sophisticated and frequent marketplaces to engage in credit card fraud, identity theft, and rental of DDoS attack capability.[56] These cyber actors may also have links to nation-state actors[57] and non-state actors, such as resistance movements, which can derive much needed funding from cyber-crime.

- *Espionage:* With the growing and near universal use of computer networks to transmit and store information, both non-state and state-sponsored cyber actors have incentives to use cyberspace operations to obtaining that information by intercepting emails, stealing files from databases, or obtaining files and documents from hard drives and external drives.

- *Warfare:* Both state and non-state actors may engage in cyberspace operations during hostilities for the purpose of denying, degrading, disrupting, destroying, or manipulating adversary information or information systems, or, conversely, countering such attempts.[58] For example, cyberspace operators, presumably acting on behalf of or in support of the Russian government, shut down electrical power plants in Western Ukraine in late December 2015.[59]

> *3.2 Key purposes: hactivism, crime, espionage, and warfare. In general, the first three drive a legal response, while the last are more likely to involve a military response.*

ACTORS

It is useful to characterize cyber adversaries, or actors, to understand the necessary resources to apply to defending against them and responding to them. The key attributes to consider when characterizing cyber actors are the size of the adversary organization, which provides a measure of the resources it can apply, the capabilities they possess, and their motivations.

Because cyber technology is accessible to everyone from children to nation states, there is a corresponding range of sizes of cyber actor organizations. Typical scales include individuals (1), small cells (10s), groups (100s), and complete enterprises (1000s). The size of the organization determines to some degree the group's accomplishments based on the resources and skills applied to each stage of executing an attack. However, automation can compensate to some extent: a single *bot-herder*[60] can command tens of thousands of machines for conducting attacks.

In cyber warfare, not all adversaries are equally capable. To differentiate, there are several models, all of which characterize actors in a range of tiers. The Defense Science Board's model[61] is widely referenced and summarized in Figure 3-4. In this model, a key discriminator among the tiers is the source of the vulnerabilities in targets attacked that are exploited by the actor. The kinds of vulnerabilities available to an actor are, in turn, related to their resources, primarily time, funding, number of people, and expertise. While there is a general relationship between the size of a group with the type of threat actor described earlier, the middle tiers can be ambiguous.

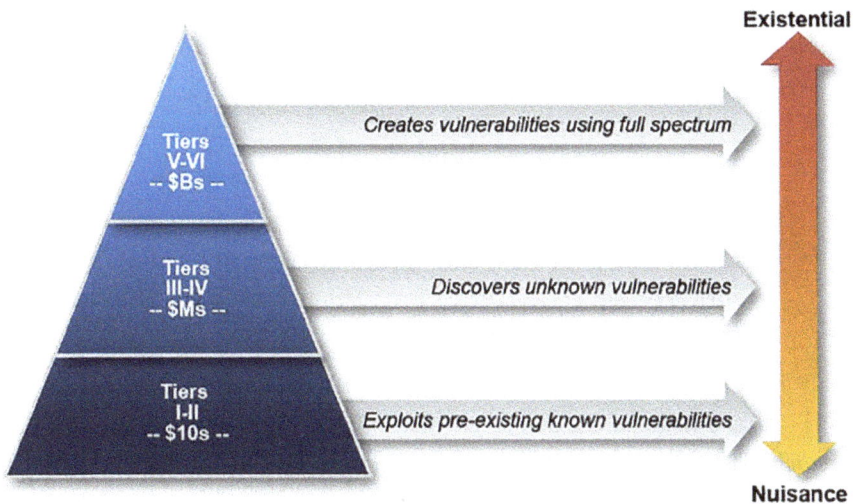

Figure 3-4. Cyber threat taxonomy.

These categorizations cannot be treated rigidly, and determining which category of attacker is responsible for an intrusion, let alone full attribution to a specific attacker, can be a difficult problem. Advanced attackers can easily mimic the techniques and behaviors of less skilled actors, using only the minimal amount of tradecraft necessary to achieve their goals. Advanced adversaries may also employ lesser skilled organizations as proxies, giving the proxies access to more advanced capabilities.

Actors can sometimes be identified by their tradecraft, or the specific tactics, techniques, and procedures they use (behavioral characterization), as well as the particular tools (malware, etc.) and vulnerabilities they exploit. Some actors also tend to target certain classes of victims. Capturing this evidence over a series of attacks allows analysts to identify an *intrusion set*, which they can also attribute to a specific actor, assuming the actor has not already taken credit for the attack. The sophistication of the actor can determine the effectiveness of identifying them this way; actors that can afford many different tools develop many different tactics and carefully test them before deploying them, which reduces the similarities in attacks that lead to association with a named intrusion set. Likewise, more sophisticated means of forensic analysis can uncover even smaller digital tracks; such means would be limited to better resourced defenders.

There are also varying motivations for why different actors undertake cyber operations, and these motivations can shape the size and structure of the resulting hacking organizations. JP 3-12 identifies four types of threat actor organizations:[62]

- Individual actors: anyone acting alone or in a small group, usually with political motivations or for bragging rights.

- Criminal organizations: financially motivated, these groups generally steal information that they can sell or use for extortion.

- Transnational groups: loose-knit organizations such as the hacking group Anonymous or various terrorist groups. They typically pursue some social or political agenda through messaging and fundraising (often illicitly) via cyber.

- Nation states: well-funded, state-sponsored actors working on behalf of a country.

3.3   Types of actors: individuals, criminal organizations, transnational groups, and nation states. The primary objective in any response is to take the actor off the field, and in this sense, it is important to recognize that at all of these levels, individuals are the ultimate actors, just as in physical space.

While a cyber operation may be undertaken by one of these kinds of groups toward a particular end, we also see instances where different groups work toward the same end, sometimes collaboratively. For example, during fighting in Ukraine, not only were state-sponsored cyber units of the two nations involved, but also transnational groups such as Anonymous and individual "hacktivists," most of whom seemed to align with Ukraine in attacking targets, including Russian state media.[63]

TARGETS

IT Networks

Most users and devices interact with the Internet through IT networks. In home and office environments, these networks consist of endpoint or host devices, such as computers—both individual desktop and shared servers—and routers and switches that interconnect them. A local area network (LAN) is connected to a wide area network through a gateway, usually a router configured to pass traffic into and out of the network. These networks are typically defended by firewalls set up to only allow certain traffic to pass into the network, and in more advanced networks (typically corporate networks), intrusion detection devices monitor network traffic for known attack signatures or patterns in the incoming data that indicate an attack. These LANs may also be configured to collect activity logs from the various networks and host devices for further examination by cyber defenders.

IT networks are the traditional targets of hackers. Any network beyond a very small one quickly becomes complex, and unless network administrators and defenders are persistent in updating all of the devices to eliminate software vulnerabilities, hackers need only find one opening to gain a foothold. Once within a network, they can pivot[64] from their first compromised system to move laterally[65] to others, with the added advantage that they are now within the perimeter defenses and look more like an authorized user.

> *3.4 IT networks have historically been the primary targets for hackers.*

Over the years, defensive tools improved, and automated software updates hindered the challenge of entering an IT network directly. However, the explosion of devices connected via local wireless networks from phones to children's toys increased the number of internal targets. Behind most of the network hacks is still the softest target—the human user—and hackers regularly attempt to exploit them through social

71

engineering attacks, such as spearphishing. Once able to compromise the user, they can use that individual's credentials on the network to move freely. Cyber awareness training now focuses on this weakness, as well as new tools to detect these kinds of attacks and the resulting compromises. However, the combination of the desire for efficiency and lack of technical awareness continue to make IT networks and their users a primary point of attack.

> *3.5 Increased network security led hackers to focus on exploiting human users via attacks, such as spearphishing.*

Internet Exchange Points and Backbone

Much of what we refer to as the Internet is represented in the interconnections among Internet service providers, which provide access to the Internet from IT networks, and content delivery networks that form the Internet's backbone. Each provider operates an autonomous system—an independent, large network—that meets at Internet exchange points (IXPs) that represent both logical and physical connections between the providers' networks. This hierarchical structure (illustrated in Figure 3-5, a snapshot of the Internet) makes routing traffic between points more efficient and provides redundancy. However, this design can also be vulnerable to attack; because so much traffic routes through the physical location of the IXP, it is a lucrative target for an adversary who wishes to significantly affect the Internet.

Figure 3-5. Internet structure.

Industrial Control Systems and Supervisory Control and Data Acquisition Systems

An increasingly important target set encompasses the many industrial control systems (ICS) and SCADA systems employed in industry. These systems automate monitoring and interaction with electrical (including nuclear), water, traffic control, and almost every other infrastructural system. Many of these "legacy" computer networks were built as a matter of convenience and efficiency and without the intention of connecting to—or even before the widespread use of—the Internet. As

73

a result, many of these legacy infrastructure computer networks contain few, if any, security controls. ICS network operators tend to focus on the industrial aspects of the systems and may not even know they are vulnerable to cyber attack. ICS/SCADA systems will be more thoroughly discussed in chapter 6.

Internet of Things

More devices in daily life contain Internet-connected computers, leading to the exponential growth of potential targets for hackers. Because these so-called Internet of Things (IoT) devices contain sensing, computing, communication, and actuation capabilities,[66] they bring new opportunities to attackers, and their increasingly ubiquitous nature provides access to places not previously accessible. IoT devices, such as printers, routers, video cameras, thermostats, refrigerators, and televisions, typically have the ability to sense the physical environment and take action (actuation) based on the sensor inputs, and adversaries seeking direct effects on the physical environment may choose to attack them, such as through an integrity attack. Additionally, much like ICS/SCADA technology, the technical security capabilities for the IoT are often substandard, and coupled with the computing and communication capabilities of the IoT, they have been used in DDoS attacks. The IoT will also be discussed in more detail in chapter 6.

OPERATIONAL CHALLENGES

As warfare in cyberspace is a new concept, it is not yet well understood, which creates challenges for the warfighter. Among the most significant challenges are attribution, policies and authorities, and "cyber thinking."

The section on actors discussed the difficulties in attributing an attack to a specific adversary. Cyberspace provides the opportunity to

attack with anonymity. In addition to technical means such as using multiple hops and anonymous routing such as the Tor network, actors may also employ others as proxies, or even mimic another actor, perhaps another adversary to the target. Attributing an attack to a sophisticated actor that does not want to be identified is extremely difficult. Further, unlike a physical attack, a cyber attack may not be discovered for months. A recent report suggests that typical time to discovery of cyber attacks ranges from months to years.[67]

Operationally, the impact is clear: if the attacker's identity is not known, or if the attack occurred some time ago, how should a nation respond? Forensic investigations initiated upon the discovery of an attack can take a long time and often do not result in a high-confidence conclusion. Attribution is discussed further in chapter 8.

> 3.6 *Attribution of a cyber attack to an attacker is one of the most significant challenges in effectively responding.*

A second significant challenge is the constantly evolving legal and policy regime. Unclear policies result in indecision as to whether an attack should be mitigated legally or militarily. In either case, there are also questions of escalation and proportionality, as mentioned earlier with respect to the Tallinn Manual assertion of a legal right to respond with physical force during a time of war. Without clear statements of policy, countries cannot be certain which level of action will lead to certain kinds of response. Adversaries may intentionally leverage this ambiguity through the use of proxies and deception.

Another policy challenge that hampers both response and action concerns equities. As previously discussed, there is a close relationship between intelligence activities in cyberspace and operational (combat) activities. In some cases, the ability to more specifically attribute an attack to an adversary is based on intelligence, and acting on that intelligence may expose those sources and methods. Similarly, cyber capabilities are single-shot weapons; once used and observed by the adversary, they can often be easily countered, and they may also reveal the sophistication of US cyber capabilities. This combination may make responding to certain attacks in cyberspace a much less attractive option than other means available among the instruments of national power, for example, legal measures to indict five Chinese military personnel for conducting cyber espionage.[68] The legal environment around cyber is discussed further in chapter 9.

Finally, one of the biggest hurdles yet to overcome is the newness of cyber as a domain of military operations. We do not refer to "cyber warfare" here, as this is exactly the problem at hand[69]—conflicts are not fought exclusively in cyberspace, so there is no such concept as "cyber war." Many warfighters, especially those in senior officer and non-commissioned officer leadership roles, did not grow up with cyber capabilities. The technical nature of the cyber domain can be intimidating, but perhaps more importantly, there is not yet a battle-tested doctrine of warfare that includes the cyber domain.

> *3.7   Legal, policy, and doctrine for cyber are all immature, leading to challenges in fighting and defending in cyberspace.*

This lack of integrated thinking about cyber is difficult to correct. Training, and more importantly, indoctrination, are essential parts of any potential solution. However, because we lack significant experience in conducting a war that includes the cyber domain, we question whether current training and the nascent doctrine in development are effective. One historical trend suggests that the introduction of new technologies (e.g., mechanized warfare) and domains (e.g., aerial warfare) substantially changed warfighting; cyber and cyberspace represent *both* new technologies and a new domain.

KEY TAKEAWAYS

> 3.1   The kill chain is a common model for a cyber operation. The objective of a defender is to detect and disrupt the kill chain at the earliest phase of an operation.

3.2 Key purposes: hactivism, crime, espionage, and warfare. In general, the first three drive a legal response, while the last are more likely to involve a military response.

3.3 Types of actors: individuals, criminal organizations, transnational groups, and nation states. The primary objective in any response is to take the actor off the field, and in this sense, it is important to recognize that at all of these levels, individuals are the ultimate actors, just as in physical space.

3.4 IT networks have historically been the primary targets for hackers.

3.5 Increased network security led hackers to focus on exploiting human users via attacks, such as spearphishing.

3.6 Attribution of a cyber attack to an attacker is one of the most significant challenges in effectively responding.

3.7 Legal, policy, and doctrine for cyber are all immature, leading to challenges in fighting and defending in cyberspace.

ENDNOTES

1 Both definitions are taken from P. W. Singer and Allan Friedman, *Cybersecurity and Cyberwar: What Everyone Needs to Know* (New York, NY: Oxford University Press, 2014), 12–13. The latter definition is used by the authors themselves (p. 13).

2 See US Joint Chiefs of Staff, "Cyberspace Operations," Joint Publication 3-12 (JP 3-12), June 8, 2018, GL-4.

3 As defined by the Merriam-Webster dictionary.

4 Barry M. Leiner et al., "Brief History of the Internet," *Internet Society*, 1997, 2.

5 US Joint Chiefs of Staff, "Information Operations," Joint Publication 3-13 (JP 3-13), November 20, 2014, ix–x.

6 US Joint Chiefs of Staff, JP 3-12, I-7.

7 Ibid., I-2–I-4.

8 Ibid., I-4.

9 Information-related capabilities are the tools, techniques, or activities that affect any of the three dimensions of the information environment. See JP 3-13, "Information Operations."

10 Both JP 3-12, "Cyberspace Operations," and JP 3-13, "Information Operations" contain information on the relationship between cyberspace operations and information operations. See JP 3-12, I-7, and JP 3-13, II-9.

11 US Joint Chiefs of Staff, JP 3-12, I-7.

12 Headquarters, Department of the Army, "Cyberspace and Electronic Warfare Operations," Field Manual 3-12, April 11, 2017.

13 US Air Force Doctrine, "Annex 3-12, Cyberspace Operations," to Joint Publication 3-12, "Cyberspace Operations," November 30, 2011, https://doctrine.af.mil/DTM/dtmcyberspaceops.htm.

14 Navy Doctrine Library System, "Cyberspace Operations," Naval Warfare Publication (NWP) 3-12, July 2011.

15 United States Marine Corps, "Marine Corps Cyberspace Operations," MCIP 3-40.02, October 6, 2014.

16 Eric M. Hutchins, Michael J. Cloppert, and Rohan M. Amin, "Intelligence-Driven Computer Network Defense Informed by Analysis of Adversary Campaigns and Intrusion Kill Chains," Paper presented at the 6th Annual International Conference on Information Warfare and Security, Washington, DC, 2011.

17 Per JP3-13, cyberspace is a domain within the information environment, and one of five domains in total.

18 Michael N. Schmitt, ed., *Tallinn Manual on the International Law Applicable to Cyber Warfare* (New York, NY: Cambridge University Press, 2013).

19 The Tallinn Manual, and its successor, Tallinn Manual 2.0, was created by a group of cyber legal and policy experts between 2009 and 2012 (Tallinn 2.0 was released in 2017). While the work was sponsored by the NATO Cooperative Cyber Defence Centre of Excellence, it is not considered a NATO manual. Among the more controversial legal findings was that nations could legally respond with physical force to cyber attacks under certain conditions, on the principle that hackers are participants in a wartime effort.

20 Stephen Jackson, "NATO Article 5 and Cyber Warfare: NATO's Ambiguous and Outdated Procedure for Determining When Cyber Aggression Qualifies as an Armed

Attack," Center for Infrastructure Protection and Homeland Security, George Mason University, last modified August 18, 2016, http://cip.gmu.edu/2016/08/16/nato-article-5-cyber-warfare-natos-ambiguous-outdated-procedure-determining-cyber-aggression-qualifies-armed-attack/.

21   North Atlantic Treaty Organization, "Wales Summit Declaration," Press Release (2014) 120, September 5, 2014, last modified September 26, 2016, http://www.nato.int/cps/en/natohq/official_texts_112964.htm.

22   Ashton B. Carter, "Information Operations (IO)," Department of Defense Directive 3600.01, Under Secretary of Defense (Policy), May 2, 2013.

23   Joint Task Force Transformation Initiative, "Guide for Applying the Risk Management Framework to Federal Information Systems: A Security Life Cycle Approach," National Institute of Standards and Technology (NIST) Special Publication 800-37 Revision 1, February 2010, updates as of June 2014, 1.

24   National Institute of Standards and Technology, "Standards for Security Categorization of Federal Information and Information Systems," Federal Information Processing Standard Publication 199 (FIPS 199), February 2004, 7.

25   Non-repudiation describes means to prevent an entity from conducting an action and subsequently denying responsibility for that action. A signature on a document is one way to provide non-repudiation.

26   National Institute of Standards and Technology, FIPS 199, 7.

27   US Department of Homeland Security, Cybersecurity and Infrastructure Security Agency (CISA) Cyber + Infrastructure, "Crypto Ransomware," Alert (TA14-295A), October 22, 2014, last revised September 30, 2016.

28   National Institute of Standards and Technology, FIPS 199, 7.

29   US Department of Homeland Security, Cybersecurity and Infrastructure Security Agency (CISA) Cyber + Infrastructure, "Understanding Denial-of-Service Attacks," Security Tip (ST04-015), November 4, 2009, last revised June 28, 2018.

30   US Department of Homeland Security, Cybersecurity and Infrastructure Security Agency (CISA) Cyber + Infrastructure, "Heightened DDoS Threat Posed by Mirai and Other Botnets," Alert (TA16-288A), October 14, 2016, last revised October 17, 2017.

31   A firewall is a device designed to filter network traffic and typically serves as a boundary between different networks (e.g., between a friendly network and the Internet itself). See Sean-Philip Oriyano, "Evasion," Chapter 17 in *CEH Certified Ethical Hacker Study Guide*, Version 9 (Indianapolis, IN: John Wiley & Sons, 2016).

32   Sean-Philip Oriyano, "Physical Security," Chapter 19 in *CEH Certified Ethical Hacker Study Guide*, Version 9 (Indianapolis, IN: John Wiley & Sons, 2016).

33   Phishing is sending a large number of emails to an arbitrary group of users, hoping that some small percentage of them will click a malicious link. Since these are general, they are relatively easy to block. Spearphishing is sending a carefully crafted email to a specific user, employing knowledge that user will find familiar, to trick the user into trusting the email. Whaling is a spearphishing attack against a high-level target, such as a senior political, military or business personality.

34   US Department of Homeland Security, Cybersecurity and Infrastructure Security Agency (CISA) Cyber + Infrastructure, "Avoiding Social Engineering and Phishing Attacks," Security Tip (ST04-014), October 22, 2009, last revised August 22, 2019.

35   US Joint Chiefs of Staff, JP 3-12, I-4.

36 Sean-Philip Oriyano, "Social Engineering," Chapter 10 in *CEH Certified Ethical Hacker Study Guide*, Version 9 (Indianapolis, IN: John Wiley & Sons, 2016).

37 Mat Honan, "How Apple and Amazon Security Flaws Led to My Epic Hacking," *Wired*, August 06, 2012, https://www.wired.com/2012/08/apple-amazon-mat-honan-hacking/.

38 Microsoft Corporation, "Mitigating Pass-the-Hash and Other Credential Theft, Version 2," *Microsoft Trustworthy Computing*, July 7, 2014, https://technet.microsoft.com/en-us/dn785092.aspx.

39 For more information, see Carnegie Mellon University, Software Engineering Institute, "Common Sense Guide to Mitigating Insider Threats, 5th Edition," Technical Note CMU/SEI 2015-TR-010, December 2016.

40 US Joint Chiefs of Staff, JP 3-12, IV-20.

41 Sean-Philip Oriyano, "Malware," Chapter 8 in *CEH Certified Ethical Hacker Study Guide*, Version 9 (Indianapolis, IN: John Wiley & Sons, 2016).

42 Facebook, for example, has 1.2 billion users who log on every day. See Mathew Ingram, "Facebook's Growth Appears to Be Unstoppable, at Least for Now," *Fortune*, February 1, 2017, fortune.com/2017/02/01/facebook-growth.

43 US Joint Chiefs of Staff, JP 3-12, I-6.

44 J. M. Berger and Jonathon Morgan, "The ISIS Twitter Census: Defining and Describing the Population of ISIS Supporters on Twitter," The Brookings Project on US Relations with the Islamic World, Analysis Paper, no. 20, March 5, 2015, https://www.brookings.edu/wp-content/uploads/2016/06/isis_twitter_census_berger_morgan.pdf, 2.

45 Ibid.

46 Social media mining is the process of representing, analyzing, and extracting actionable patterns from social media data. See Reza Zafarani, Mohammad Ali Abbasi, and Huan Liu, *Social Media Mining: An Introduction* (New York: Cambridge University Press, Draft Version April 20, 2014).

47 Machine learning is the science of getting computers to act without being explicitly programmed. Refer to Coursera, "Machine Learning," course offered by Stanford University, https://www.coursera.org/learn/machine-learning (accessed August 1, 2017).

48 Zafarani et al., *Social Media Mining*, Chapter 1.3.

49 Roger Dingledine, Nick Mathewson, and Paul Syverson, "Tor: The Second-Generation Onion Router," Paper presented at the 13th USENIX Security Symposium, San Diego, CA, 2004, 1.

50 Emin Caliskan, Tomas Minarik, and Anna-Maria Osula, "Technical and Legal Overview of the Tor Anonymity Network," Tallinn, Estonia: NATO Cooperative Cyber Defence Center of Excellence, 2015), 24.

51 Caliskan et al., "Tor Anonymity Network," 3.

52 Merriam-Webster dictionary defines "cryptography" as "the enciphering and deciphering of messages in secret code or cipher; also: the computerized encoding and decoding of information." In this case, the cryptography enforces integrity and non-repudiation, preventing the transaction from later denial or copy.

53 Neil B. Barnas, "Blockchains in National Defense: Trustworthy Systems in a Trustless World," Air University Academic Research Report, June 2016, http://www.dtic.mil/doctrine/education/jpme_papers/barnas_n.pdf, 19.

54 Lillian Ablon, Martin C. Libicki, and Andrea A. Golay, *Markets for Cybercrime Tools and Stolen Data: Hacker's Bazaar* (Santa Monica, CA: RAND Corporation, 2014), 11–12.

55 Francois Paget, "Hacktivism: Cyberspace Has Become the New Medium for Political Voices," McAfee Labs White Paper, 2012, https://www.mcafee.com/us/resources/white-papers/wp-hacktivism.pdf, 3–4.

56 Ablon et al., *Markets for Cybercrime Tools and Stolen Data*, ix.

57 Jen Weedon, "Beyond 'Cyber War': Russia's Use of Strategic Cyber Espionage and Information Operations in Ukraine," Chapter 8 in *Cyber War in Perspective: Russian Aggression Against Ukraine*, ed. Kenneth Geers, Tallinn, Estonia: NATO Cooperative Cyber Defence Center of Excellence, 2015, 73.

58 US Joint Chiefs of Staff, JP 3-12, II-7.

59 Kim Zetter, "Inside the Cunning, Unprecedented Attack of Ukraine's Power Grid," *Wired*, March 3, 2016, https://www.wired.com/2016/03/inside-cunning-unprecedented-hack-ukraines-power-grid/.

60 A bot-herder is a cyber hacker who creates a virtual army of "bots" – computers that he has hacked and taken control of – and then manages ("herds") them to perform actions on his command, such as sending large quantities of junk ("spam") email, or participating in distributed denial of service (DDoS) attacks. Because these "bots" are usual widely distributed personal computers, blocking them is very difficult for defenders.

61 Department of Defense, Defense Science Board, "Task Force Report: Resilient Military Systems and the Advanced Cyber Threat," Office of the Under Secretary of Defense for Acquisitions, Technology and Logistics, January 2013.

62 US Joint Chiefs of Staff, JP 3-12, I-11.

63 Pierluigi Paganini, "Crimea – The Russian Cyber Strategy to Hit Ukraine," Infosec Institute, March 11, 2014, http://resources.infosecinstitute.com/crimea-russian-cyber-strategy-hit-ukraine/.

64 Pivot: use a now compromised computer as a launching point for subsequent attacks.

65 Lateral movement: move within a network from one machine to another, as opposed to movement originating from outside the network, which is generally more visible to defenders who are watching traffic crossing the boundary, but not within the network.

66 Jeffrey Voas, "Networks of 'Things,'" National Institute of Standards and Technology (NIST) Special Publication 800-183, July 2016, ii.

67 Verizon Enterprise, "2017 Data Breach Investigations Report," 10th Edition, 2017, http://www.verizonenterprise.com/verizon-insights-lab/dbir/2017/#report.

68 US Department of Justice, Office of Public Affairs, "U.S. Charges Five Chinese Military Hackers for Cyber Espionage Against U.S. Corporations and a Labor Organization for Commercial Advantage: First Time Criminal Charges Are Filed Against Known State Actors for Hacking," *Justice News*, May 19, 2014, https://www.justice.gov/opa/pr/us-charges-five-chinese-military-hackers-cyber-espionage-against-us-corporations-and-labor.

69 Brett T. Williams, "The Joint Force Commander's Guide to Cyberspace Operations," *Joint Forces Quarterly* 73, April 2014. 12–19.

CHAPTER 4.
RESISTANCE SECURITY IN THE CYBER DOMAIN

INTRODUCTION

The first criterion for success for both the resistance movement and the force authority is survival. For example, resistance movements need to operate under a cellular organizational structure so that the whole movement is not incapacitated due to a single compromise. Precluding infiltration of the organization is critical to both the resistance movement and the force authority.

In the past, precluding infiltration relied primarily on keeping opposing human agents from penetrating the organization. Today, however, the heavy reliance on cyber capabilities leads to new and multi-dimensioned threats of infiltration. Moreover, cyber infiltration can cause a loyal member to become an unwitting agent of the opposing side if cyber capabilities are compromised. Therefore, maintaining the security of their cyber assets and communications needs to be a top priority for both the resistance movement and the force authority.

This chapter describes ten key takeaways of cybersecurity useful to Army special operations forces. While there are many more aspects to cybersecurity than described here, these are high-level principles that, if not followed, could lead to the demise of either the resistance movement or the force authority.

Seven of the ten principles of cybersecurity apply equally to the resistance movement and the force authority. The remaining three principles, highlighted throughout the chapter, apply differently to each component.

THE IMPORTANCE OF CYBERSPACE AND CYBERSECURITY TO THE RESISTANCE MOVEMENT AND STATE SECURITY SERVICES

Cyber capabilities, and in particular social media, can be leveraged by resistance movements to pursue their tactical and strategic ends. While the cyber domain provides many new opportunities to resistance movement leaders, these new cyber mechanisms also involve inherent security risks. For example, resistance movements have always needed to avoid being compromised. Now, however, the wide array of cyber channels of communications allows for many different methods of infiltration, from chat sessions to air-gap[1] jumping malware that can reach networks isolated from the Internet. As described in *ARIS Undergrounds in Insurgent, Revolutionary and Resistance Warfare*, resistance movements need to grow and expand, but expansion risks increased opportunities of compromise.[2]

The importance of cybersecurity to the resistance movement can be described on a spectrum based on the consequences of cyber compromise. For example, in extremely repressive environments, a cyber compromise can lead to physical torture and death to individuals and eventually the collapse of the resistance movement. Conversely, in less repressive environments, cyber compromise can lead to embarrassment for the resistance movement or the state security service, thereby eroding its political influence.

Figure 4-1 presents a visual representation of this spectrum of consequences of cyber compromise depending upon the degree of repression in the environment. An example of an extreme consequence is the identification of Syrian resistance fighters against the Assad government via compromised chat sessions that likely led to their deaths, as described in more detail later. In a similar manner, force authorities facing violent resistance may also lose lives through poor cybersecurity. An example of an embarrassing consequence is the hacking of the Democratic National Convention emails in 2016 just before the presidential election. Many commentators blame the democratic candidate's loss of the election to the release of the hacked emails.[3]

> 4.1 *The magnitude of the consequences of cyber compromise determines the degree of importance that should be assigned to cybersecurity of the resistance movement or the state security services.*

US military personnel advising either a resistance movement or a force authority needs to be able to help assess the level of consequence that can result from a cyber compromise and help prioritize resources toward increasing cybersecurity.

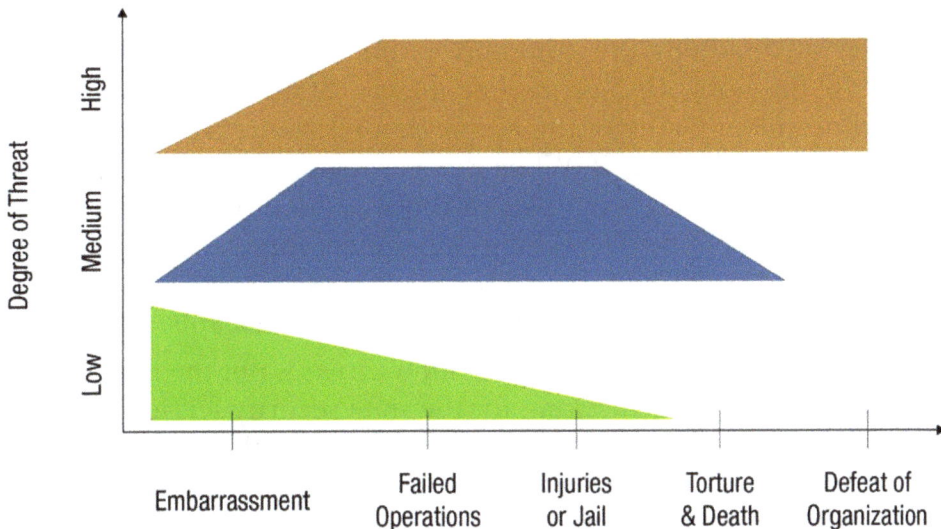

Figure 4-1. The spectrum of consequences for cyber compromise.

Most of the examples in this chapter address situations where the consequences of cyber compromise are extreme and usually lethal. Even if a resistance movement avowed nonviolent means to achieve change, repressive governments often use violence or the threat of violence as the primary mechanism to suppress the movement. Depending on the level of openness in a regime, even nonviolent movements might be illegal and considered dangerous to a state's authority.[4] Conversely, resistance movements such as al Qaeda, a group whose primary tool was violence against almost everyone in Iraq in the mid-2000s, was also a target of cyber compromise by the state security services. As a result, cybersecurity is essential to protecting resistance movements, whether they are violent or nonviolent, and of greater importance when the consequences of cyber compromise are lethal.

Cyberspace as a New Vector of Compromise

Resistance movements in lethally repressive environments have always needed to protect the identities of their members. As described in *ARIS Human Factors Considerations of Undergrounds in Insurgencies*, dividing the resistance movement into *cells* (whether sequential or parallel) can help protect the resistance movement from complete collapse by localizing the effects of the compromise.[5]

A resistance movement's use of enforcement squads and other enforcement methods are common within violent resistance movements. In the book *Of Spies and Stratagems*, Stanley Lovell describes such enforcement tactics of the Norwegian resistance against the Nazi occupation in World War II. When the Norwegian resistance identified informants, they made examples of those so identified. Rather than killing the informants, the resistance members surgically removed the tongue and released the former informant back into society. This had an extremely chilling effect on other would-be informants.[6]

In the cyber realm, the most common threat is not the lone insider informant intentionally working for the state security services (though that is still a threat[7]). The most common threat is the compromise of cyber capabilities used by resistance movements by loyal resistance members. In a similar manner, the primary threat to the state security services is also cyber compromise of its loyal members through state-owned cyber assets. In both cases, compromise of cyber capabilities can either lead directly to physical compromise or in turning an otherwise loyal member into a witting or unwitting insider for the other side.

When opponents compromise cyber assets of loyal members of the resistance movement or the state security services, this often creates a much larger source of information leakage than human informants achieved through traditional infiltration of the organization.

> 4.2   *The largest source of information leakage from resistance movements or state security services is likely via loyal members whose cyber assets are compromised, rather than the relatively small number of disloyal human informants.*

Infiltration may now be accomplished by compromising a non-human cyber asset, such as computers or cell phones via malicious code or a bot, rather than compromising a human. This reality of cyber operations gives the perpetrator of a cyber compromise a much larger space in which to operate to find vulnerabilities in the target cyber assets. When the perpetrator compromises an opposing side's asset, the perpetrator is usually confident that the compromise is real and that the information obtained via that compromise is valid. This makes the information obtained via a cyber compromise usually more reliable than information from an informant who may be a double agent.

A cyber compromise may be against the cyber capabilities used for internal communications, external communications, or both. Internal communications within a resistance movement, for example, may be facilitated by emails to members or by removable media (such as thumb drives and memory sticks). If any one member in the cell has their email compromised, then the whole cell is at risk. If any one member transfers malware by removable media, then everyone who uses that media will also be compromised. In that case, the US military in Afghanistan was infected by malware that was carried to many cyber assets via removable media.[8]

External communications for a resistance movement include websites or chat rooms used, for example, to drum up international support and generate recruits. However, if the cyber assets are compromised when using chat or websites, the resistance movement members can also be exposed. In a similar manner, state security services may have their websites compromised or other cyber assets infected via chat room communications. Cameras and microphones installed on personal computers can be activated surreptitiously by an adversary, allowing the adversary to identify not only the person but also the location and identities of others in the view of the camera or the range of the microphone. Emails sent across open communication channels can also be intercepted and read by an adversary.

US advisors of either a resistance movement or a state security force need to help identify which types of cyber assets could be leveraged for the greatest effect against them, as well as advise on methods to better secure these cyber assets.

Why the Resistance Use of Cyberspace is Different

Cyber capabilities provide resistance movements with great reach in their external and international communications and a wide range of internal communications options. For example, state security service oversight of the landline phone system, as in Russia, no longer sufficiently controls internal communications. The plethora of Internet-based communications methods combined with cross-media connections[9] to the Internet greatly expanded the range of options available to the resistance movement.

Meanwhile, every cyber capability is vulnerable to eventual compromise. Just as the resistance movement must contend with an inherent tradeoff between expansion of the movement versus the risk of compromise, there is an inherent tradeoff between the use of a wide range of cyber capabilities versus the security of these capabilities. While it is very difficult for the state security service to effectively monitor every cyber communications capability, it is extremely difficult for the resistance movement to adequately secure every cyber communications capability. Achieving cybersecurity on each type of cyber capability requires specialized software and skills that are often not widely available to a resistance movement.

> *4.3   There is an inherent tradeoff between the benefits of the large internal and external reach and variety of cyber capabilities versus the <u>need to secure every type</u> of cyber capability used.*

In general, the larger the number and variety of cyber assets, the greater the number of opportunities for compromise. Traditionally, the larger the variety of resources, the greater the chance of survival of the organization. However, in cyberspace, the more types of cyber assets used, the more types of vulnerabilities and the greater need for resources arise to successfully secure cyber assets.

Since the Internet was made available to the public, both individuals and societies as a whole became significantly dependent on its capabilities. These capabilities extend beyond the original PC-based access of the Internet and now include a wide range of communications mechanisms that integrate their functions with and through the Internet. The Internet can connect with smartphones and tablets for browsing, texting, tweeting, and a variety of other communications. The cross-media flow between the Internet and cellular phones became ubiquitous not only in the developed world but also in the developing world. Places with no landline infrastructure are now connected by cellular and, in some cases, satellite communications that can reach the Internet and therefore the rest of the world.[10]

Resistance movements adopted and exploited these new capabilities before the force authorities in many repressive nations were aware of these capabilities. The success of the Arab Spring was dependent not only on social media to recruit, organize, and message but also on the regimes' lack of monitoring of social media and its use as an organizing avenue for resistance movements.[11,12]

Social media was instrumental in organizing and mobilizing the massive protests that occurred in Tunisia, Egypt, and many others. While the repressive regimes monitored and controlled the traditional media outlets, they were not watching social media. However, other nongovernment sources watched social media. The information disseminated provided a ground truth that could be spread quickly across the world. "Mostly what we got was people on the ground -- participants, dissidents -- because the Egyptian government was clueless."[13] After many of the resistance activities were successful, the regimes' state security services learned to use, monitor, and manipulate social media, thereby precluding similar mass mobilization activities in their countries.

State security services educated themselves about social media and other cyber capabilities and have taken steps to make sure another Arab Spring type of event does not happen. In addition to monitoring social media and other Internet communications, many state security services use much more active measures to identify members of resistance movements (such as through implanting malware) and to counter messages using their own bloggers, as will be described further in chapter 5.[14,15]

For example, many resistance movements use chat rooms to recruit members, organize themselves, and plan events. The state security services will then enter the chat rooms pretending to be someone supportive of the resistance movement when they are in fact trying to weaken or destroy it, as described in the next section.

Examples of Resistance Movement Compromise via Cyberspace

FireEye, a cybersecurity company, published a report, titled "Behind the Syrian Conflict's Digital Front Lines," that describes fake female avatars fooled Syrian resistance members on Skype into believing they were supportive of the resistance. Using simply a common name and image, the avatar "would develop a rapport with the victim before sending a malicious file."[16] Asking whether the resistance member was on a computer or a phone helped to determine the type of malware sent. The avatar then requested a photo of the resistance member and sent a photo in return. However, the return photo included malware, which then yielded complete control of the resistance member's computer or cellphone to an organization supporting the state security service.[17]

Honey traps—where a person of the opposite sex,[18] almost always a "young and attractive" female persona, infiltrates an adversary—is as old as written history. Sayings from the time of Sun Tzu and later include the Chinese maxim to "use beauty to ensnare a man."[19] Getting the target sexually and even romantically involved traditionally leads to blackmail opportunities and turning a previously loyal member into working for the opposing side. While an old ploy, it still works.

What makes this type of threat even more effective than ever, however, is the fact that the adversary does not even need to find an attractive member of the opposite sex to achieve the same results. As seen in Figure 4-2, a fake avatar on the Internet can be created by anyone of any gender, while the target is often unaware of this capability.

To further spread the malware among the resistance, the fake avatar previously described hosted a Facebook profile with photos of the same person to deepen the fabricated background of the avatar. While the profile comprised pro-opposition content, many posts contained malicious links which, when clicked, delivered malware to the computer or network of whomever clicked on the link. To continue with the ruse, the links with malicious content invited visitors to install cybersecurity tools such as a virtual private network (VPN)[20] and Tor,[21] giving the impression that they added to their cybersecurity when in fact they substantially decreased it.[22]

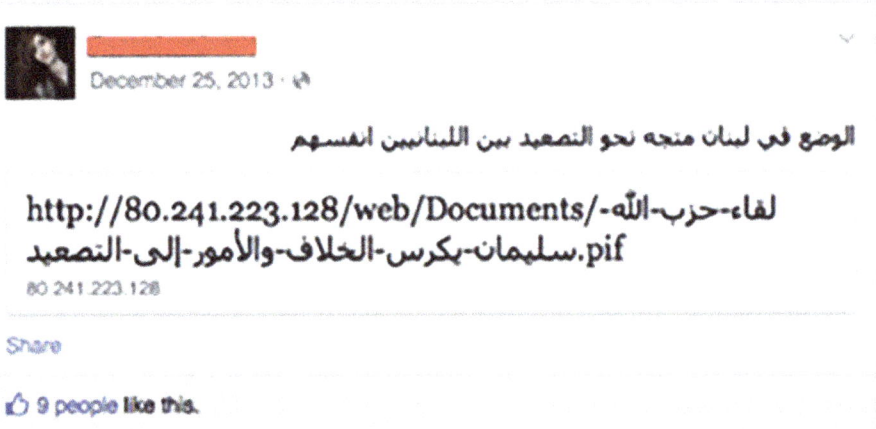

Courtesy of FireEye.

Figure 4-2. Sample female avatar posting malware links on Facebook.

In addition to fake Facebook pages purportedly supportive of the Syrian resistance, an organization supportive of the state security service set up a fake website pretending to align with the resistance to target, infect, and identify resistance members seeking news about the conflict. While the news content was real—obtained from legitimate sources advocating democracy in Syria—the website actually inserted malware in the form of a Flash Player[23] upgrade necessary to view the content. Malware embedded in video chat software was also made available on the website.[24]

Extending the honey trap approach to include online matchmaking, the phony opposition website included a number of women's fake profiles and a "LiveCamID." Clicking on the LiveCamID led to a page where an infected version of otherwise legitimate live video software was available for download. Moreover, the fake profile pages were fronted by a fake Facebook login page that collected credentials of the resistance members who logged in.[25]

The state security services use of the fake avatars, profiles, and websites "was likely able to acquire large collections of data by breaching only a relatively small number of systems due to the opposition's use of shared computers for satellite-based Internet access."[26] Many resistance movements, like the Syrian resistance, tend to be fairly resource constrained. Therefore, resource sharing of cyber capabilities is a common vector of cyber compromise. A state security service that can successfully compromise one "shared device can easily steal the Skype databases and stored documents of several targeted individuals or organizations as well."[27]

In a similar manner, the state security service must also train their personnel to beware of similar ploys aimed at them. One example includes the 2017 Hamas penetration of the Israeli Defense Force (IDF) networks. This penetration was accomplished through social media requests to IDF soldiers with the requesting profiles displaying pictures of attractive females, similar to the example against the Syrian Resistance. In this case, the fake profiles contained avatars created by Hamas. After accepting the requests, the avatars exchanged dialogue with the IDF soldiers (including the use of Hebrew slang) and extracted military operational details from them, as well as requesting to download a file.[28]

These downloaded files allowed malicious code within the application to penetrate the terminals used by the IDF soldiers and the hosting network.[29] In this case, some of the affected IDF soldiers recognized they were exploited, which enabled IDF network security technicians to discover the embedded malicious code and attribute it to Hamas.[30]

As with honey traps used against the Syrian resistance, the IDF soldiers were sent images of attractive females who were not even supportive of Hamas. Their images were stolen from various websites and used without their owners' permission or knowledge.[31] It is much easier to set a honey trap in cyberspace where the perpetrator *pretends* to be a young and attractive female than it is to find a physical human willing to be a live honey trap.

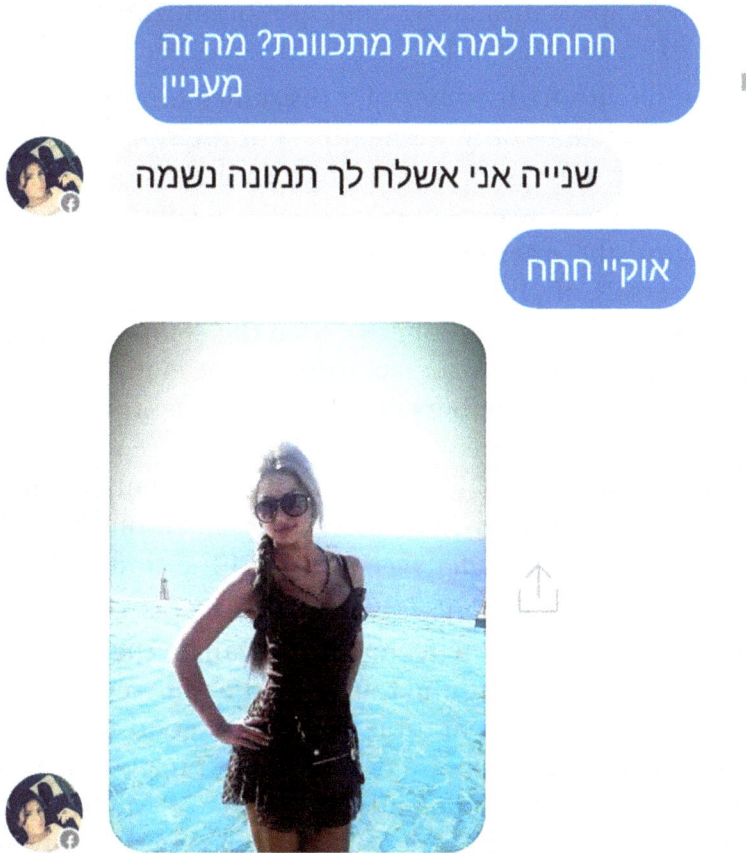

חחחח למה את מתכוונת? מה זה מעניין

שנייה אני אשלח לך תמונה נשמה

אוקיי חחח

Figure 4-3. Sample Hamas female avatar to trick IDF soldiers.

Examples of Resistance Mitigation Options to Avoid Compromise via Cyberspace

The preceding examples of a resistance movement's shared cyber assets via cyber compromise leads to the need for a separation of cyber capabilities to avoid compromising multiple cells.

> *4.4   Separation of cyber capabilities should follow the sequential or parallel cellular organization of the resistance movement to avoid compromising multiple cells via single shared cyber assets.*

Just as cellular organization helps the resistance movement survive if one cell becomes compromised, cellular separation of cyber capabilities avoids a cyber event that compromises the whole organization.

95

Sharing cyber capabilities across multiple organizational cells should be avoided to ensure that only one compromised cell is lost at a time.

For example, if one member of a resistance movement is arrested, then only the identities of those members of the resistance in that specific cell are at risk of compromise. In a similar manner, if the cyber assets of a resistance movement of one cell are compromised, only the individuals and the operations associated with that cell are compromised. Conversely, if the cyber assets of a resistance movement are shared across a large number of cells, then the compromise of one central cyber asset can lead to the compromise of many members across many cells. Even if the members do not know each other, if they use the same compromised cyber asset, it is as though there was an informant that knew the identities of all of the various cell members. In the previous example of the Syrian resistance, its reliance on shared computers for satellite-based Internet access allowed the compromise of a few assets to likely expose large numbers of resistance members.[32]

While resistance movements need to organize their cyber assets in cell structures for their continued survival, the state security services do not. Even though some separation of access is good for security, state security service segmentation for cybersecurity purposes can be accomplished at a much larger scale and still be secure. Major state security service organizations can each have their own cybersecurity assets and the ability to monitor them and enforce compliance from central locations. This capability provides state security services the benefits of economy of scale in cybersecurity practices that are not available to the resistance movement. Resistance movements will need to perform cybersecurity practices and enforcement at the cell level, which is much more difficult to accomplish than state security services with centralized cybersecurity monitoring.

> 4.5 Cybersecurity education and training for its members are essential to the survival of the resistance movement, as well as to the security of the state security service.

Just as a resistance movement should train its members in the traditional physical and operational security procedures, it should also train its members in the basics of *cybersecurity*. While it may seem obvious to some audiences that fake identities, websites, and software are common threats, many people remain unaware of them in this context. As

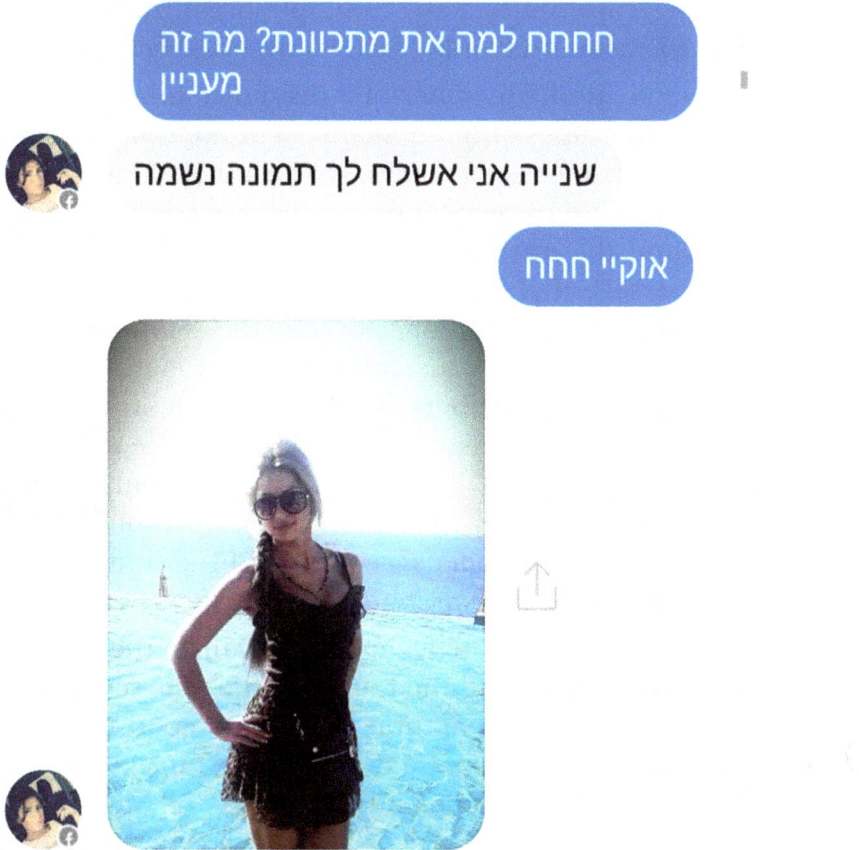

Figure 4-3. Sample Hamas female avatar to trick IDF soldiers.

Examples of Resistance Mitigation Options to Avoid Compromise via Cyberspace

The preceding examples of a resistance movement's shared cyber assets via cyber compromise leads to the need for a separation of cyber capabilities to avoid compromising multiple cells.

> 4.4 *Separation of cyber capabilities should follow the sequential or parallel cellular organization of the resistance movement to avoid compromising multiple cells via single shared cyber assets.*

Just as cellular organization helps the resistance movement survive if one cell becomes compromised, cellular separation of cyber capabilities avoids a cyber event that compromises the whole organization.

Sharing cyber capabilities across multiple organizational cells should be avoided to ensure that only one compromised cell is lost at a time.

For example, if one member of a resistance movement is arrested, then only the identities of those members of the resistance in that specific cell are at risk of compromise. In a similar manner, if the cyber assets of a resistance movement of one cell are compromised, only the individuals and the operations associated with that cell are compromised. Conversely, if the cyber assets of a resistance movement are shared across a large number of cells, then the compromise of one central cyber asset can lead to the compromise of many members across many cells. Even if the members do not know each other, if they use the same compromised cyber asset, it is as though there was an informant that knew the identities of all of the various cell members. In the previous example of the Syrian resistance, its reliance on shared computers for satellite-based Internet access allowed the compromise of a few assets to likely expose large numbers of resistance members.[32]

While resistance movements need to organize their cyber assets in cell structures for their continued survival, the state security services do not. Even though some separation of access is good for security, state security service segmentation for cybersecurity purposes can be accomplished at a much larger scale and still be secure. Major state security service organizations can each have their own cybersecurity assets and the ability to monitor them and enforce compliance from central locations. This capability provides state security services the benefits of economy of scale in cybersecurity practices that are not available to the resistance movement. Resistance movements will need to perform cybersecurity practices and enforcement at the cell level, which is much more difficult to accomplish than state security services with centralized cybersecurity monitoring.

> *4.5   Cybersecurity education and training for its members are essential to the survival of the resistance movement, as well as to the security of the state security service.*

Just as a resistance movement should train its members in the traditional physical and operational security procedures, it should also train its members in the basics of *cybersecurity*. While it may seem obvious to some audiences that fake identities, websites, and software are common threats, many people remain unaware of them in this context. As a result, education and training of resistance movement members in basic cybersecurity "hygiene"[33] will be essential to both the cyber and the physical security of the resistance movement and its members.

While such an effort may sound easy, it is not. Efforts as simple as teaching employees to not click on links, open attachments, or download software from email phishing attempts have proven extremely difficult in governments and large corporations in the Western world, where most educated employees are very familiar with computers and computer networks.[34] Human beings are typically seen as the weakest link in the security chain regardless of the advancement level of network security measures.[35] Regardless of cyber defense personnel hardening network settings, these measures can be bypassed when a member of the workforce demonstrates weak cybersecurity hygiene practices, allowing for effective exploitation by a cyber intruder.

The lack of basic cyber hygiene in developed nations is also seen in developing nations. However, the very real threats to the personal safety of resistance members (and their families) provide incentives to learn and practice basic cyber hygiene techniques.

In addition to learning such techniques, the resistance movement should monitor and enforce the use of cyber hygiene practices. The resistance movement can attempt "Red Teaming"[36] or "penetration testing"[37] of its members' cyber capabilities to ensure they follow cyber hygiene. In a similar manner, state security services also benefit from preventative cyber hygiene practices to protect both their cyber assets and any physical assets connected to their cyber assets from cyber attacks. US military advisors to both resistance movements and state security services need to encourage the establishment of testing and enforcement of cybersecurity practices.

> 4.6 *Testing and enforcement of the security of the cyber capabilities used by a resistance movement or a state security service are as important as physical testing and enforcement of member loyalty.*

Testing the loyalty of resistance movement members is a historically-based, effective technique. Elaborating on the previous Norwegian World War II resistance example, all members of a cell were told of a meeting at a certain time and place. Then all but one member (the suspected mole) were told it changed. If the Nazis showed up at the original place, the one not told of the change was identified as the informant.[38]

Testing of cybersecurity of resistance movement members and the organization as a whole should also be undertaken by the resistance movement. Sending spear phishing emails as tests of resistance movement members is one way to determine which members practice good cybersecurity hygiene. Basic penetration testing of member computers and cell phones and checking for weak or default passwords of hardware devices and software applications could also be useful in identifying vulnerable cyber assets.

All of these cybersecurity hygiene checks require time and expertise that the resistance movement probably does not have in abundance. If a resistance movement only has a few members who perform at Red Team cybersecurity testing well, it will be difficult for them to reach the cyber assets of all of the individual cells because they are distributed in small cells.

The preferred course of action for the resistance movement is to have all of their cyber assets in one place or to allow the cybersecurity experts of the resistance movement have access to each of the cell's cyber assets. However, that much information about the resistance movement's whole organization in one place is itself a poor security practice—both physically and in terms of cybersecurity. Physically, if a resistance movement cyber expert is compromised, then the whole resistance movement is at risk. In the same vein, if the tools used by the resistance movement's cybersecurity experts are compromised, then all of the organization's cyber assets will likely be compromised.

An alternative approach is for the resistance movement to distribute cybersecurity "kits" of software to help each cell to "self-check" its own cyber assets for good or bad cybersecurity practices.[39] For example, a small resistance cell has a set of cyber assets that need to be checked for security. The kit could include programs that check for default passwords, weak passwords, and basic security settings for multiple types of cyber assets, such as personal computers, smartphones, or websites. The kit could also include sample spear phishing and chat room tests that one member of the cell sends to other members to make sure good cyber hygiene practice is performed. For example, fake "friending" messages or links to malicious websites could also be included in the kit. Lastly, the kit should include cyber hygiene training and education materials to keep the resistance members alert and able to identify common cyber threats. While these examples represent only a few of the ways in which a cell's cyber assets might be compromised, these basic checks could help preclude the cell from being readily compromised by an adversary's first attempt.

The requirement to broadly ensure cybersecurity across all friendly cyber assets is one area where the state security service has a striking advantage over resistance movements. As previously described, state security services can perform centralized cyber testing and enforcement because their organizations do not require the division into small cells to survive. Moreover, the scarce human resources with extensive cybersecurity expertise can be shared across a wide range of the state's organizations, which allows for economies of scale when securing the state's cyber assets. In contrast, the resistance movement cells must remain intentionally decentralized, requiring distributed mechanisms to test and enforce cybersecurity.

CYBERSECURITY AND INTERNAL COMMUNICATIONS

Resistance movements and state security services both use a wide range of cyber mechanisms for *internal* communication.40 For resistance movements, these internal communications support the functions of:

- Creating and maintaining organization
- Performing C2 of organization and operations
- Training members
- Planning and preparation
- Executing of plans

Each of these internal communication functions can be supported by a range of cyber capabilities. Table 4-1 shows some of the cyber mechanisms commonly used to support internal communications, both within the resistance movement and a state security service.[41]

Table 4-1. Internal communications functions by cyber mechanisms.

	Mobile Phones	Email	Removable Media	Cloud
Organization	X	X	X	X
C2	X	X	X	X
Training	X	X	X	X
Planning and Preparation	X	X	X	X
Execution of Plans	X	X	X	

Traditional communication channels and methods include clandestine face-to-face meetings, dead drops, and radio transmissions, to name a few. Today and for the foreseeable future, cyberspace provides many new realms in which to communicate with less exposure to physical observation. However, the use of these varied cyber communications channels also exposes the resistance movement (and the state security service) to many opportunities for identification, compromise, and infiltration via cyber means.

There are many cyber mechanisms available and more appear each year. This section only focuses on some of the more common and emerging cyber mechanisms used by resistance movements for internal communications. (Chat is also used for internal and external communications, but the risks of chat were previously described.) The mechanisms are:

- Mobile phones (including both cell phones and smartphones)
- Email
- Removable media (e.g., USB sticks, DVDs, external drives)
- The Cloud

Mobile Phones

Mobile phone communications are not as private as they appear. For example, where encrypted cell phone traffic is available, it provides better security. If, however, the state security service owns the cellular infrastructure, the phone may already be compromised. Cell towers regularly push "baseband" updates to the cell phones contacting the cell towers. These updates can include malware. As a result, cell phones used in regions where the state security service control the cellular infrastructure may be readily compromised by this technique.[42]

In addition, commercially available Lawful Intercept[43] software can be used by the state security service to spy on cell phone conversations.[44,45,46] Another threat vector against cell phones occurs when hackers opportunistically hack cell phones.[47] Opportunistic hacking is not aimed at a particular target, but phones are hacked because they are vulnerable. The target of this hack may be used for nefarious purposes by the hacker, including blackmailing the owner with incriminating materials on the phone. Deutsche Telekom hosts a "dashboard" website display that maps phone hacks it detects around the world.[48] Deutsche Telekom uses a suite of honeynets to detect cell phone hacking attempts distributed across the world where it and its partners own cell towers. These hacking attempts are then reported on a dashboard at www.sicherheitstacho.eu that displays these hacking attempts in real time.[49]

Some governments have fairly sophisticated cell phone eavesdropping capabilities they deployed abroad. For example, during the 2006 Israeli incursion into Lebanon, Iran and Syria provided Hizbollah with sophisticated code cracking capabilities against Israeli-encrypted communications, as well as extensive monitoring of Israeli military personnel's personal cellular communications. These capabilities against Israeli-encrypted and -unencrypted cellular communications were used to great effect in terms of military intelligence and propaganda coups.

For example, the Iranian's tracked the private cell phones of individual reservist soldiers to identify Israeli units' movements and reported them to Hizbollah fighters in the area. When the Israelis took casualties, Hizbollah announced those casualties over its television station before the Israeli Armed Forces could announce them.[50]

As a mitigation technique, resistance organizations use "burner phones," which are disposable phones used for a small number of calls. They also use multiple Subscriber Identity Module (SIM) cards swapped into their phones, where each SIM card has a different phone number and identity. Both of these techniques make tracking the phones more difficult. Another ploy is for high-ranking members of a resistance to pass their phones to lower-level operatives who move in a significantly different direction than the previous cell phone owner. This "cell phone swap" was one of the techniques Osama Bin Laden used to escape from Tora Bora.[51]

Lastly, smartphones have many settings used to track the movements and other activities of the phone's user. Probably the most common threat vector entails the use of Global Positioning System (GPS) to track the owner's movements, highlight places of interest in the user's vicinity, and to mark photos with one's GPS location. Turning off the GPS to remove the ability for a phone to announce a GPS location is a good security practice. It is important to ensure that any apps that use GPS on the phone are also disabled.

Note that even manually disabling GPS tracking does not necessarily mean that the smartphone's movements are not trackable. It was recently discovered that Google tracks smartphone locations even when geolocation has supposedly been turned off.[52] In the wake of such a discovery, many have been wondering how to prevent such tracking, but there are a number of complicated steps that one must perform to actually keep smartphones from being tracked by Google.[53]

Email

Email is one of the most common sources of cyber compromise because it is so often successful. The two most common methods of compromise are via infected attachments or links to malicious sites embedded within the email. Clicking on either the attachment or the link results in malware downloaded to the user's host machine.

Education and training of resistance members of the risk of email attachments and embedded links is essential to the survival of the resistance movement. Applying the enforcement technique described in Principle #6 will help ensure that the resistance member's use of email follows proper cybersecurity hygiene practices.

Another good email practice for resistance movement members is to use many email addresses and personas. Similar to the use of multiple cell phones or SIM cards (as previously described), multiple email addresses used for only short periods of time also aids in cybersecurity. Such precautionary measures hinder the state security services' to track the various addresses and attempt to compromise them.

Removable Media

Resistance movement organizations need to store and share information about the organization and to plan activities. A commonly used type of cyber device for storing and sharing electronic data are removable media, such as the USB devices, also known as thumb drives or memory sticks. Because such small devices can now store terabytes of data and are yet relatively inexpensive, they are commonly used as cyber communications mechanisms.

However, USB devices can also carry malware. Worse, USBs and other removable media can cross the "air gap" often used to separate more sensitive networks from the Internet. In a now classic example described by former US Deputy Secretary of Defense William Lynn, USB devices with malware were left in restrooms of US military bases in Afghanistan. These devices were then used by US service members to carry data between unclassified and classified networks. This situation resulted in the infection of classified networks previously considered "isolated" from the Internet by USB-borne malware.[54] Moreover, criminal elements now use similar air-gap jumping malware, increasing the risk against resistance movements that use such devices.[55]

One mitigation technique is to use USB devices with "write blocks" to preclude malware from being placed onto the USB device. At the same time, one must be aware that the USB devices with write-block capability may already have been compromised in the supply chain before reaching the user.

Lastly, while USB devices are convenient in their ability to store and transfer large amounts of data, the loss or compromise of such a device can be a boon for the state security service. For example, when al-Zarqawi was killed, intelligence operatives recovered a thumb drive in his pocket (along with other memory devices) that provided the US forces substantial actionable information about al Qaeda in Iraq:[56]

> Within a week of discovering computer equipment in the bombed-out safe house of slain terrorist Abu Musab al-Zarqawi, U.S. and Iraqi forces carried out more than 450 raids targeting followers of al-Qaida's leader in Iraq.[57]

In a similar manner, 2.7 terabytes of data were found on Bin Laden's computers and media during the raid in which he was killed.[58]

Although not all was stored on removable media, the same principle of encrypting data at rest applies.

One of the main tradeoffs of encryption is the need for one or more encryption keys to be securely stored. For example, if users choose to encrypt data files, they should uniquely encrypt each file with a large number of keys, or the same key is applicable to many files. The more keys used, the better the security, but the more difficult it is to work with the data. Due to the need to manage many encryption keys, vendors are now offering encryption key management services for their customers.

Resistance movements need to encrypt their data at rest, compartmentalize the amount of data stored on any single device, as well as plan to react when such a compromise occurs. The state security service also needs to prepare similar plans in the event of compromise. US advisors to both resistance movements and state security services need to advise and assist on the creation of contingency planning and the recovery from a cyber-security compromise.

> 4.7   *Resistance movements and state security services need to have contingency plans in place to recover quickly and effectively from a cyber-security compromise.*

Lack of encrypting data at rest can result in exposing large amounts of data about resistance movement or state security service operations, personnel, organization, and plans when obtained by the opposing side. Time is of the essence when such a compromise of information at rest occurs. The side that captured the data must quickly sift through it to identify actionable information to exploit and harm the opposing side. Conversely, the side that lost the data must assume that all of the data lost will be used by the opposing side and alert all affected parties before they are identified and attacked. Without a clear and comprehensive plan by which to respond when such a compromise of substantial data occurs, the result mirrors the type of "roll up" of the resistance movement as described in the case of al Qaeda in Iraq.

The Cloud

The Cloud is a form of Internet-based computing resources that allows users to lease and connect to cyber assets and services provided by cloud service providers (CSPs) while paying for only the time and assets used. While clouds can be small (such as a few computing nodes in a laboratory), major CSPs commonly provide many thousands of computers (or computing nodes) and terabytes of memory leased to remote users for a price. The price paid by the user depends on the number and type of computing nodes, memory used, and bandwidth used to transmit data to or from the cloud.

There are generally three types of clouds, as shown in Figure 4-4:[59]

- *Storage clouds*, where the CSP only provides memory and the user only pays for the amount of memory used per month

- *Utility clouds*, where the CSP provides a range of computing power using virtual machine, and the user only pays for the number and type of computing nodes or other computing services used per month

- *Data-focused clouds* (or data clouds), optimized for parallel programming to perform big data analysis.

Storage Clouds	Lots of Memory				
Utility Clouds	Lots of Memory	+	Virtual Machines		
Data Clouds	Lots of Memory	+	Distributed File System	+	Parallel Programs

Adapted from Cloud Computing 101: A Primer for Project Managers

Figure 4-4. Types of clouds and their key elements.

Due to the large number of CSPs around the world, and the ability to access most public clouds from anywhere in the world, the cloud provides another place for resistance movements in which to communicate and store information. Storage clouds are likely the least expensive and easiest to use, but the opportunity to buy a long-term lease for persistent use of a utility cloud or even a data-focused cloud should not be overlooked. Data stored in clouds can be used by resistance movements as locations for dead drops to exchange information. They can also be used as places of temporary storage by a single member who might be travelling or otherwise at risk and does not want to carry the sensitive data with them.[60] Placing the encrypted data in the cloud as a temporary holding place until the immediate threat passes can allow a resistance movement member to retrieve the stored data once conditions are sufficiently safe.

Unfortunately, not all cloud storage is secure. For example, if someone "backs up" their iPhone data on iCloud, the data are encrypted. However, Apple can access information stored within a backup, including photos, videos, device settings, application data, iMessage,[61] SMS,[62] MMS[63] messages, and voicemail.[64]

If the CSP provides the encryption, then the service provider also has access to the encryption keys. If the user provides the encryption for data stored in the cloud, then only the user can access that data (assuming use of strong encryption techniques). Resistance movements, therefore, should opt for providing their own keys to secure their data at rest in the cloud.

Note that the weak link for data stored in the cloud is not the encryption mechanism but the access mechanism. When a user accesses a cloud, even by secure means, a user name and password must be used. Unless there are additional protections, such as "two-factor authentication",[65] then an adversary using brute force guessing of the user name and password will eventually break the code and access the network. Once the adversary achieves access, malware can be installed to track the encryption and decryption processes, thus exposing the encryption keys to the adversary.[66]

CYBERSECURITY AND EXTERNAL COMMUNICATIONS

Resistance movements and state security services both use a wide range of cyber mechanisms for *external* communications. For resistance movements, these external communications support the functions of:

- Recruitment and mobilization
- Subversion of state security service members and assets
- Propaganda
- International support
- Operation equipment and readiness

Each of these functions reaches beyond the boundaries of the existing resistance movement to transmit information to external audiences and gain membership, resources, and support. These external communications can be either public or clandestine. Because clandestine communications tend to be primarily accomplished one on one, this section focuses on primarily cyber-based, public, external communications.

External communications via public channels may be openly associated with the resistance movement or may be anonymous, obfuscated, or hidden. Cyber mechanisms for external communications covered in this section are:

- Websites (both created or hijacked, but excluding social media)
- Social media
- Pirate and Internet radio
- Cyber-related supply chains

Table 4-2 illustrates some of the cyber mechanisms commonly used to support external communications, both within the resistance movement and within the state security service. Supply chain risks exist for both cyber assets and physical assets obtained via cyber assets, including browsing for, ordering, shipping, receiving, and operating physical objects and software.

Table 4-2. External communications functions by cyber mechanisms.

	Websites	Social Media	Pirate and Internet Radio	Cyber Supply Chain
Recruitment and mobilization	✔	✔	✔	
Subversion of state security service	✔	✔	✔	
Propaganda	✔	✔	✔	
External support	✔	✔	✔	
Operation readiness				✔

Websites

Websites can reach a very large number of people at the same time. If the website can remain up and not be blocked, the messages will likely reach their intended target audience. Reaching that audience may be for purposes of recruiting new members, subverting members of the state security service, gaining international support for the movement, and supporting propaganda. New websites are constantly created by resistance movements, but each one is often short lived due to the state security service or web-hosting organization blockage or removal.

For example, "As far back as 2001, Alneda, an Arabic Web site used by al-Qaida to send messages to its followers, was a known entity. US authorities tried several times to shut down the site, but it is nearly impossible to prevent such an operation from popping up on another server."[67] While each website has a limited lifespan, the sequence of websites over time provides a persistent online presence of the resistance movement. In addition to search engines, online chat rooms, personal emails, and tweets can guide interested parties to the new location of the site.

Public messages from the resistance movement can be posted on websites of others via hacking the site. The message carries even more weight if the targeted website is owned by the state security service. For example, hackers supportive of ISIS hacked the US Central Command's (CENTCOM's) Twitter and YouTube accounts and then tweeted pro-ISIS messages and uploaded pro-ISIS videos. While there was very little damage to the sites, the attention gained for ISIS was a propaganda victory.[68] An example of a jihadist website can be seen in Figure 4-5.

Courtesy of Memri.org.

Figure 4-5. Sample jihadist website.

In more permissive, less repressive environments, websites generate mass mobilization in relatively short periods of time if conditions are right.[69] For example, in the Occupy Movement, the digital media acted as a "stitching technology" that helped facilitate flows of information and action across interconnected but dispersed networks:[70]

> It [the movement] initially appeared on a Tumblr microblog, begun by a single Occupy sympathizer, that invited visitors to post their own experiences with economic injustice and related issues. The frame then migrated to Occupy sympathizers on Twitter. Eventually, the frame became a fixture at the physical Occupy campsites in New York and around the globe.

Conversely, public websites with messages aimed against a more repressive and lethal state security service will often be hosted outside the control of the state security service (such as in another country). This often allows the website a longer lifespan, even if the state security service blocks access to it from the citizens of the target nation. In 2005, al Qaeda posted job advertisements on the Internet site of a London-based Asharq al-Awsat "asking for supporters to help put together Web statements and video montages.... Al-Qaida-linked groups also set up their own sites, which frequently have to move after being shut by Internet service providers."[71]

The security of these websites needs to be of paramount importance to the resistance movement. If a state security service can hack into the resistance movement website, then malware can be distributed to supporters of the resistance movement very quickly.

Furthermore, a hacked resistance website can reveal identities of members, and anonymity can be lost. For example, once a resistance movement's website is hacked, users of the site may be redirected to similar-looking websites run by the state security service, which allows the state security service to not only monitor a username and password but also a user's browser.[72]

Even viewing content on the site can be used to collect information against the resistance movement. If a state security service also monitors who accesses certain sites by compromising them, the identities of potential members or supporters of the resistance movement are at risk.

Just as important, if the state security service compromised a website known to be owned by a resistance movement, then the state security service can post particularly offensive material on the site, claiming condonement by the resistance movement. This deception technique creates severe political repercussions against the resistance movement even as it tries to deny ownership of the post. For example, hackers were able to hack Twitter administrator accounts by testing simple passwords for specific logins. Once the hackers achieved administrative privileges, they sent phony Twitter messages from President Obama as well as from staff at Fox News.[73] Although the hoax was quickly identified, it would be much more difficult for a resistance movement to overcome phony messages from its legitimate accounts if a hacker were to gain administrative privileges over its sites.

Education and training of current members of the resistance movement can help reduce compromise via phony websites. However, potential recruits with no training will likely be the most susceptible to identification by the state security service prior to joining the resistance movement.[74] If the new recruit previously visited such a site, aimed at collecting data on pro-resistance visitors, then that recruit and his/her family are already at risk due to previous consideration of potentially supporting the resistance. In more permissive environments, this risk of compromise is not a significant problem, but in environments in which the state security services often uses torture and lethal force, such a compromise could be a significant problem for both the recruit and the resistance movement. Principle #8 usually only applies to resistance movements.

> *4.8 Resistance movements need to interview and screen new recruits to determine whether they may already be identified by the state security service via phony pro-resistance websites.*

While there is not a preferred mitigation method, multiple alternatives exist for the resistance movement aware of any new recruit's likely compromise via cyber means even before joining the resistance movement. Just as physical screening of new recruits often involves testing individuals to see if they divulge vital information to the opposing side, initially providing new recruits with cyber assets and monitoring software is one way to determine whether they attempt to make clandestine contact with the opposing side. At the same time, the monitoring software should not remain in place permanently because the state security services may exploit the same monitoring channel over time.

Social Media

Chapter 5 will describe many uses of social media by resistance movements, as well as some of the problems social media presents to resistance movements.

An earlier section of this chapter described state security service use of fake Facebook pages to collect profile data on resistance members, as well as providing links to sources of malware to compromise resistance movement cyber assets. This section describes additional security issues associated with the use of social media by either the resistance movement or the state security service.

One common security risk is that social media sites can be monitored by both sides: the state security service and the resistance movement members of each other. A significant amount of personal information is often posted on social media, which can be used by an opponent to identify, threaten, or co-opt a user.

Much of the data intentionally posted by a user can result in an unintentional level of exposure. For example, the opposing side can leverage users' posts about personal relationships or personal history (e.g., schools attended or places frequently visited). Pretending to be a graduate of the same school is a common ploy to initiate contact.

In other cases, information can be unintentionally posted due to a human procedural or configuration error. For example, one of Verizon's vendors exposed the names, addresses, and phone numbers of approximately six million customers.[75] In a similar event, data on approximately 198 million voters was left on an open online database by an analytics contractor employed by the Republican National Committee. It was only taken down when a cybersecurity analyst discovered it.[76]

The greater the number and variety of information someone posts about themselves, the more an opponent can learn about them. Where one lives, went to school, previous employment, friends' identities, and personal appearance are often posted with little regard for an adversary collecting data to cause harm or collect future leverage. Even if a resistance member employs online operational security (OPSEC) on social media, if friends or relatives do not practice OPSEC, the resistance member is still at risk. Sharing photos with tags to a person's name exposes both the resistance member and his/her online friends and relatives to the attention of the state security services.

For example, if someone posts support of the resistance movement, then the state security service can "connect the dots" of the rest of the information on the page to create an accurate description of that person and relationships with family and friends. Because online connections are often related to physical connections, the state security services can connect the resistance member's online contacts with physical identities. Information on the resistance member's friends and families could be used to coopt or coerce that person to become a witting or unwitting tool of the state security service against the resistance movement. Even if the resistance member practices proper cyber hygiene, but the member's family, friends, and followers do not, the resistance member's identity could be identified by the state security services.

In a similar manner, the resistance movement can monitor social media posts of known members of the state security service to identify patterns of life, family, and friends and take coercive action against that state security service member or his/her associates.

The need to preclude operations security leaks via cyber assets leads to the need to scan social media to identify members that violate OPSEC. For example, US military services each host offices dedicated to identifying OPSEC leaks posted by service members on social media sites.[77] After the bin Laden take down, the name of the Navy Seals unit involved in the raid was leaked to the press, which allowed individuals and their family members to be identified from open source materials.[78] Resistance movements need to perform similar functions for ensuring the OPSEC of their members and operations that might appear on social media.

> 4.9 *The resistance movement and the state security service each needs to scan social media to identify members that violate OPSEC.*

One of the most serious sources of information unintentionally posted to social media is photographs. Modern cameras and smartphones often include GPS; the location where the photo was taken (as well as its time) is often included in the photo's metadata. This type of data exposure can compromise both resistance and state security service members' physical locations and may lead to member identification and tracking over time.

Photos posted by family or friends of a resistance member can provide the state security services (and vice versa) with an image that would otherwise not be available via public means. Worse yet, some social media sites use facial recognition software to automatically tag individuals in a photo with their name and links to their profiles.[79] As long as a photo was tagged with a name once on the social media site, there is a good chance that the facial recognition software automatically tags subsequent images.

This photo-matching capability may be particularly trouble-some for a resistance movement member using a *nom de guerre* while another photo associated with the member's real name is posted on a friend's site:

> Facebook's facial recognition research project, Deep-Face (yes really), is now very nearly as accurate as the human brain. DeepFace can look at two photos, and irrespective of lighting or angle, can say with 97.25% accuracy whether the photos contain the same face. Humans can perform the same task with 97.53% accuracy. DeepFace is currently just a research project, but in the future it will likely be used to help with facial recognition on the Facebook website. It would also be irresponsible if we didn't mention the true power of facial recognition, which Facebook is surely investigating: Tracking your face across the entirety of the web, and in real life, as you move from shop to shop, producing some very lucrative behavioral tracking data indeed.[80]

If Facebook can track a person's presence across the Internet now, then state security services can as well. At the same time, resistance movements could also track the movements of key state security service members. Both resistance movement and state security service members need to check their individual privacy settings for all social media channels in use. In addition, they inform friends and relatives that posting their images on the Internet is a threat to their personal safety. Unfortunately, it is becoming more difficult for someone to remain anonymous or perform anonymous browsing, as will be described in chapter 8 on attribution.

Figure 4-6 illustrates the concept of facial recognition software techniques where key facial features are identified and the distances between them are mapped.[81]

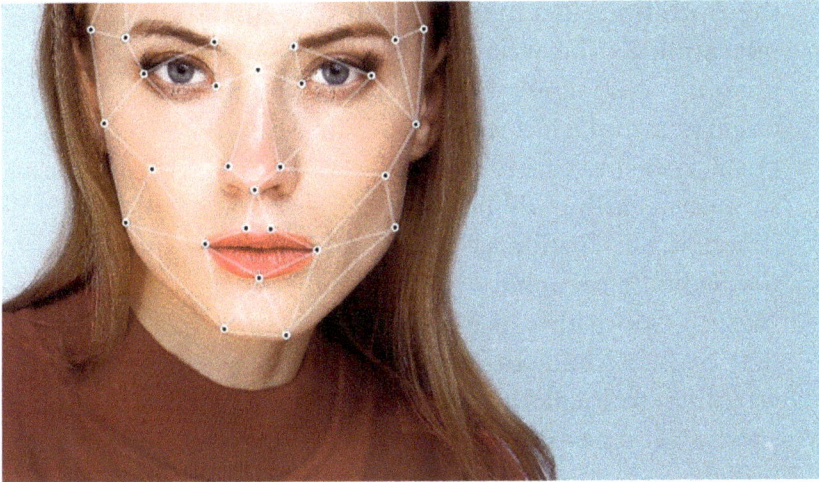

Figure 4-6. Sample facial recognition software face mapping.

The state of the art of facial recognition software continues to expand, and China advanced it into the physical realm. Some Chinese police are now equipped with "smart glasses" that take an image of a person in their view, which is then matched to a database of the person's identity. As long as an image of that person is available, modern surveillance techniques in the physical world can be used to quickly identify a person of interest.[82]

Pirate Radio and Internet Radio

One of the more traditional external communications mechanisms for resistance movements is pirate radio, which is a radio station that broadcasts without a valid license. Pirate radio was used extensively throughout World War II and the Cold War. For example, during WWII, allied radio stations from London, Moscow, and America broadcast live Hungarian news. In Hungary, Prime Minister Miklos Kallay-sanctioned the operation of an illegal radio station. From August 3, 1942, the Voice of America broadcast in Hungarian from a boat that sailed near Salonika in the waters of Greece.[83]

The Taliban used pirate radio extensively in Pakistan:

> Through their pirate FM transmitters, the Taliban have demanded that local parliamentarians, security forces and other government officials resign from

their positions as a mark of protest against the military operations; otherwise they should be prepared for a jihad directed against them. The Taliban radio broadcasters, popularly known as "FM Mullahs," continuously transmit anti-American and anti-government sermons, calling democracy "un-Islamic" and those practicing it "infidels." In their fiery radio speeches, the Taliban preachers have demanded that the non-Muslim minorities of Malakand pay jizya (protection tax) or face jihad. In the same tone, they have issued warnings to local NGOs, musicians and anybody else involved in "un-Islamic" activities. Those defying their orders are butchered, and daily announcements of the details of their deaths are broadcast on FM channels.[84]

The cyber version of pirate radio is Internet radio. Anyone with an Internet connection can listen to broadcasts streamed live via the Internet. Unlike normal web postings or YouTube, these broadcasts are live, just like a normal radio station. Being a live broadcast, this sets a time window for the reach of Internet radio to be similar to physical pirate radio. In a similar manner, these fleeting broadcasts can be sent via the Internet from different Internet addresses at different broadcast times, providing a longer term survivability to Internet radio than a simple web posting.

Pirate radio was used in 2017, transmitting to Mosul as part of the fight against ISIS in Syria.[85] The "Alghad" (meaning "tomorrow") pirate radio station in Iraq broadcasted online as well as in the physical airwaves. It was founded by a refugee from Mosul after it was overrun by ISIS. The station broadcast news and music to the population of Mosul and the surrounding Nineveh province to counter the isolation and propaganda imposed by ISIS on the city. The staff use *nom de guerres* like Al Mawsily, which means "from Mosul." Individuals trapped in Mosul often called into the radio station to have their voices heard even though they risked being discovered and killed by ISIS.[86]

One of the advantages of Internet radio is that if the operator keeps changing Internet addresses, it is less susceptible to hacking by the opposition. Even so, those providing the Internet radio need to secure their servers and cyber office equipment to make sure that their software applications upon which their broadcasts rely are not compromised.

Supply Chain

Many nations are concerned about the potential of another nation state to compromise or damage the nation's supply chain. For example, in October 2016, the US DoD Joint Staff issued a warning about Lenovo computers and handheld devices made in China due to cybersecurity concerns:

> Bill Gertz writes a J-2 intelligence directorate report stated cybersecurity officials have discussed how Lenovo devices could lead to the integration of compromised hardware into the DoD supply chain and bring cyber espionage risks. The report also tackled alleged attempts from Lenovo to acquire U.S. information technology companies in a push to gain access to classified DoD and military information networks... The state research institute Chinese Academy of Science has a 27-percent stake in Lenovo Group. The National Security Agency has previously linked China to cyber spying reports against the Pentagon as well as U.S. and foreign defense contractors, the report noted. [87]

Resistance movements face a similar problem. How do they know the equipment and software they use are safe and uncompromised? As described earlier, USBs may be pre-infected before use by the intended target. Similar concerns apply to any cyber equipment or software. What are the sources of the personal computers, cell phones, removable drives, chat software, security software, and smartphone applications? Hardware and pre-installed firmware are also potential targets for adversary compromise. If these cyber assets are already compromised before they are purchased in shrink-wrapped containers, it will be very difficult to determine whether or not they have been altered without extensive forensic testing. Any of these pieces of equipment or software could be an adversary attack vector against the cyber equipment of the resistance movement or the state security service.

Just as nations need to investigate and monitor their sources of cyber equipment and software, so too must resistance movements. Is their cyber equipment stolen from the state security service? If so, has it been checked for software that alerts the previous owner to the location of the stolen equipment? Is the equipment purchased on the black market? If purchased and shipped from overseas, how long was it held in customs? Was the equipment smuggled into the country for the resistance movement?

There are no simple answers to resolve all of the potential threats possible via the supply chain. This section emphasizes that the resistance movement should be aware their supply chain may include pre-loaded malware to compromise their cyber assets. That awareness can then lead to decisions to vary the sources of cyber assets and test the equipment, such as the Red Team checks when the equipment or software is first obtained. Such testing may include operating the cyber assets in a safe location (possibly overseas) and identifying whether the assets attempt to establish a C2 connection to an adversary. Once tested, the cyber assets might be wrapped in tamper-detecting seals to indicate tampering after such tests.

Supply chain vulnerabilities require an awareness of supply chain risks. Identifying reliable sources of cyber assets is crucial to a resistance movement in particular. If the resistance movement allies with a nation state, the equipment could be more thoroughly checked by the nation state prior to its deployment to the resistance movement.

> *4.10 Both the resistance movement and the state security service need to be aware of supply chain risks, vary sources of cyber equipment, and test cyber equipment and software when first acquired.*

KEY TAKEAWAYS

The previous chapters described how cyber could be used to support the achievement of resistance movement and state security service objectives. The various cyber and Internet-connected communications capabilities provide a wide range of internal and external communications methods, and combinations of these capabilities are even more powerful when used together. There are, however, many risks associated with using cyber capabilities. As described in Key Takeaway 4.2, the broadest threat to both resistance movements and state security services is not from disloyal informants but from loyal members with compromised cyber assets.

Unless the resistance movement follows principles and practices that lead to cybersecurity, all of these advantages can become serious disadvantages. What was once a boon to influence reach and anonymity has now become a potential liability if each cyber capability used by a resistance movement is not adequately secured.

To recap the principles for cybersecurity that US advisors to resistance movements or state security forces should find useful:

4.1	The magnitude of the consequences of cyber compromise determines the degree of importance that should be assigned to cybersecurity of the resistance movement or the state security services.
4.2	The largest source of information leakage from resistance movements or state security services is likely via <u>loyal</u> members whose cyber assets are compromised, rather than the relatively small number of disloyal human informants.
4.3	There is an inherent tradeoff between the benefits of the large internal and external reach and variety of cyber capabilities versus the <u>need to secure every type</u> of cyber capability used.

4.4 Separation of cyber capabilities should follow the sequential or parallel cellular organization of the resistance movement to avoid compromising multiple cells via single shared cyber assets.

4.5 Cybersecurity education and training for its members are essential to the survival of the resistance movement, as well as to the security of the state security service.

4.6 Testing and enforcement of the security of the cyber capabilities used by a resistance movement or a state security service are just as important as physical testing and enforcement of member loyalty.

4.7 Resistance movements and state security services need to have contingency plans in place to recover quickly and effectively from a cybersecurity compromise.

4.8 Resistance movements need to interview and screen new recruits to determine whether they may already be identified by the state security service via phony pro-resistance websites.

4.9 The resistance movement and the state security service each needs to scan social media to identify members that violate OPSEC.

4.10 Both the resistance movement and the state security service need to be aware of supply chain risks, vary sources of cyber equipment, and test cyber equipment and software when first acquired.

All of these principles apply equally to both the resistance movement and the state security service except for principles #4, #6, and #8.

Key Takeaway 4.4 is unique to the resistance movement due to its need to maintain a cellular structure for physical security. To stay physically secure, the small cells need to keep their persons and their cyber assets distributed and "unknown" to each other. For cyber assets, no cell should share its cyber assets with another cell. When cyber assets are a scarce resource, it is difficult for a resistance movement to always follow the small cell structure if they need to share limited cyber assets and scare cybersecurity personnel.

Testing and enforcement (Key Takeaway 4.6) are more easily accomplished by the state security service due to its ability to centrally manage and test cyber assets. The scarce human resources with sufficient cyber expertise, as well as the cyber assets themselves, can be shared within a state security service organization. Unlike a resistance cell structure, the state's organization does not need to divide itself further below the organizational level or major blocks within the organization. Any compromise of a state security service cyber asset is not likely to result in the disintegration the state's organization.

Unfortunately for the resistance movement, the scarce resources available to test and enforce cybersecurity should not have knowledge of the whole cyber asset suite, whichresults in a single point of failure for the resistance movement. Instead, the resistance movement should distribute "kits" of software that help each cell test and enforce its cybersecurity status. These kits should also be checked for malware before distribution. Because resistance movements need to retain their cell structure, the true benefit rests in the distributed capability to test and enforce security for the cyber assets in each cell.

Screening recruits (Key Takeaway 4.8 is one measure for the resistance movement to be aware that state security services already identified resistance movement recruits who visited pro-resistance websites. Additional care must be taken regarding well-meaning and loyal recruits who do not know they were identified as potential recruits by the state security services.

Chapter 10 presents a fictional example of a resistance movement (called the Red Berets) and its use of some of the preceding cyber principles in its operations.

ENDNOTES

1 Air-gap jumping malware allows malware to reach cyber assets that are disconnected from the Internet. The malware rides on removable media, such as CDs or memory sticks, to reach their intended target separated from the Internet by an air gap.

2 Paul J. Tompkins Jr., "Nonviolent Resistance," Chapter 3 in *Undergrounds in Insurgent, Revolutionary, and Resistance Warfare*, Second Edition, Assessing Revolution and Insurgent Strategies (ARIS) series, January 25, 2013, 12.

3 Richard Greene, "The Russian Hack Absolutely Affected the Outcome of the 2016 Election," *HuffPost*, December 15, 2016.

4 See Paul J. Tompkins Jr., "Nonviolent Resistance," Chapter 3 in *Legal Implications of the Status of Persons in Resistance*, Assessing Revolution and Insurgent Strategies (ARIS) series, November 2012, for a discussion of the legality of nonviolent resistance movements.

5 Paul J. Tompkins Jr., *Human Factors Considerations of Undergrounds in Insurgencies*, Second Edition, Assessing Revolution and Insurgent Strategies (ARIS) series, January 25, 2013, 40–43.

6 Stanley Lovell, *Of Spies & Stratagems: Incredible Secrets of World War II Revealed By a Master Spy* (New York: Prentice-Hall Pocket Books, 1964), 55.

7 Evan Perez, "How Celebrity Hacker 'Sabu' Helped Feds Thwart 300 Cyber-attacks," *CNN Business*, May 27, 2014.

8 Noah Shachtman, "Under Worm Assault, Military Bans Disks, USB Drives," *Wired*, November 19, 2008.

9 Cross media communications are media that communicate directly with other types of media. For example, Short Message Service (SMS) can be sent from a smartphone via the Internet to another phone or to a personal computer. Voice conversations can be transmitted from a legacy telephone via the Internet using Voice Over IP (VOIP) to other telephones, mobile phones, or personal computers.

10 Grace Dobush, "How Mobile Phones are Changing the Developing World," *Consumer Technology Association Blog*, July 27, 2015.

11 Emily Banks, "Egyptian President Steps Down Amidst Groundbreaking Digital Revolution," *Mashable*, February 11, 2012.

12 John D. Sutter, "Will Twitter War Become the New Norm?" *CNN online*, November 15, 2012.

13 Ibid.

14 Sarah Cook, "China's Growing Army of Paid Internet Commentators," *Freedom House*, October 11, 2011.

15 Ai Weiwei, "China's Paid Trolls: Meet the 50-cent Party," *NewStatesman*, October 17, 2012.

16 Daniel Regalado, Nart Villeneuve, and John Scott Railton, "Behind the Syrian Conflict's Digital Front Lines," *FireEye Threat Intelligence Special Report*, February 2015, 11.

17 Ibid., 12.

18 Men have also been used to trap females, and members of the same sex have been used to trap their targets as well.

19 John Barkai, "Cultural Dimension Interests, the Dance of Negotiation, and Weather Forecasting: A Perspective on Cross-Cultural Negotiation and Dispute Resolution," *Pepperdine Dispute Resolution Law Journal* 8, no. 3 (April 2008).

20 Virtual private networks employ encryption to provide secure access to a remote computer over the Internet.

21 Tor stands for "The Onion Router," which is free software enabling anonymous communication on the Internet. The term "onion" comes from the software adding multiple layers of encryption to messages on the path when leaving the source, and peeling off these three layers of encryption by the time it arrives at its destination. See chapter 8 for details.

22 Regalado et al., "Behind the Syrian Conflict's Digital Front Lines," 13.

23 Adobe® Flash Player freeware software is a lightweight browser plug-in and rich Internet application runtime that includes allowing users to view multimedia and streaming video and audio.

24 Regalado et al., "Behind the Syrian Conflict's Digital Front Lines," 13.

25 Ibid., 14.

26 Ibid.

27 Ibid.

28 Paul Goldman and Alistair Jamieson, "Hamas Used Fake Social Media Accounts to Hack Israeli Soldiers' Phones: IDF," *NBC News*, January 12, 2017.

29 Petronella Technology Group, "Israeli Soldiers Hacked by Fake Social Media Profiles," https://petronellatech.com/israeli-soldiers-hacked-by-fake-social-media-profiles/.

30 Goldman and Jamieson, "Hamas Used Fake Social Media Accounts."

31 Ibid.

32 Regalado et al., "Behind the Syrian Conflict's Digital Front Lines," 14.

33 Cybersecurity hygiene or cyber hygiene refers to following basic cybersecurity practices that are expected of anyone using cyber assets in order to preclude easy compromise. For example, making your password "Password" is not following good cyber hygiene. Using strong passwords and changing them often is part of good cybersecurity hygiene.

34 Stephen Northcutt, "Spear Phishing," *SANS Method of Attack Series*, 2017.

35 Bradley Fulton, "The Weakest Link: The Human Factor Lessons Learned from the German WWII Enigma Cryptosystem," *Sans Institute Information Security Reading Room*, 2 (2001).

36 Red Teaming is a process designed to detect system and network vulnerabilities and test security by taking an attacker-like approach to system/network/or data access See Wikipedia, "Tor (anonymity network)," https://en.wikipedia.org/wiki/Tor_(anonymity_network).

37 Penetration testing (also called pen testing) is the practice of testing a computer system, network or Web application to find vulnerabilities that an attacker could exploit (http://searchsoftwarequality.techtarget.com/definition/penetration-testing).

38 Lovell, *Of Spies & Stratagems*, 54.

39 These kits need to be free from malware in the first place. Otherwise, the whole organization could be compromised via the kits designed to provide cybersecurity. See section on supply chain cyber risks.

40 See Paul J. Tompkins Jr., "Communications," Chapter 7 in *Undergrounds in Insurgent, Revolutionary, and Resistance Warfare*, Second Edition, Assessing Revolution and Insurgent Strategies (ARIS) series, January 25, 2013, for an in-depth discussion of resistance movements' communication, including both technical and non-technical means.

41 Note that chat can be used for both internal and external communications.

42 Christos Xenakis and Christoforos Ntantogian, "Attacking the Baseband Modem of Mobile Phones to Breach the Users' Privacy and Network Security," 7th International Conference on Cyber Conflict: Architectures in Cyberspace, NATO CCD COE Publications, Tallinn, May 26–29, 2015.

43 Lawful interception is obtaining communications network data pursuant to lawful authority for the purpose of analysis or evidence. Such data generally consist of signaling or network management information or, in fewer instances, the content of the communications (Wikipedia: https://en.wikipedia.org/wiki/Lawful_interception).

44 Cora Currier and Morgan Marquis-Boire, "Secret Manuals show the spyware sold to despots and cops worldwide," *Intercept*, October 30, 2014.

45 Romanidis Everipidis, "Lawful Interception and Countermeasures In the era of Internet Telephony," COS/CCS 2008-20 (Master of Science thesis, School of Information and Communication Technology, Royal Institute of Technology, Stockholm, Sweden, 2008).

46 Institute for Human Rights and Business, "Human Rights Challenges for Telecommunications Vendors: Addressing the Possible Misuse of Telecommunications Systems. Case Study: Ericsson," Case Study No. 2, November 2014.

47 Bill Ray, "How I Hacked SIM Cards with a Single Text – and the Networks DON'T CARE: US and Euro Telcos Won't Act Until Crims Do, White Hat Sniffs," *Register*, September 23, 2013.

48 Deutsche Telekom AG, "Security Dashboard Shows Cyber Attacks in Real Time," *Press Release*, June 3, 2013.

49 Ibid.

50 Iason Athanasiadis, "How Hi-Tech Hezbollah Called the Shots," *Asia Times*, September 9, 2006.

51 Jarrett Murphy, "Osama's Satellite Phone Switcheroo," *CBS News*, January 21, 2003.

52 Ryan Nakashima, "AP Exclusive: Google Tracks Your Movements, Like It or Not," *Associated Press*, August 13, 2018, https://www.apnews.com/828aefab64d4411bac257a07c1af0ecb.

53 Ibid.

54 William J. Lynn, III, "Defending a New Domain: The Pentagon's Cyberstrategy," *Foreign Affairs*, September/October 2010.

55 Dan Goodin, "Meet 'badBIOS,' the Mysterious Mac and PC Malware That Jumps Airgaps," *Ars Technica*, October 31, 2013.

56 Jim Miklaszewski, "US ID's al-Qaida Iraq Boss, Launches Raid," *NBC News*, June 15, 2006.

57 Patience Wait, "Special Report: The New DNA: DOD Lab Excavates Bits, Bytes to Dig Out Information," *Government Computer News*, July 31, 2006.

58 Emily Rand, "Source: 2.7 terabytes of Data Recovered from bin Laden Compound," *CBS Evening News*, May 6, 2011.

59 Patrick D. Allen, *Cloud Computing 101: A Primer for Project Managers* (CreateSpace Independent Publishing Platform, 2015), 23–26.

60 Note that IoT devices with sufficient memory can also be used for temporary information storage for later retrieval. See chapter 6 on information infrastructure and chapter 10 for a fictional example.

61 iMessage is an Apple instant messaging service capable of transmitting text documents, photos, videos, contact information, and group messages over the Internet.

[62] SMS stands for Short Message Service and is the basis of text messaging services via telephone, Internet, and mobile telephony services.

[63] MMS stands for multimedia messaging service and is used to transfer pictures and multimedia messages via the Internet or mobile telephony services.

[64] Jason Cipriani, "What You Need to Know About Encryption on Your Phone," *CNET*, March 10, 2016.

[65] There are many types of two-factor authentication. In general, two-factor authentication includes a password, or personal identification number (PIN), and a token (such as a CAC card), a bank card, or a smartphone to which a second PIN is texted. An example of two-factor authentication is a login process that includes sending the user a one-time PIN to his/her cell phone that must be used in 10 minutes or the login is not allowed. Three-factor authentication includes the first two components, plus some physical characteristic of the user, such as a fingerprint, eye iris, or voice print. These are biometric factors in authentication. Note that two-factor authentication may also entail a password or PIN as well as a biometric feature without using a token or a physical card.

[66] Allen, *Cloud Computing 101*, 93–97.

[67] Tony Maciulis, "Al-Qaida Recruitment in Shadowy Net World: As Terror Network Becomes Increasingly Tech-Savvy, Is the US Prepared?" *MSNBC*, March 17, 2006.

[68] Spencer Ackerman, "US Central Command Twitter Account Hacked to Read 'I love you Isis,'" *Guardian*, January 12, 2015.

[69] For example, in 2002, Danny Wallace in the United Kingdom posted an advertisement in a local London paper that simply said "Join Me!" and gave a link. In a couple of months, thousands signed up, not knowing what they joined, with whom they joined, or why. Neither did Danny Wallace. So he turned it into an organization for performing random acts of kindness. A happy ending, but why would thousands sign up not knowing for the purpose or intention? See "Ten Years of Join Me," *Leader*, March 21, 2011, http://www.join-me.co.uk.

[70] Bennett and Segerberg, *The Logic of Connective Action*, 160–164.

[71] "Al-Qaeda Puts Job Ads on Internet – Arab Paper," *Irish Times*, October 6, 2005.

[72] Nick Nikiforakis and Gnes Acar, "Browser Fingerprinting and the Online-Tracking Arms Race," *IEEE Spectrum*, July 25, 2014.

[73] Ryan Singel, "Twitter Settles with Feds Over '09 Obama Hack," *Wired*, March 11, 2011.

[74] In highly repressive environments, resistance movements utilize lengthy vetting processes to ensure operational security. Under these conditions, recruitment may take months to years. Individuals recruited under these circumstances report not being aware that they were targeted for recruitment for some time. See Paul J. Tompkins Jr., *Undergrounds in Insurgent, Revolutionary, and Resistance Warfare*, Second Edition, Assessing Revolution and Insurgent Strategies (ARIS) series, January 25. 2013, 33 and Paul J. Tompkins Jr., *Case Studies in Insurgency and Revolutionary Warfare – Colombia (1964–2009)*, Assessing Revolution and Insurgent Strategies (ARIS) series, 2014, 247–249.

[75] Jessica Guynn, "Verizon Data from 6 Million Users Leaked Online," *USA Today*, July 12, 2017, updated July 13, 2017.

[76] Elizabeth Weise, "Republican Party Data on 198M Voters Exposed Online," *USA Today*, June 19, 2017, updated June 20, 2017.

[77] "Army 'Big Brother' Unit Targets Bloggers," *Defense Tech*, October 13, 2006.

[78] Nate Hale, "Whatever Happened to OPSEC?" *In From the Cold* blog, May 14, 2011.

[79] Facebook, "What is Tagging and How Does It Work?" *Facebook.com*, 2017.

80 Sebastian Anthony, "Facebook's Facial Recognition Software is Now as Accurate as the Human Brain, but What Now?" *Extreme Tech*, March 19, 2014.

81 This is not an example of the Deep Face software technique.

82 Jon Russell, "Chinese Police are Using Smart Glasses to Identify Potential Suspects," *Tech Crunch*, February 8, 2018.

83 Wikipedia, "Hungarian Radio," translated from Hungarian via Google translator.

84 Mukhtar A. Kahn, "The FM Mullahs and the Taliban's Propaganda War in Pakistan," *Terrorism Monitor* 7, issue 14 (May 26, 2009).

85 Moni Basu, "Good Morning, Mosul: Pirate Radio Risks Death to Fight ISIS on Airwaves," *CNN*, October 22, 2016.

86 Ibid.

87 Scott Nicholas, "DoD Joint Staff Issues Cybersecurity Warning Against Lenovo Computers, Handheld Devices," *ExecutiveGov*, October 25, 2016.

CHAPTER 5.
NARRATIVES, SOCIAL MEDIA, AND SOCIAL NETWORKS IN INFORMATION OPERATIONS

INTRODUCTION

Information can be accessed or communicated in many ways. People can communicate face to face and with friends, family, colleagues, or community members. They can obtain information through books, newspapers, and periodicals, or other mass media such as radio and television. Increasingly, especially in the developed world, the online information environment is a source of news, information, and opinions on issues large and small. Figure 5-1 shows the percentage of people from each country in the world with Internet access. While Africa, Latin America, and parts of Asia lag behind the global average, these regions are experiencing considerable growth in Internet penetration.

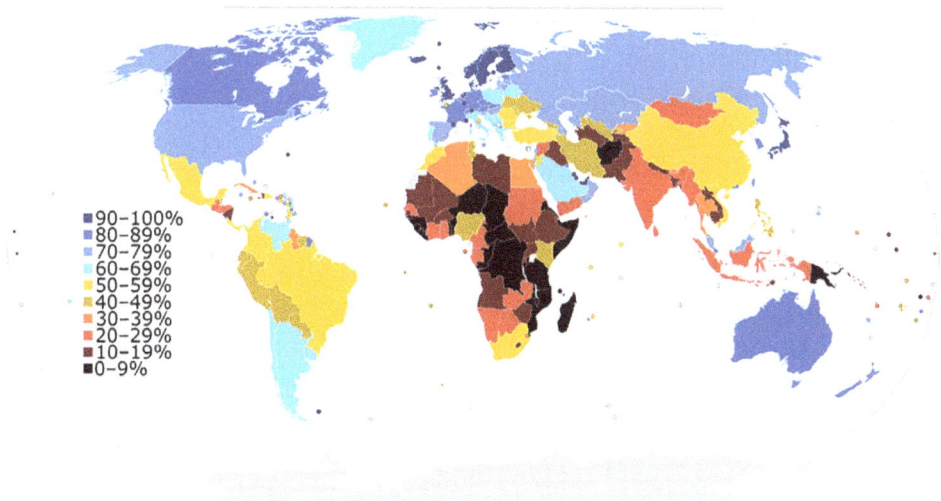

Figure 5-1. Internet users in 2015 as a percentage of a country's population.

Any given social media message has the potential to reach a very large audience. Popular users may have tens of millions of followers who see posts they share. Even obscure users can have a post go "viral" and reach an enormous audience, if enough followers repost or share the message, and their followers do the same, and so on. Thus, social media can rival, with the potential to supplant, other traditional sources of news and information such as print newspapers, television, and radio. As it is essentially free to publish on social media, it provides an opportunity for voices or stories, such as those of a resistance or other non-mainstream group, to propagate quickly and easily. It also bypasses the filters of traditional editorial processes. Modern adversaries of the United States transitioned from the woodline and moved online, where they find asymmetric advantage. This advantage is largely due to the fact that influence is counter-intuitive. The cyber environment lowered costs and increased accessibility to audiences, and influence is more effective than kinetic targeting for achieving strategic goals. Carl von Clausewitz said that war is [influence] by other means. The low cost and easy access to the Internet makes this asymmetric advantage available to resistance movements. Leaders, especially senior leaders, must therefore study and learn the changes to modern warfare in this new environment to set strategy and effectively make risk and resource decisions.

It is important to remember that the online population is not representative of the general population. They is a strategic population, however, especially for resistance movements. Online personas are more likely to hold activist and anti-establishment views, making them a ripe audience for social movements.

> *5.1   Online populations are not representative of the
> general population.*

Social network analysis (SNA) provides an important tool for understanding modern threats. SNA provides an analytic method for identifying influential online communities, understanding their goals and motives, and identifying influential opinion leaders, either for online or offline unconventional warfare development.

When assessing social media data and tools, it is important to recognize that high follower counts, number of tweets, and volume of activity is not equal to influence. Influence must be understood within the context of an appropriate and valid online social network. Most vendor-provided tools do not provide this data because the calculations are time intensive and costly to implement, even for a computer. Commanders must understand and demand appropriate metrics to understand the environment.

The network structure of the online environment affects behaviors, attitudes, beliefs, and norms. When adversaries or resistance movements deliberately shape the network, such as the example from Da'esh, it creates cognitive obstacles to influence. These obstacles take the form of conformity, majority illusion, and echo chambers, among others. Combatting propaganda in this environment with facts and counterarguments is akin to conducting a frontal assault against an enemy with clear fields of fire and well-developed obstacles. At the very least, operators need to breach the obstacles. Commanders must learn to leverage measures of influence to develop effective maneuver strategies in the information environment. The US Army Special Operations Command (USASOC) G9 describes this concept as "expanded maneuver."

WHAT ARE NARRATIVES?

Definition and Description

At their simplest, narratives are kinds of stories that individuals, groups, cultures, and nations tell themselves, and others, about themselves and the world. These stories shape our understandings of who we are and of the cultures and societies of which we identify. Narratives provide a framework for how we think about and act in the world. The human mind itself is designed to learn from and remember stories.[1]

The term narrative has also been used to characterize:

> the reservoir of cultural material that forms over generations in every society. At an individual level, narratives are the stories that, strung together, form the autobiographies of our lives. At a group level, such as in a culture or society, histories, myths, literature, and other stories help explain and solidify group identity. They help bring to life the norms, values, and beliefs that separate the group from others. The ability of narratives to help form group identity and cohesion is an important part of mobilizing a target audience to participate in social or political activities, such as in a resistance movement.[2]

In the cyber resistance context, we can consider narratives more specifically, in terms of the elements and aspects of existing cultural narrative material that a group draws upon to advance their strategic purpose. These are not mundane, everyday stories. These narratives are intended to be compelling and persuasive. They are stories with components of conflict and suspense, emotional intensity, narrative arcs and archetypal characters.[3]

Narratives must "hang together" to function.[4] They must be coherent and present compelling reasons, arguments, or explanations that resonate with their audiences. They must fit into broader patterns supplied by history, biography, and culture. To be effective, narratives should be compatible with the existing worldview of the interpreter.[5]

Narrative can be used to help frame how people view an event or issue. A frame is a lens that shapes identification, label, and interpretation.[6] Any sociopolitical issue can be understood in many different ways. However, a group or society is likely to focus on a small subset of dimensions to manage complexity. These frames help influence opinions in the issue. How an individual responds to a frame often depends on the individual's prior attitudes and beliefs,[7] though a new frame can break through prior positions.

In work on framing the death penalty issue in the United States,[8] exposure to either a pro-death-penalty story framed along the moral dimension, or an anti-death-penalty story framed along the moral dimension, significantly swayed people from their previously held beliefs. When a story was framed, however, along the innocence dimension, in which an innocent person was wrongfully sentenced to death (a fundamentally anti-death-penalty frame), it was more positively received, even by those who were in favor of the death penalty.

Noted sociologist David Snow describes three types of framing that aid social movements in making progress toward their goals in constructing compelling narratives and generating collective action.[9] *Diagnostic* framing identifies the "problem" ailing society, who is victimized by this problem, and who or what is responsible for causing the problem. Thus, it defines what or whom the social movement must stand against. For example, diagnostic framing might assert that the suffering of the peasant class is the fault of the ruling elite. *Prognostic* framing, the second type, proposes solutions and tactics for resolving a problem. These are actions the social movement must perform or changes that authorities must carry out to alleviate the problem. Finally, *motivating* framing rallies and inspires members of the movement to take action. Frames thus can help link grievances and resentments about perceived problems to specific goals, actions that can be taken, and justifications for those actions.[10]

The rhetoric of Boko Haram, a terrorist group, in Nigeria provides examples of each of these types of framing. Boko Haram, which means "Western education is forbidden" in Hausa, is an Islamic fundamentalist insurgency in northern Nigeria. It is perhaps most widely known for kidnapping nearly three hundred school girls in Chibok, Nigeria. Boko Haram had been conducting violent and terrorist attacks against the government and the population since 2002. Boko Haram defined the "problem" of Nigeria as a corrupt, secular leadership against the Muslim population.[11] The Western education system, according to Boko Harma, was the source of this secularism and violated the Quran. This diagnostic framing, that Nigeria was contaminated by secularism and corrupt leaders, helped set the stage for its prognostic framing. Boko Haram's self-defined purpose was to cleanse Nigeria of its current corrupt, un-Islamic government and create a caliphate. Its motivational frame exhorted Muslims to take action, calling for the destruction of the Nigerian state as a religious obligation, consistent with sharia. This framing helped Boko Haram attract adherents while discrediting the government and those who worked with and supported it.

Narratives and Collective Action, Activism, and Resistance

Narratives can do much more than instill traditional norms, values, and beliefs in people. They can persuade people of new goals, bring about changes in their sense of identity, and spur them to organized, collective action.

In our context, collective action describes action taken by groups of regular people to confront opponents, elites, or authorities.[12] Protests represent a type of collective action. Depending on whom you confront and where, such actions can be risky, with the potential for injury, arrest, or even death. It is often easier and safer to stay on the sidelines as a free rider because collective action often produces public goods, such as political reforms, available to anyone regardless of whether or not they personally participate in the risk. As a result, motivating participation in collective action can pose a significant challenge.

Narratives may hold part of the key to participation in collective action. These compelling stories can help shape the basis of one's personal identity, thus encouraging action and limiting the appeal of free riding. A shared narrative helps build bonds and deepen trust with others, supporting cooperation.[13]

> 5.2   A shared narrative helps build bonds and deepen trust with others, supporting cooperation.[14]

Social and other digital media provide protest organizations with an opportunity to tailor messages to reach and mobilize a broader audience. However, such personalization presents risks, including the potential trade-off between flexibility and effectiveness. If an organization attempts to mobilize participants who desire greater personalization in affiliation, definition, and expression, this could decrease their effectiveness at collective action because they would no longer be united through common collective action frames[15] or narratives.

In their study of protests of a G20 meeting, academics W. Lance Bennett and Alexandra Segerberg found that a social movement that used a diversity of social media technologies and variety of frames was more effective at mobilization than a movement pushing a single collective action frame.[16]

The Put People First coalition used technology to allow individuals to send personalized protest messages to the G20 and was able to mobilize roughly thirty-five thousand people for its march. It avoided a single, specific narrative framing of the problem symbolized in the G20, instead broadly describing economic crisis and urging reforms of banking, finance, or trade systems.

The Meltdown coalition, whose online presence unilaterally presented its information and framing to users without opportunities for personalization, mobilized roughly five thousand people for its protest a few days later. The group's framing asserted that bankers were causing economic catastrophe and that the solution was to overthrow capitalism.

NARRATIVE EXPRESSION IN SOCIAL MEDIA

Memes and "Fake News"

The rise of the Internet changed the dynamics of popular resistance. This includes protests, movements, and other forms of activism designed to change the status quo.[17] The ability of movements to instantly diffuse tactics, strategies, and information magnifies through social media.

Hashtags are words, acronyms, or phrases that label a social media post, such as *#bringbackourgirls*. This hashtag was originally used in Nigeria after Boko Haram kidnapped hundreds of Nigerian girls from their schools. *#Bringbackourgirls* first gained traction among communities in Nigeria, before breaking out and becoming popular worldwide.[18] It attracted attention to the plight of the kidnapped girls, a story that news media did not paid much attention, and spurred international pressure to secure their release. Figure 5-2 shows an example social media post incorporating this hashtag.

Figure 5-2. First Lady Michelle Obama holding a sign with the hashtag "#bring-backourgirls" in support of the 2014 Chibok kidnapping.

A hashtag can be used to associate a post, and the user who sent it, with a social movement or activist cause (*#occupywallstreet*), reference an event (*#sandyhook*), or link the tweet with some existing meme or running joke (*#whatnottosay*).

The Occupy Wall Street movement, which began in the fall of 2011, initially gained little traction or coverage from mainstream media in the United States.[19] It was originally framed by the media dismissively, as a frivolous, disorganized group with no clear goals. However, through social media postings on Twitter and YouTube, *#occupywallstreet* messaging quickly reached millions within the first month of the movement and pushed forward its own framing as a movement concerned with issues of economic inequality, Wall Street abuses, and the corrupting influence of corporate money on politics. This framing was picked up by portions of the blogosphere and eventually by the mainstream media as well.

Memes are short, distinctive phrases,[20] or catchy images, gifs (short, animated clips), or other media. Memes express ideas, concepts, opinions in a way that spreads rapidly, like a contagion, through a group or population. Memes commonly use emotion or humor to express their

136

message, rather than relying on logical argument. Thus, memes can be ideal for conveying and reinforcing narratives through social media or any computer-mediated communication.

If a meme resonates with existing narratives, members of a group are more likely to propagate it within their social network.[21] While external, societal factors, such as news media and government institutions, may ignore or try to counter a meme, social media provides a ready outlet for its continued spread in populations with access to it. Social media posts can also go viral, spreading rapidly throughout the social media platform on which they originate and even "jumping" onto other platforms where they are widely discussed.

When the amount of content to be conveyed exceeds the space available for a social media post, users may include links to external resources, such as online articles, in their posts. These linked items can contain content that ranges from basic factual background material, to persuasive stories and narratives, to total fabrications and falsehoods, often in the guise of "news."

The phenomenon of online "fake news" began receiving substantial attention prior to the 2016 US presidential election. These false articles are designed to look like credible journalistic reports from genuine news outlets, although they are not based in facts or real events or written by actual journalists. They are easily spread online to vast audiences willing to believe the fiction and spread the word via social media. Some fake news stories have been shared on Facebook millions of times.[22] Fake news websites are often motivated by advertising revenue, though others have political or ideological goals, or may reflect a nation state conducting influence operations.[23]

> *5.3 Fake news stories that are framed consistently with a target audience's identity, cultural values, or accepted narratives are more likely to be believed than real news that does not conform to those expectations.*

The use of social media and the Internet more broadly have been associated with the potential for increasing political fragmentation and social polarization, as people are selectively exposed to much more information that echoes their pre-existing views and opinions.[24] Many major social media platforms serve as conduits for fake news. The algorithms these platforms use to promote content, as well as their advertising models, enable fake news stories to spread faster and wider than accurate stories on the same topics, or fact checking or debunking of their fake news content.[25] An astute resistance movement or adversary could exploit this to his/her advantage by developing or seeding fake news stories designed to increase support of the cause or to fragment and weaken their opposition.

FACTORS INVOLVED IN ENABLING OR INHIBITING SOCIAL MEDIA TRANSMISSION, SPREAD, OR ADOPTION OF NARRATIVES

Transmission

To be effective, a narrative must reach, and resonate with, as many people as possible. While a social media post or meme may spread organically, through an existing online social network, there are a number of additional ways it can transmit and amplify.

Social networks at their most basic are composed of people and the ties that link them together. These can be ties of friendship, kinship, or affection.[26] They can be working or even negative relationships (dislike or fear). In a digital environment, social media, particularly social networking sites, help people maintain and cultivate their existing physical world social ties to keep up with friends and family, while forming new connections with others.[27]

Through social media platforms, a person can easily share content, spreading it rapidly across his/her social network, and engage in discussions about it, to further reinforce and amplify the message through network members who can transmit it to their own networks in turn. Social movements, or others wishing to further their agenda, may strategically cultivate their social networks, attempting to build connections with others who could be influential in amplifying their messaging. These could be thought leaders on a relevant issue, public figures, bloggers or journalists, or appropriate celebrities.

In some cases, new technology has been developed to amplify the transmission of a message. For example, the Islamic State of Iraq and Syria (ISIS, also known as Da'esh or ISIL) created an Android application (app) called The Dawn of Glad Tidings to enable the group to spread information via social media.[28] This app allowed Twitter users to make their accounts available to ISIS, so ISIS could send tweets that looked like they originated with that user. The app was later removed from the Android store.

Bots (short for "software robots") can also be used to amplify social media messaging as well. Bots can be described as:

> Social media identities that use automated scripts to rapidly or strategically disseminate content [that] are rapidly becoming an important element of online politics, and seem to blur the lines between political marketing, algorithmic manipulation, and propaganda.[29]

Bots are becoming more sophisticated in their behavior, mimicking human patterns of online activity and becoming harder to detect.[30] While bots can disseminate or amplify innocuous or legitimate content, they can also be programmed with more malicious intent. Bots can be designed to deceive, exploit, or manipulate social media discourse with malware, misinformation, rumors, or false accusations. For example, bots have been observed to artificially inflate apparent support for a political candidate, giving a false impression about his popularity with the general public.[31] Bots have been used in efforts to manipulate public opinion on various policy issues and the stock market.[32] Bots can fool humans into thinking they are genuine. When acting en masse, bots also have the capacity to shape narratives and framing, the flavor and emphasis of news reporting, and public discourse itself.

"Sock puppet" accounts, similar to bots, are also intended to deceive others into believing that they are the accounts of actual users, but sock puppets are not as fully automated as bots. Instead, users control accounts that reinforce posts and narratives of another account, creating an appearance of broader, grassroots, independent support for that account or the issues and positions on which it posts. Sock puppet accounts will retweet or positively comment on the social media posts of a puppet master account, inflating its influence. Some sock puppet accounts have been reportedly used to help infiltrate groups such as ISIS, exposing members who are then arrested by the authorities.[33]

"Trolls" are users who deliberately post offensive or provocative content with a goal of upsetting others or derailing the social media conversation. Trolling is another way that motivated content can be spread or disrupted. Trolls can work independently, choosing the individuals or organizations they wish to victimize or torment. This deceptive, disruptive, or harmful behavior when committed by individuals can be motivated by a desire to damage a particular community or group, or it can be driven by joy taken in hurting or humiliating others.[34] Trolling is typically done from accounts that trolls create expressly for this purpose, and they may maintain several false profiles. These profiles may falsely present the gender, race, ethnicity, or nationality of the actual troll to increase their disruptive effectiveness. Trolls can organize themselves into groups, to "swarm" a target with their abusive or deceptive content.[35] Trolls can also be organized into "troll armies," groups of individuals commonly employed by a nation state actor to deliberately interfere with conversations or groups critical of the regime and undermine the regime's opponents. Further discussion of troll armies can be found in later sections of this chapter.

Adoption

There are a number of factors that can have an impact on adoption of narratives presented through social media. Certain figures have more credibility or influence in a community, and thus their statements and framing of events hold more weight and thus are more likely to be accepted. Such figures include those more prominent or central in a social network with many connections to others. People in powerful positions in a social network are particularly able to reinforce existing norms and behaviors, framing, and narratives. Perceptions among the majority of a network or social group create conformity pressures for individual members of the group.[36] Outside members are more likely to adopt beliefs consistent with their perception of what the rest of the network thinks.

WHO IS ON SOCIAL MEDIA?

Generalization

It is important to understand the segments of a population that use social media. Many people assume that because they use social media and their friends use social media, that essentially everyone uses social media. They then make the logical leap that social media data can be used to understand the broader information environment and provide accurate insights regarding the attitudes, sentiment, and beliefs of people in a given area. This approach is fundamentally flawed.

Personas on social media are not generalizable to the offline population, nor are they representative of the general population in a given area. A person's presence on social media and the manner in which they interact in social media will vary with age, gender, socio-economic status, personality, and whether they live in a developed or developing country, among other factors. In many cases, the population of social media users may hold strategic value. In those cases, collecting data online, conducting online activities, and assessing the impact of those activities may present a highly effective approach.

Developed countries have higher rates of Internet users than developing countries.[37] In the United States, for example, 89 percent of adults reported using the Internet in contrast to a median of 45 percent across twenty-one emerging and developing countries.[38] Social media use in the United States varies based on platform. Platform use across the developing world varies widely from country to country. Any assumptions about platform use, without recent empirical data, are possibly incorrect. It is generally the case, however, that Internet users in developing countries use the Internet for social networking at higher rates than in the developed world.[39]

Many people access the Internet through smartphones. Around 72 percent of Americans use a smart phone, compared to only 28 percent in Nigeria or Ukraine.[40] It should be noted that even in countries with low smart phone adoption rates, the size of that population may represent a large Internet market and high volume of data. For example, only 17 percent of the Indian population report smart phone use, but in a country with over one billion people, this equates to still hundreds of millions of people.

Social media can provide an important avenue to collect relevant population data, to include attitudes and beliefs. It is important to keep in mind, however, that social media does not represent the general population. Different platforms represent different segments of a population based on a wide range of variables. Careful planning is required to understand which platforms are relevant for specific population segments. Effective social media campaigns conducted on one platform, intended for a specific audience, may not be effective on a different platform, region, or audience.

Personality

Personality is an important variable that affects an individual's social media use and warrants special attention. While all people are likely to use social media, regardless of personality, their personality will affect the extent to which they use social media, the platforms they prefer, and manner in which they interact with others online. From the perspective of resistance, some of these personality traits may be aligned with traits that drive people to be more engaged in activism or have a greater need to affiliate with online groups. More on this topic can be found in chapter 7.

CHARACTERISTICS OF REAL SOCIAL NETWORKS

Network Topology

There are several structural tendencies that are often found in networks.[41] Some of these network structures imply potential properties that drive social relationships. Cellular networks are sometimes thought to be clandestine. A core-periphery network implies it may be a volunteer network. Scale-free networks are either evolved with a preferential attachment model or optimized for efficiency. Small-world networks attempt to explain the average, short distance among actors. A lattice network usually implies some kind of physical constraint on link formation within a network. This section will briefly overview these pertinent social networks.

An Erdos-Renyi Random Network (ER-random network) is defined as a network where all nodes have the same probability of connection.[42] In an ER-random network, the density of the network is equivalent to the probability of two connected nodes.

Another classic network structure is the lattice network. A lattice network is one in which all nodes have the same degree, or same number, of connections. Lattice networks often have a high diameter (longest, shortest path in the network), meaning it may take a large number of steps for knowledge or resources to transfer from one end of the network to the other.

The small-world network is a hybrid between the lattice and the ER-random networks.[43] A small-world network is defined by its method of construction. The links in a lattice network are randomly rewired. If the number of rewiring connections in the lattice is sufficiently large, relative to the number of links in the network, then the network will become ER-random.

The scale-free network is a network that consists of a few hub, or central, nodes with many connections, while most nodes have relatively few connections. A scale-free network can be constructed through a process called preferential attachment.[44] In a preferential attachment process, new nodes joining a network form links with highly connected nodes with more likelihood than with the minimally connected nodes. In this sense, the rich get richer, in terms of network connections. It should be noted that just because a preferential attachment process creates a scale-free network, a scale-free network does not imply a preferential attachment evolution process for the network. Networks optimized for social exchange will also exhibit a scale-free structure.

The core-periphery network consists of a core of a few densely connected actors. Most nodes have more connections to the core and fewer connections between other nodes in the periphery. This structure is common among religious and volunteer organizations. There often exists a core nucleus of leadership and those more dedicated to the organizations and a periphery of those that affiliate or attend organization events.

The cellular network is a network with multiple, highly connected clusters, the cells, and several links between cluster "leaders." This hypothetical structure is meant to describe clandestine networks. It would be a mistake to believe that all clandestine networks behave in this manner, however. During the US military intervention in Afghanistan, for example, the United States was very effective in targeting Taliban cell leaders and removing them from the network. The cell remained operationally ineffective for a long time until new leadership could reconnect to the larger network. The Taliban adapted its tactics, making it more ER-random than cellular.

Some of these network structures imply potential properties, driving social relationships. Cellular networks are can be clandestine with a core-periphery network structure that implies it may be a volunteer network. Scale-free networks are either evolved with a preferential attachment model or optimized for efficiency. Small-world networks attempt to explain the average, surprisingly short, distance among actors. A lattice network usually implies a physical constraint on link formation within a network.

The ER-random network implies that actors form relationships at random. However, relationships are often not formed at random. Why then do many networks observe an ER-random-like structure? The answers lie within two additional concepts: the Dunbar number and network horizon.

Dunbar Number

Robin Dunbar is an anthropologist that discovered an interesting correlation between the surface area of a primate's cerebral cortex and the maximum size of its social group.[45] For humans, this equates to an estimate of one hundred and fifty people (known as the Dunbar number). This means that when a social group exceeds this number, it is no longer possible to know everyone in the network.

There are several people who argue on the exact value of the Dunbar number. Some think it is slightly larger or smaller than one hundred and fifty, or that it varies from person to person. It is widely accepted, however, that there exists an upper limit of the number of meaningful connections a person can maintain.

> 5.4   *When organizations exceed roughly one hundred and fifty nodes, there is a meaningful change in the behavior and dynamics of the network.*

It is at roughly one hundred and fifty nodes or more at which knowledge and resource exchange within a network becomes a challenge. Some describe this value as the threshold between small and medium-size networks. This becomes important when looking at online networks. Many online networks exceed one hundred and fifty nodes. Closer inspection, however, may reveal many small sub-networks under this threshold. Obviously, the definition of a link is important in determining whether this limitation applies. It will, however, apply for real social networks.

NETWORK EFFECTS ON SOCIAL PSYCHOLOGY

Social psychology allows us to understand how social groups impact an individual's psychology and his or her conformity to social norms. It is critical for understanding influence and persuasion. The Internet affects the dynamics of social psychology differently from face-to-face interaction. Online influence, interaction, mobilization, and group dynamics are typically counter-intuitive. This section will describe some of the emerging science for understanding the impact of the Internet to change online group dynamics.

Majority Illusion

The majority illusion describes an important concept that contributes to counter-intuitive online behavior.[46] People are unable to observe all actors in a network or understand their knowledge, attitudes, beliefs, or intention. Their only view of the network is of those with whom they are directly connected and to a lesser extent one degree separated from them.

Perhaps this affects attitudes toward Da'esh. Multiple population polls find that salifist, takfiri, or extremist ideology is very rare (approximately 1 percent) among Middle Eastern populations.[47] Da'esh has high exposure, however, in online and traditional media, which increases the majority illusion effect. When asking the same people who report no affinity toward Da'esh if others support the organization, they say yes, even though they cannot name anyone.[48] Surveyed individuals cannot know the attitudes of the entire population, but that does not stop them from developing the perception that more people support Da-'esh than actually do. It is therefore possible that people will misperceive social norms based on their view of the network.

The majority illusion can occur based on two key network properties: exposure and assortativity. Exposure describes the network degree centrality of those holding the minority view. When minority node exposure is high, people are more likely to observe the minority view. The assortativity describes the extent to which high degree nodes are connected to high or low degree nodes. When high degree nodes are connected to other high degree nodes, assortativity is low. When high degree nodes are connected to low degree nodes, assortativity is high. High assortativity also creates a majority illusion effect. In the Da'esh example, the exposure of the minority who hold Da'esh beliefs is exaggerated by the media and their online messaging, allowing them to reach otherwise low degree nodes who are normal people. This may create a majority illusion effect. Network effects on the Internet can create conditions of assortativity and exposure that may magnify the majority illusion.

> 5.5 *The majority illusion is important because when people misperceive social norms, their behavior changes. Social conformity does not drive people to change attitudes, beliefs, and intention toward the true social norm. Rather, it drives people to change toward the perceived social norm.*

Pluralistic Ignorance

The importance of the majority illusion is further illustrated in a related concept called "pluralistic ignorance."[49] Pluralistic ignorance describes an individual's misperception of a social norm that drives behavior. For example, if a typical American college student is asked how many alcoholic beverages he/she consumes at a party, the response might be three to four on average. If you ask students how many alcoholic beverages their friends consume, they might, in theory, report six to eight. It is the misperception of the true social norms that lead American students to drink more, use drugs, and engage in more promiscuous sex than is common in the general population. Once the student leaves college and joins the general population, his or her perceptions of social norms change once again, and they resume more moderate behavior. While the phenomenon of pluralistic ignorance is well documented, the underlying mechanisms are not. The majority illusions provide a relatively new framework for studying influence within this context.

Social Conformity

One of the most robust findings in social psychology was pioneered by Solomon Asch in 1956.[50] Using simple experiments, Asch studied the conditions under which people conformed to group pressure. In Asch's experiments, a test subject was asked to orally report on the length of a line in a group of several others' lines. The other respondents were research actors of the experiment, however, and would unanimously report an obviously wrong answer. The focus of the experiment was to measure whether the unsuspecting participant would conform to group pressure and report the wrong answer.

Asch found that people conformed on approximately 37 percent of trials. One-third of conforming respondents said they did so because they felt that their personal perceptions must be wrong and that the group must be right. This is known as *informational conformity*. The other two-thirds did not feel comfortable disagreeing with the group. This is known as *normative conformity*.[51]

In one version of the experiment, one of the research actors reported the right answer. Asch found that conformity dropped from 37 percent to approximately 5 percent. Apparently, the reinforcement

147

of one dissenting view can disrupt the normative effect. While respondents reported warmth and good feeling toward the group member that agreed with them, they denied that it impacted their own decision to conform. This point is important because it shows that people refuse to believe they are personally affected by conformity.

Asch conformity can be observed in online communities. Echo chambers often develop increasingly polarized views. An echo chamber is a network cluster with high internal communication or connections and low external communication and connections. When members are exposed to high volumes of biased content, they develop informational and normative conformity and are likely to believe misinformation that is accepted and advocated by the group. It is important to note, however, that misinformation correction campaigns, where allied forces use true facts and logic to "correct the record" or "contest the space," are ineffective tactics and backfire in what is known as the "boomerang effect."[52] This concept will be discussed in a later section.

Asch's experiments allow us to understand conditions of social conformity. They have been repeated with many different variations. One variation, in particular, investigates social network effects on conformity.

Network Conformity

People interact online primarily through social networks. Social network scholar Ian McCulloh conducted a variant of Asch's conformity experiment among platoons (twenty to thirty people) in the US Army.[53] In McCulloh's variant, he collected social network data of friendship and respect among members of the platoon. After a period of thirty to forty-five days, he conducted a conformity experiment, similar in structure to Asch's study. Respondents were evenly divided between central and peripheral actors. The remaining people were the research actors of the experiment.

McCulloh modified the experiment by using military knowledge questions. All enlisted soldiers in the Army are required to memorize basic military knowledge for promotion boards held by senior non-commissioned officers (NCO). There was a slight modification to the logo on the Microsoft PowerPoint slides for each question to queue the research actors to report a wrong answer.

148

The findings from McCulloh's experiment were significant. In his results, people that were central in the network did not conform. People on the periphery of the network conformed over 80 percent of the time. In a subsequent study, McCulloh and colleagues conducted psychometric evaluation of an entire infantry brigade combat team prior to its deployment to Afghanistan, two to three months into their deployment, and again upon redeployment.[54]

This study measured mental health, such as depression, post-traumatic stress disorder (PTSD), social support, prior combat experience, and social network information. The study team found that the most significant factor predicting PTSD, depression, and unfortunately suicide was social isolation in the friendship network. In fact, when controlling for social isolation, direct combat had no effect on PTSD unless it involved the social loss of a friend.[55]

People need social acceptance. Without acceptance, most people develop significant mental health issues. When they occupy positions on the periphery of a social network, they are much more likely to conform to group norms to gain acceptance. Without this acceptance, they are likely to develop mental health problems and are thus biologically driven to conform. While Western culture views conformity with a negative connotation, Eastern cultures, making up two-thirds of the world's population, view the concept positively and would translate the term as harmony instead of conformity. Socially isolated people are much more susceptible to online social mobilization and are more likely to join social movements. This is referred to as *network conformity*. For this reason, analysis of online social movements, resistance, and activism must consider the social networks in which they occur.

> *5.6 Socially isolated people are much more susceptible to online social mobilization and are more likely to join social movements.*

INFLUENCE UNDER CONDITIONS OF RESISTANCE

Many people believe the Internet facilitates influence and propaganda that contributes to resistance, social mobilization, and even radicalization. It is important to understand some basic principles of influence to better understand how the Internet and social media interact with an influence process. A thorough explanation of models of behavior change are well beyond the scope of this chapter. The neurocognitive model of influence is presented because it integrates aspects of the three classic models of behavior change, which are social judgment theory, cognitive dissonance, and the theory of reasoned action.

Neurocognitive Model of Influence

The neurocognitive model of influence is useful for understanding influence under conditions of cognitive resistance. A message is not always intended for a receptive audience. In many cases, messages are intended to change someone's behavior or beliefs. If the message recipient holds differing opinions and does not behave as desired, they are likely to resist the message. This is cognitive resistance. It can be challenging to assess the effectiveness of messages in these circumstances.

Rhetorical persuasion consists of logical appeals. When the logical appeal falls within an individual's latitude of acceptance,[56] it can be effective at increasing self-integration and subsequent behavior change. When the logical appeal introduces facts inconsistent with a person's existing views and the appeal falls outside of the person's latitude of acceptance, it invokes counter-arguing. Counter-arguing not only prevents self-integration, but people can become more practiced at rationalizing their existing views and become more polarized in the opposite direction from the message intent.

Impact of Online Influence

Social media is full of messages, debates, logical arguments, and narrative appeals. Many people make the mistaken assumption that because a message is communicated that it will be believed. The neuro-cognitive influence model shows that this is simply not true. Many messages can actually illicit a boomerang effect and achieve the opposite results from those intended.

Logical debates are rarely effective on social media. They simply invoke counter-arguing. An excellent example is found in Greenpeace's social media campaign against Nestlé.[57] Greenpeace wanted Nestlé to stop buying palm oil from Sinjar Mas, a company accused of destroying rain forests. Greenpeace took its case to court in 2008, where the facts were evaluated, and Sinjar Mas and Nestlé were found to be ethical companies taking proper environmental precautions.

In 2010, Greenpeace produced a parody of a Kit Kat commercial and posted it to YouTube. It received only a few hundred views. Nestlé cited a copyright violation and had the parody video removed from YouTube. Greenpeace immediately posted the video on Vimeo and its own webpage, and within a week, the video had been viewed four hundred million times. People "liked" Nestlé's Facebook page to post hate messages. At the height of this social media firestorm, Nestlé received a negative message every ninety seconds on average for three weeks. The more Nestlé explained facts in the case, the more people became outraged. Eventually, the mainstream media started to report on the social media campaign, resulting in a drop in stock prices. At that point, Nestlé stopped buying palm oil from Sinjar Mas. Perhaps the most interesting part of the story is that, during this time, sales were up five percent from the previous year, so Nestlé actually sold more chocolate.

There are several lessons that we can learn from the Nestlé-Greenpeace social media battle. First, censorship is an excellent way to increase attention and views on social media. If Nestlé had not attempted to sensor the parody video, the media campaign would have received less attention.

The second lesson is that facts and logic are ineffective methods of persuasion. Even though Nestlé had proven its case in an objective court of law, people remain unconvinced. Instead, Nestlé's logical appeals promoted counter-arguing and increased negative opinion and sentiment. A more effective strategy would be to provide alternative content that is more interesting and compelling.

The third lesson is that it is possible to achieve a strategic objective entirely online. Greenpeace was ineffective in its attempts at a logical appeal. Greenpeace learned that it is much easier to advance causes on social media where people tend to exhibit personality traits that make them more likely to believe conspiracy theories and join in activist movements. To succeed, Greenpeace did not need to prove a point. The organization simply needed to play to people's emotions, create a large online movement, and show the data to fearful decision makers.

The fourth lesson we can learn from this case is that social media analysis focused on counting the volume of posts or identifying trending hashtags is insufficient to inform decisions. Most social media analysis solutions cannot relate social media metrics to other metrics that matter, such as sales or sentiment within a consumer base.

This behavior of choosing forums and friends can lead to echo-chambers, where clusters of actors tend to be like-minded and agree with each other. It is quite possible, however, that the Internet and echo-chambers do not create polarized groups. Polarized people join common groups to voice their concerns. In this manner, polarization creates the echo chamber, and social media simply makes this behavior more observable.

Lieutenant Colonel Matt Benigni explored this phenomenon in the context of Da'esh information warfare.[58] Da'esh established an online charity that was a financial front company supposedly supporting the children of Syria. It used "benefactor bots" to allegedly donate funds to the site. It used "support bots" to mention the benefactor bots in posts with an online target audience. This behavior led Twitter to make friend recommendations to the online target audience, which in turn led to real people friending benefactor bots. These bots delivered malware that hijacked the target audience bots, allowing Da'esh to communicate to a target online community from trusted sources. The support bots flooded the media feed during times the real people used their accounts, based on pattern of life analysis, to mask their activities. By drawing real people into an artificially constructed echo chamber, Da'esh effectively radicalized audiences.

Influence and persuasion is a somewhat counter-intuitive phenomenon. People prefer to believe that they make their decisions independent of others, despite the overwhelming evidence to the contrary. The Internet compounds this problem; online influence and persuasion is even more counter-intuitive. It is critical that special operators learn the science of influence and persuasion to conduct missions ranging from PSYOP to unconventional warfare.

SOCIAL MEDIA FIRESTORMS

Definition

A firestorm is defined as a large, negative, word–of-mouth discussion conducted over social media.[59] Examples include the Greenpeace-Nestlé campaign, public response to Chipotle's 2015 E. coli outbreak, or the 2017 online outcry to cancel the Stephen Colbert show in America. Firestorms are interesting because they provide a clear example of large numbers of people joining together to protest a specific issue.

Observed Behaviors

Firestorm behavior is counter-intuitive as well. Hemank Lamba's 2015 study collected data for eighty major Twitter firestorms. They found that a small number of users were responsible for the majority of tweets. The firestorm was driven more by discussion and responses than by actual events. They tended to be self-reinforcing. Most interestingly, they occurred over a very short duration. The typical firestorm only lasted a couple days. Those firestorms that persist past a couple days tend to involve an opponent, usually a large corporation or government agency, using logical appeals in an attempt to refute the statements of the online movement. Firestorms that are able to persist beyond a few days can be more effective in achieving their goals. They can only achieve this kind of effectiveness, however, by drawing the large corporation or government into an online debate.

Russia demonstrated a different, more effective, approach. Whenever online firestorms are initiated against Russia, its government does not engage in an online fight.[60] Instead, it introduces more compelling alternate content. An example of this was the incident where Russian air defense shot down a civilian airline. Rather than argue, they flooded social media with tabloid-like information, drawing international and activist attention away from the incident. In this manner, the Russian government distracted social media users, which allowed the firestorm to die quickly.

Online users also tend to distrust governments or large organizations. The planned and coordinated strategic communications of these entities do not appear authentic and are therefore seen as untrustworthy.

Firestorms are fueled by debate. Facts and logic are not helpful in shaping these online movements. Large companies or governments are not trusted in the conversation. It is far more effective to either distract users with alternate, more compelling content or to facilitate online communities that can carry an uncoordinated, yet more authentic campaign.

SOCIAL MEDIA SPREAD OF INFORMATION AND COUNTERING TECHNIQUES

Social media provides previously unprecedented speed and reach for the promulgation of messages. This section discusses the following topics related to the spread of ideas in social media and techniques to counter that spread as used by both resistance movements and state security services:

- Virality
- Counter messaging
- Attacking, refuting, and distracting
- Poisoning the well
- False claims

- Astroturfing
- Censorship

Virality

In 2010 and 2011, the world watched the Arab Spring revolutions unfold across the Middle East and North Africa. Social media was a key factor in generating the widespread support and mobilizing huge, national-level protests that were previously unprecedented. As a result, observers from around the world concluded that all was required to organize a revolution was social media.[61]

In reality, one major reason that the Arab Spring occurred was that the state security services were not watching social media channels. "Mostly what we got was people on the ground -- participants, dissidents -- because the Egyptian government was clueless."[62] At the time, they were unaware of the reach and anonymity of social media and thus were taken by surprise.

The Internet has made a big difference in Egypt. For years, the country's secret police and state-controlled media very effectively suppressed most dissident activities. Without the relatively free arena of online social networking sites and tools like Facebook, Twitter, and YouTube, young Egyptians like Ghonim[63] could not have built the resilient and creative force that finally toppled Hosni Mubarak.[64]

When a particular message is quickly promulgated and magnified in social media, it is called "going viral." Anyone who has a message to share wants it to go viral, to reach as many people as possible, and have them pass it along to many others. "As [Ghonim] told CNN's Wolf Blitzer on Friday, 'We would post a video on Facebook and it would be shared by 50,000 people on their walls in hours.'"[65]

The mass mobilizations generated during the Arab Spring created a cascading effect, wherein the growth in the number of people mobilized increased exponentially. A few individuals started a message on Twitter or Facebook, which was then shared by others, who shared with their friends, who in turn shared with their friends, etc. As a result, a viral message mobilized very large numbers of people in a short period of time.[66] Use of automated retweet apps, like the ISIS "Dawn of Glad Tidings" can quickly spread automatically without the need for human intervention.

These messages to mobilize were tremendously successful, partly because the state security services provided no counter messages to preclude the cascade effect. It was a relatively free arena, which allowed mass mobilization demonstrations to become organized even when other channels of communications were watched and censored. This is notable because the free arena of social media changes the reach and speed of the messages.[67]

Unfortunately for resistance movements fighting oppressive regimes, most if not all of the social media channels are now watched. Repressive regimes cannot afford an Arab Spring-like event in their nations and monitor social media attentively.

Not only are state security services now watching social media, they also participate in social media, both overtly and covertly. Overtly, the Egyptian police created a presence on Facebook to connect better with the Egyptian populace.[68] The Egyptian Army also joined Facebook later that week.[69] Covertly, the state security services sponsor counter messaging in the same social media to preclude the cascade effect from forming unchallenged, as described in the next section on counter-messaging.

Counter Messaging

More than simply watching, regimes participate in social media. Because the greatest threat to a repressive regime is a massive mobilization of the populace, state security services must present counter messages to the messages of dissent, reform, revolution, or mobilization. For example, China hired large numbers of "50-Cent bloggers" to support the government and to counter any and all messages that may be critical of it.[70]

Russia hired similar "troll armies," called "web brigades," to support the party line and discredit any negative messages online.[71] Besides questioning the patriotism of a message critical of the regime, other techniques to attack the messenger and the message include questioning the religious "purity" of the messenger, or even questioning his/her motives. Sowing doubts about the messenger is a common technique to reduce public support for the message.[72]

Both Chinese and Russian state security services use "trolls" to guide and manipulate online dialogues. Initially, most counter messaging focused on discrediting the messenger, which was often sufficiently effective at suppressing dissent within a nation. However, China's 50Cent bloggers have become more sophisticated because the online audience (known as netizens) has become more sophisticated. As described by an anonymous 50-Cent blogger in a 2011 interview by Ai Weiwei:

> The netizens are used to seeing unskilled comments that simply say the government is great or so and so is a traitor. They know what is behind it at a glance. The principle I observe is: don't directly praise the government or criticize negative news. Moreover, the tone of speech, identity and stance of speech must look as if it's an unsuspecting member of public; only then can it resonate with netizens. To sum up, you want to guide netizens obliquely and let them change their focus without realising it.
>
> *[Weiwei] How big a role do you think this industry plays in guiding public opinion in China?*
>
> Truthfully speaking, I think the role is quite big. The majority of netizens in China are actually very stupid. Sometimes, if you don't guide them, they really will believe in rumors.
>
> *[Weiwei] Because their information is limited to begin with. So, with limited information, it's very difficult for them to express a political view.*
>
> I think they can be incited very easily. I can control them very easily. Depending on how I want them to be, I use a little bit of thought and that's enough. It's very easy. So I think the effect should be quite significant.[73]

The number of 50-Cent bloggers is unknown, but each is prolific. Continuing with the interview of the 50-Cent blogger:

> The process has three steps – receive task, search for topic, post comments to guide public opinion. Receiving a task mainly involves ensuring you open your email box every day. Usually after an event has happened, or even before the news has come out, we'll receive

an email telling us what the event is, then instructions on which direction to guide the netizens' thoughts, to blur their focus, or to fan their enthusiasm for certain ideas. After we've found the relevant articles or news on a website, according to the overall direction given by our superiors we start to write articles, post or reply to comments. This requires a lot of skill. You can't write in a very official manner, you must conceal your identity, write articles in many different styles, sometimes even have a dialogue with yourself, argue, debate. In sum, you want to create illusions to attract the attention and comments of netizens... I go online for six to eight hours nearly every day. I'm mainly active on our local BBS and some large mainstream internet media and microblogs. I don't work over weekends, but I'll sign in to my email account and see if there's any important instruction.[74]

Moreover, the interviewed blogger estimated the percentage of online dialog produced by the bloggers group to be substantial:

Because I do this, I can tell at a glance that about 10 to 20 percent out of the tens of thousands of comments posted on a forum are made by online commentators.[75]

Through the use of these trolls, the state security services are able to monitor social media sites for burgeoning messages that could create dissention among the public. In response, they create counter messages that are then promulgated by other trolls or by automated bots when a large response is required. As described in the next section, pretending that certain messages are generated by grassroots citizens as opposed to the government or another organization is a common technique to sway public opinion on social media.

These troll armies appear to be sufficiently effective to prevent a cascading effect of mass mobilization and support for the resistance movement to preclude them from reaching the level of the Arab Spring. Although often clumsy and apparent, even these artificial voices in social media have a dampening effect on the opportunities for mass mobilization by a resistance movement against repressive state security services. Social media is now watched by the state security service, and the anonymity of social media is no longer guaranteed.[76]

Attacking, Refuting, and Distracting

Social media has extensive reach. With that reach, a single message can become widespread if it gains popularity, as described in the earlier section on virality. The messages spread more quickly when there is little to block or refute the original message.

One of the strongest types of messages to express refutation is the *accusation*. An accusation is easy to make and does not require proof or evidence when exposed to the public. Moreover, an accusation gives the accuser the initiative and defines the arena in which the argument occurs. The accusation may frame the issue to the disadvantage of its target. The target of the accusation is immediately on the defensive and must choose to address the accusation, remain silent, or distract the populace from the accusation.

> 5.7 *An accusation gives the accuser the initiative and defines the arena in which the argument occurs. The target of the accusation is immediately on the defensive and must choose to address the accusation, remain silent, or distract the populace from the accusation.*

False accusations are a common ploy used in social media and other public communication channels. Although US law assumes that a party is innocent until proven guilty, most of the rest of the world does not, nor does the media or the general public:

> Some audiences naively accept all accusations on the assumption that "if it weren't true, why would they say so?" Critical thinking and considering the source and its motivations are rare commodities in the modern world (particularly if the audience is already pre-inclined to believe the source of an accusation rather than the target). During the Korean War, for example, any successful military action against Chinese forces generated a Chinese accusation of war crimes against civilians. Although ludicrously false, such accusations played well to Chinese citizens and allies, and caused the U.S. discomfort in the international community.[77]

In the Ukraine conflict, Russia-driven social media frequently portrayed pro-Kiev factions as fascists or Nazis due to the support of many Ukrainians for Germany in WWII.[78] Another theme in Russian disinformation was purported brutality against pro-Russian separatists. They disseminated pictures of atrocities from other conflicts and presented them as events that occurred in Ukraine. To support these accusations, Russian trolls posted gruesome pictures from the Syrian war of dead women and children and led users to believe the event occurred in Ukraine:[79]

> Just how much of the Russian TV and print media's 24/7 coverage of rampaging Ukrainian extremists; swastika-bearing neo-Nazis; pro-Ukrainian thugs beating Russian speakers; "Right Sector" extremists gunning down unarmed civilians at a checkpoint, and joyous Crimeans welcoming their Russian saviors was fabricated? How much of it was real? The answer: Very little, if any. But it was successful nonetheless.
>
> Vladimir Putin's invisible social-media campaign included fiction writers posting on fake Facebook (or the Russian version *Vkontakte*) accounts, pretending to have witnessed some horrendous crime committed by Ukrainian extremists. A second wing handled the shadowy distribution of photo-shopped or staged photos, again featuring Ukrainian atrocities. A third wing spreads rumors to destabilize entire communities and districts. No one really knows from where the fake photo came or who originated the rumor, but they continue to spread through the targeted population.
>
> Even when exposed, they have already done their damage. It is difficult to wipe out a graphic image imbedded in a viewer's brain, no matter how false.[80]

Once an accusation is made in social media, it takes the initiative in the competition of ideas and may become viral. For example, the Guardian "reports 40,000 comments a day by an 'orchestrated pro-Kremlin campaign" of pro-Russian trolling on Ukraine stories.[81] Once it becomes viral, it becomes very difficult to counter completely (see the later section on poisoning the well). For example, the "Pizzagate" shooting incident of 2016 was initiated by a person trying to verify for himself a set of false stories and faked evidence that the Comet Ping

Pong pizzeria and other nearby businesses were involved in a pedophile ring associated with Democratic politicians. The faked evidence included edited photos of accused persons as well as genuine photos taken at different locations, which were intended to provide proof of a pedophile ring. Based on all of this purported evidence, an individual tried to investigate the accusations himself with a loaded weapon and fired three shots. Fortunately, no one was injured. Even though the accusation was thoroughly debunked by many independent sources, a poll taken just two weeks after the shooting event still demonstrated high response rates for those who believed the original accusation was true.[82]

The target of a false accusation can choose to remain silent because sometimes silence is the best response. As mentioned in the "Combatting Fake News" conference report:

> An important implication of this point is that any repetition of misinformation, even in the context of refuting it, can be harmful (Thorson, 2015, Greenhill and Oppenheim, forthcoming). This persistence is due to familiarity and fluency biases in our cognitive processing: the more an individual hears a story, the more familiar it becomes, and the more likely the individual is to believe it as true (Hasher et al 1977; Schwartz et al, 2007; Pennycook et al., n.d.).[83]

In the book *LikeWar*, the authors concluded that, "What counted most was *familiarity*." As long as people see a similar headline as one they had seen before, they are significantly more likely to believe it.[84]

Conversely, others may view silence as consent and presume that not refuting an accusation validates its accuracy. As a result, a rapid response in the same media channel can effectively attempt to refute the accusation and maintain the issue as an argument rather than a *fait accompli*. Social media is a battleground of ideas, and refutation of accusations is an essential part of precluding one-sided dominance in that battlespace.

In general, the speed and reach of social media requires speed in response. Participants need members who can predict likely accusations and prepare responses to remain ahead of the competition. Significant effort must be expended to quickly disprove false accusations whenever possible. Delays in responding to an accusation can be interpreted as resulting from the time required to prepare a lie.

An alternative to direct refutation is distraction. Rather than engage in a debate against an accusation, simply distract the audience away from the topic. As described by Alex Tabarrok:[85]

> We estimate that the government fabricates and posts about 448 million social media comments a year. In contrast to prior claims, we show that the Chinese regime's strategy is to avoid arguing with skeptics of the party and the government, and to not even discuss controversial issues. We infer that the goal of this massive secretive operation is instead to distract the public and change the subject, as most of these posts involve cheerleading for China, the revolutionary history of the Communist Party, or other symbols of the regime.[86]

An example of social media used in China by the 50-Cent bloggers was described as the following:

> For example, each time the oil price is about to go up, we'll receive a notification to "stabilise the emotions of netizens and divert public attention." The next day, when news of the rise comes out, netizens will definitely be condemning the state, CNPC and Sinopec. At this point, I register an ID and post a comment: "Rise, rise however you want, I don't care. Best if it rises to 50 yuan per litre: it serves you right if you're too poor to drive. Only those with money should be allowed to drive on the roads . . ."
>
> This sounds like I'm inviting attacks but the aim is to anger netizens and divert the anger and attention on oil prices to me. I would then change my identity several times and start to condemn myself. This will attract more attention. After many people have seen it, they start to attack me directly. Slowly, the content of the whole page has also changed from oil price to what I've said. It is very effective.[87]

Distraction adds benefits of avoiding any repetition of the accusation, as well as avoiding risking upsetting an audience via a fumbled response. Distraction serves as a powerful tool, even more so since the advent of social media because of its speed and reach.

> 5.8 *Any repetition of misinformation, even in the context of refuting it, can be harmful. If a rapid, successful refutation is not feasible, silence may be a better response. However, distraction away from the accusation appears to be even more successful than silence in countering a false accusation.*

Poisoning the Well

One particularly disturbing effect of the accusation is that even when clearly disproven, damage is still done to the target of such accusation. In a paper by Australian psychologist Stephan Lewandowsky *et al*, the authors noted that:

> The wealth of studies on this [misinformation] phenomenon have documented its pervasive effects, showing that it is extremely difficult to return the beliefs of people who have been exposed to misinformation to a baseline similar to those of people who were never exposed to it.

> For example, Green and Donahue (2011) first presented people with a report that was found to change people's attitudes about an issue (e.g., a report about a heroin addicted child changed people's attitudes toward the effectiveness of social youth-assistance programs). Participants then received a retraction stating that the report was inaccurate, either because of a mix-up (error condition) or because the author had made up most of the "facts" in order to sensationalize the report (deception condition). The results showed that participants were motivated to undo their attitudinal changes, especially in the deception condition, but that the effects of misinformation could not be undone in either condition. The misinformation had a

> continuing effect on participants' attitudes even after
> a retraction established the author had made it up.[88]

Humans tend to place greater emphasis on the first information received, even if that information is later refuted. In one study, test subjects were exposed to false information about the negligent owner of a building that caught fire due to the storage of oil paints and gas cylinders. Even after the retraction, the test subjects believed negligence was still a cause of the fire.

It follows that when people later re-encounter the misinformation (e.g., "oil paints and gas cylinders were present"), it may be more familiar to them than without the retraction, leading them to think, "I've heard that before, so there's probably something to it." This impairs the effectiveness of public information campaigns intended to correct misinformation.

Even the repetition of the information in negated form reinforces a false weight of evidence argument. The many accusations appearing in the media carry significant long-term impact, even after retractions are made and the accusation proven false.[89]

> *5.9* *False accusations appear to result in political damage to the target in spite of any refutations or disproval, which is why "poisoning the well" is such an effective attack.*

False Claims

Many false claims can be made and promulgated simply due to the wide reach and influence of social media. A few individuals can claim to speak for many, the number of posts in different social media channels can give a false impression of independent corroboration, and many accusations from supposedly independent sources can provide a false weight of evidence. Each of these three manipulative uses of social media is described in this subsection.

False Claim of Speaking for the Masses

As described in the preceding sections, social media can provide a voice for large groups of people. This has become such a common phenomenon that there is now an assumption that if someone claims to speak for the masses that the claim is true. For example, a reporter may select a representative tweet from Twitter claiming or at least implying that the contents represent public opinion.

During the 2012 presidential debates, reporters focused not so much on what candidates said but on tweets about the debate. The content of the tweets became the news, rather than the content presented by the candidates:

> Journalists should get off Twitter and watch the debate, without being influenced by their pals, and then report on what they actually saw instead of what the emerging narrative is. I imagine none of them wanted to dispute the narrative their colleagues were pumping out and risk being ostracized for having an independent thought. But Americans and their bosses are paying them to cover events and not to tweet it.[90]

Focusing on the content of tweets rather than on the event is bad enough, but the anonymity of social media makes it possible to manipulate the news itself. For example, a selected tweet can be handpicked by a reporter to place the desired spin on the message to better align with journalistic bias. Worse, rather than waiting for the "right" tweet to appear, the reporter can write (or have a colleague write) the desired tweet and report on the self-created tweet as though it represents the large corpus of tweets posted by the masses. The anonymity of social media lends itself to manipulation by organized groups, or even a small number of people backed by social bot software.[91]

Besides tweets, a similar approach of generating false representative social media content entails video of a person on the street, who is really a plant by the person creating the video. The supposedly spontaneous interview is well rehearsed, and all the key points desired by the reporter are addressed by the supposedly random interviewee. Once posted to YouTube or on a news website, the impression is that this one video depicts mass opinion, as opposed to identifying it for the propaganda it is.

Many examples appeared daily from Russian troll armies generating purported atrocities by the Ukrainian government and its supporters. Many of these pretended to be first-person accounts:

> Interviews with innocent by-standers and ordinary citizens are a staple fare of the coverage. A woman shows the camera hundreds of spent cartridges she gathered after a night of violence. Extremists turn outraged local residents, on their way to visit wounded comrades, away from the hospital. A babushka, in tears, bemoans the terror in which she lives and pleads for the Russians to restore order and civilization. Pretty good stuff. I'd believe it if I did not know better.[92]

Fortunately, with so many professional influencers operating in parallel, mistakes happen:

> Three different channels have featured interviews with one Andrei Petkov, lying wounded in a hospital in the south Ukrainian city of Nikolayev. In the three interviews, he is identified by name. He is on his back in a hospital bed, describing his experiences in the previous evening's violence, which left him with serious wounds. Petkov is dressed in a black outfit, his nose bandaged. In each interview, he speaks softly, but with earnest conviction. He cuts a sympathetic and credible figure. The problem is that Andrei Petkov is a different person in each interview![93]

In one interview, this interviewee claimed to be a spy from Germany bringing weapons and fifty mercenaries to Ukraine. In the second interview, the same actor claimed to be an ordinary citizen of Ukraine at his "usual" protest against the new Ukrainian government and was attacked and injured by neo-Nazi Ukrainians. In the third interview, he said he was an innocent pediatric surgeon injured by an unprovoked attack by neo-Nazi Ukrainians.[94]

Although a blatant falsehood that was caught and exposed, it appeared that the Kremlin desired effect on the Russian populace occurred:

> Apparently Russian viewers *want* to believe these fairy tales. They want to think their country is in the right. They want to be proud of their country. Accordingly,

they make ideal subjects for Big Lie propaganda. I do not know how they will feel when they eventually learn the truth.[95]

Note that organized staging of interviews of random passersby or innocent victims is a common method of claiming to speak for the masses. The false claim of speaking for the masses is one mechanism by which a minority group, whether a resistance movement or a state security service, can create support for their position. The "majority illusion" can encourage people to support the position. Claiming to speak for the masses is one way to intentionally generate the majority illusion.

False Independent Corroboration

Independent corroboration occurs when a different observer (or set of measurements) independently confirms a claim made by the first observer. This is often required in legal proceedings and carries more weight in public opinion than a single person making a claim. When the same message appears in a wide range of apparently independent sources via different media channels, then many consider that sufficient independent confirmation.

Unfortunately, due to both the variety and anonymity of social media, one can never be sure that the corroborating statements are from independent sources. If Fred in Oklahoma posts the same information as Sue in Oregon and Charlie in Florida, then are these independent corroborations, or are Fred, Sue, and Charlie actually the same person with three online personas? Even if not the same person, are Fred, Sue, and Charlie intentionally colluding to express the same sentiments while pretending to be independent of each other?

> [Social Media] activism seems to be dominated by those who are already active in the offline environment, people associated with conventional politics, and a limited number of influential bloggers – professional influencers. It is up to them to turn a particular SM-based movement in a direction of their choice.[96]

One must be suspect of apparently independent corroboration in the media, especially social media, because the variety and anonymity of media make it highly susceptible to manipulation. What appears as independent voices cannot be readily identified as collaborative efforts to manipulate public opinion. In the case of Russian trolls supporting anti-Ukraine stories, they generated forty thousand comments a day by an orchestrated pro-Kremlin campaign.[97]

Moreover, unwitting audiences often repost stories without any critical evaluation. For example, one Russian troll army report about a dentist who was refused entry to save victims on Ukrainian atrocities from a burning building "gathered more than 5,000 shares one day after it appeared. Rosovskiy's [the dentist's] account was expeditiously translated into English, German, and Bulgarian. Why not believe the story? It seemed to make sense, after all the noise about the Kiev extremists."[98]

Repeating a lie from a different source should not count as independent corroboration, but it appears to do just that in social media, especially when it spills over into mainstream media.

False Weight of Evidence

Generating a large number of similar messages appearing from different sources over time gives the impression of a "weight of evidence" to support the claim. This tactic is especially effective when the claim contains an accusation, as previously described.

Simply making a large number of accusations against a person or a nation, regardless of how false, carries weight in the minds of many. For example, an opponent can generate many accusations of wrong doing against a politician, even though none might be true, to influence public opinion. The intent is to influence the public to believe, "If the person wasn't guilty of something, why are so many accusations from so many different people?"

The possibility that the accusations may all be part of a coordinated attack on an individual is often not considered by the general public. The principle of innocent until proven guilty does not apply beyond the courtroom and is often ignored in political debate. As long as the various accusations are similar and sufficiently regular over time, the accused suffer from the apparent weight of evidence, regardless of how unfounded they are.

Just as claiming to speak for the masses can support the intentional majority illusion, false weight of evidence is another mechanism to generate the majority illusion. By repeating the same message or accusation, using false independent confirmation, the accuser can create the desired majority illusion.

On a national level, al Qaeda repeated the same set of three accusations in its narrative against the United States since 9/11: claiming the "War on Terror" was really a "War on Islam," the West is after the Muslim world's oil, and Western men are after Muslim women. Al Qaeda and its supporters generated story after story to reinforce these same three accusations, thereby carrying sufficient weight of evidence that large populations in the Middle East believed them to be true.

To counter the al Qaeda champion of victims narrative, Western nations, such as the United Kingdom, developed and distributed counter narratives as part of a campaign. As reported in *The Guardian* in an article by Alan Travis:

> The target of the campaign - the al-Qaida narrative - is seen as linking together genuine or perceived, commonly held concerns into a "narrative of grievance" that reinforces the portrayal of Muslims as victims of western injustice. "It [the narrative] combines fact, fiction, emotion and religion and manipulates discontent about local and international issues. The narrative is simple, flexible and infinitely accommodating. It can be adapted to suit local conditions and may have a disproportionate influence on understanding and interpretation of local or global events.[99]

When al Qaeda interpreted every event to align with its narrative, it created an appearance of weight of evidence against the West and helped its claimed position of defending Muslim victims from Western aggression. Media channels that openly supported al Qaeda repeated these interpretations, further contributing to the "weight of evidence" behind the accusation.

> *5.10 Types of false claims in social media include: a) a few individuals claiming to speak for many, b) many posts in different media channels giving the impression of independent corroboration, and c) many accusations from supposedly independent sources providing a false weight of evidence.*

Astroturfing

"Astroturfing is the practice of masking the sponsor of a message or organization (e.g., political, advertising, religious or public relations) to make it appear as though it originates from and is supported by grass-roots participants."[100] Astroturfing can be used for a wide variety of purposes, such as advertising or political spin. The anonymity and reach of a variety of social media channels makes it difficult for an individual, or even governments, to identify the legitimate sources of these messages intended to manipulate public opinion through social media. While social media appears to be a bottoms up or crowd-sourced method of expression, the anonymity of these social media lend themselves to manipulation by those trying to coopt the appearance of wide-spread, bottom-up support for a particular message.

The use of bots that pretend to be real and unique people, but are actually controlled by a single individual exacerbate the problem.[101] While it is possible to distinguish social bot from human behavior, it requires time and money. In the meantime, the messages sent by the bots spread faster than detection and exposure.

In 2011, over a two-week period, researchers used "three social bots that were able to integrate themselves into the group, and gained close to 250 followers between them. They received more than 240 responses to the tweets they sent. The best performing bot was able to gain more than 100 followers and generated almost 200 responses."[102] These bots fooled human users.

Although Twitter and Facebook since improved their ability to detect bots, large social botnets still exist. In January 2017, for example, researchers accidently discovered massive collections of dormant, fake accounts on Twitter:

> The largest network ties together more than 350,000 accounts and further work suggests others may be even bigger... Some of the accounts have been used to fake follower numbers, send spam and boost interest in tending topics... "Considering all the efforts already there in detecting bots, it is amazing that we can still find so many bots, much more than previous research," Dr. Zhou told the BBC.[103]

Astroturfing can be used by both resistance movements and state security services, and social bots help magnify the reach and therefore the speed of disseminating messages.

Censorship

Now that the Internet and its connections to other communications systems (such as cell phones) are available in most of the world, traditional censorship is no longer feasible. Although North Korea still executes its citizens for simply connecting to the Internet, videos of the executions and the unrest in North Korea continue to escape from the country. If the most controlled and most repressive nation in the world cannot censor all channels, even under pain of death, there is little chance that any other nation can achieve the level of censorship previously available to the state security services of highly repressive regimes.

At the same time, censorship is not necessary for a repressive regime to survive if they learn how to manipulate social media at least as well as the resistance movement. As previously described, China's 50Cent bloggers represent an avenue to respond to negative messages in social media without the need for direct censorship. While arresting bloggers for statements against the government still occurs, much of the online "discussion" in social media involves a government-hired blogger who guides and manipulates it. Discussion is in quotes because often a single blogger argues both sides of an argument to either distract audience members away from a topic or to guide them to a particular opinion. As described by an anonymous 50Cent blogger in China:

> In a forum, there are three roles for you to play: the leader, the follower, the onlooker or unsuspecting member of the public. The leader is the relatively authoritative speaker, who usually appears after a controversy and speaks with powerful evidence. The public usually finds such users very convincing. There are two opposing groups of followers. The role they play is to continuously debate, argue, or even swear on the forum. This will attract attention from observers. At the end of the argument, the leader appears, brings out some powerful evidence, makes public opinion

align with him and the objective is achieved. The third type is the onlookers, the netizens. They are our true target "clients". We influence the third group mainly through role-playing between the other two kinds of identity. You could say we're like directors, influencing the audience through our own writing, directing and acting. Sometimes I feel like I have a split personality.[104]

When the same influencer playing different roles in the same online discussion, the dialogue becomes a part of the influence campaign that distracts audiences from genuine issues. In this case, one person can be very prolific, accomplish a particular influence objective, and act as a form of indirect censorship by pulling attention away from the real issues.

In a similar manner, Russian trolls use the "three actor format" but use three different people, one for each role:

You got a list of topics to write about. Every piece of news was taken care of by three trolls each, and the three of us would make up an act. We had to make it look like we were not trolls but real people. One of the three trolls would write something negative about the news, the other two would respond, "You are wrong," and post links and such. And the negative one would eventually act convinced. Those are the kinds of plays we had to act.[105]

State security service manipulation of social media both internally and abroad continues to become a primary tool in the global competition of ideas. Competing messages in cyber space is the new norm. When coupled with traditional terror tactics, ether by the resistance movement or state security services, the battle for hearts and minds continues against a background of more traditional lethal threats.

KEY TAKEAWAYS

Mechanisms for influence and organized resistance over social media are often counter-intuitive. Operations developed by uninformed planners often result in unintended consequences. They may even serve the interests of adversaries. Awareness of recent scientific insights into online influence, cyber resistance, and other resistance movements will equip planners to conduct effective online operations to counter adversaries that attempt to harm US interests.

5.1   The online population is not representative of the general population.

5.2   A shared narrative helps build bonds and deepen trust with others, supporting cooperation.

5.3   Fake news stories that are consistently framed with a target audiences' identity, cultural values, or accepted narratives are more likely to be believed than real news that does not conform to those expectations.

5.4   When organizations exceed roughly one hundred fifty nodes (personas), there is a meaningful change in the behavior and dynamics of the network.

5.5   The majority illusion is important because when people misperceive social norms, their behavior changes. Social conformity does not drive people to change attitudes, beliefs, and intention toward the true social norm. Rather, it drives people to change toward the perceived social norm.

5.6   Socially isolated people are much more susceptible to online social mobilization and are more likely to join social movements.

5.7 An accusation gives the accuser the initiative and defines the arena in which the argument occurs. The target of the accusation is immediately on the defensive and must choose to address the accusation, remain silent, or distract the populace from the accusation.

5.8 Any repetition of misinformation, even in the context of refuting it, can be harmful. If a rapid, successful refutation is not feasible, silence may be a better response. However, distraction away from the accusation appears to be even more successful than silence in countering a false accusation.

5.9 False accusations appear to result in political damage to the target in spite of any refutations or disproval, which is why "poisoning the well" is such an effective attack.

5.10 Types of false claims in social media include: a) a few individuals claiming to speak for many, b) many posts in different media channels giving the impression of independent corroboration, and c) many accusations from supposedly independent sources providing a "false weight of evidence."

Considering Takeaway 5.2, the information environment is a contested domain, one in which the narratives that spread through social media compete to influence public opinion and motivate collective action and resistance. The attainment of US national security objectives and foreign policy goals can no longer be effectively decoupled from understanding of, and ability to operate in, the social media landscape. As they spread online, narratives can frame what issues matter and determine the positions that populations or audiences take on them. Narratives can be powerful forces for change or for resisting change. They can persuade people of new goals, bring about changes in their sense of identity, and spur them to organized, collective action. They can also be used to evoke a glorified past that must be returned to, stemming societal progress.

Considering Takeaway 5.3 a little deeper, both resistance movements and state security services compete for influence and dominance in social media. Both sides leverage the virality of social media as other groups retransmit their messages—often regardless of whether these messages are true or false. Counter-messaging can be direct—refuting a statement made by the opposing side—or may be more oblique by distracting audiences from key issues.

Finally, regarding Takeaway 5.7, accusations give the accusing side the initiative and set the arena for subsequent discussion. Even disproven false accusations still garner significant support in spite of clear refutation. This makes it easy for an actor on social media to "poison the well" against an individual or a group.

ENDNOTES

1   Stephen John Read and Lynn Carol Miller, "Stories Are Fundamental to Meaning and Memory: For Social Creatures, Could It Be Otherwise," *Knowledge and Memory: The Real Story. Advances in Social Cognition* 8 (1995): 139–152.

2   Summer Agan, *Narratives and Competing Messages* (Fort Bragg, NC: US Special Operations Command, 2018), 7.

3   Mark A. Finlayson and Steven R. Corman, "The Military Interest in Narrative," *Sprache Und Datenverarbeitung* 37, no. 1–2 (2013), https://users.cs.fiu.edu/~markaf/doc/j2.finlayson.2013.sdv.37.173.pdf.

4   Walter R. Fisher, "Narration as a Human Communication Paradigm: The Case of Public Moral Argument," *Communication Monographs* 51, no. 1 (March 1, 1984): 1–22, https://doi.org/10.1080/03637758409390180.

5   George Lakoff, Howard Dean, and Don Hazen, *Don't Think of an Elephant! Know Your Values and Frame the Debate ; the Essential Guide for Progressives* (White River Junction, Vt: Chelsea Green Pub. Co, 2004).

6   Erving Goffman, *Frame Analysis: An Essay on the Organization of Experience* (Harvard University Press, 1974).

7   Paul R. Brewer and Kimberly Gross, "Values, Framing, and Citizens' Thoughts about Policy Issues: Effects on Content and Quantity," *Political Psychology* 26, no. 6 (December 1, 2005): 929–48, https://doi.org/10.1111/j.1467-9221.2005.00451.x.

8   Frank E. Dardis et al., "Media Framing of Capital Punishment and Its Impact on Individuals' Cognitive Responses," *Mass Communication and Society* 11, no. 2 (April 7, 2008): 115–40, https://doi.org/10.1080/15205430701580524.

9   David A. Snow et al., "Ideology, Frame Resonance, and Participant Mobilization," *International Social Movement Research* 1, no. 1 (1988): 197–217.

10 Nathan D. Bos et al., *Human Factors Considerations of Undergrounds in Insurgencies*, Second (Alexandria, VA: US Army Printing Office, 2013).

11 Temitope B. Oriola and Olabanji Akinola, "Ideational Dimensions of the Boko Haram Phenomenon," *Studies in Conflict & Terrorism* (June 1, 2017): 1–24, https://doi.org/10.1080/1057610X.2017.1338053.

12 Bos et al., *Human Factors Considerations of Undergrounds in Insurgencies*.

13 Jure Leskovec, Lars Backstrom, and Jon Kleinberg, "Meme-Tracking and the Dynamics of the News Cycle," in *Proceedings of the 15th ACM SIGKDD International Conference on Knowledge Discovery and Data Mining*, KDD '09 (New York, NY, USA: ACM, 2009), 497–506, https://doi.org/10.1145/1557019.1557077.

14 Ibid.

15 W. Lance Bennett and Alexandra Segerberg, "Digital Media and the Personalization of Collective Action," *Information, Communication & Society* 14, no. 6 (September 1, 2011): 770–99, https://doi.org/10.1080/1369118X.2011.579141.

16 W. Lance Bennett and Alexandra Segerberg, "The Logic of Connective Action: Digital Media and the Personalization of Contentious Politics," *Information, Communication & Society* 15, no. 5 (2012).

17 Jeffrey M. Ayres, "From the Streets to the Internet: The Cyber-Diffusion of Contention," *Annals of the American Academy of Political and Social Science* 566, no. 1 (1999): 132–143.

18 Clay Fink et al., "Complex Contagions and the Diffusion of Popular Twitter Hashtags in Nigeria," *Social Network Analysis and Mining* 6, no. 1 (December 1, 2016): 1, https://doi.org/10.1007/s13278-015-0311-z.

19 Kevin M. DeLuca, Sean Lawson, and Ye Sun, "Occupy Wall Street on the Public Screens of Social Media: The Many Framings of the Birth of a Protest Movement," *Communication, Culture & Critique* 5, no. 4 (December 1, 2012): 483–509, https://doi.org/10.1111/j.1753-9137.2012.01141.x.

20 Craig Silverman, "This Analysis Shows How Viral Fake Election News Stories Outperformed Real News On Facebook," *BuzzFeed*, accessed July 18, 2017, https://www.buzzfeed.com/craigsilverman/viral-fake-election-news-outperformed-real-news-on-facebook.

21 "Toward A Model of Meme Diffusion (M3D) - Spitzberg - 2014 - Communication Theory - Wiley Online Library," accessed February 28, 2017, http://onlinelibrary.wiley.com/doi/10.1111/comt.12042/full.

22 Hunt Allcott and Matthew Gentzkow, "Social Media and Fake News in the 2016 Election," National Bureau of Economic Research, January 2017, https://doi.org/10.3386/w23089.

23 Matthew Baum, David Lazer, and Nicco Mele, "Combating Fake News: An Agenda for Research and Action" (paper for Harvard Kennedy School, Shorenstein Center on Media, Politics, and Public Policy, February 18, 2017), https://shorensteincenter.org/combating-fake-news-agenda-for-research/.

24 Cass R. Sunstein, Echo Chambers: *Bush v. Gore, Impeachment, and Beyond* (Princeton University Press Princeton, NJ, 2001), https://pdfs.semanticscholar.org/4e7c/434ec3b8eaf2 6c62642dfbac56be6eef9647.pdf; Elanor Colleoni, Alessandro Rozza, and Adam Arvidsson, "Echo Chamber or Public Sphere? Predicting Political Orientation and Measuring Political Homophily in Twitter Using Big Data: Political Homophily on Twitter," *Journal of Communication* 64, no. 2 (April 2014): 317–32, https://doi.org/10.1111/jcom.12084.

25 David M. J. Lazer et al., "The Science of Fake News," *Science* 359, no. 6380 (March 9, 2018): 1094–96, https://doi.org/10.1126/science.aao2998.

26 S. P. Borgatti et al., "Network Analysis in the Social Sciences," *Science* 323, no. 5916 (2009): 892.

27 Nicole B. Ellison and Danah M. Boyd, "Sociality Through Social Network Sites," *Oxford Handbook of Internet Studies,* January 1, 2013, https://doi.org/10.1093/oxfordhb/9780199589074.013.0008.

28 "How ISIS Is Winning the Online War for Iraq," *New Scientist*, accessed May 5, 2017, https://www.newscientist.com/article/dn25788-how-isis-is-winning-the-online-war-for-iraq/.

29 Robert Gorwa, "On the Internet, Nobody Knows That You're a Russian Bot," *RealClearDefense*, accessed May 5, 2017, http://www.realcleardefense.com/articles/2017/03/21/on_the_internet_nobody_knows_that_youre_a_russian_bot_111010.html.

30 Emilio Ferrara et al., "The Rise of Social Bots," *Communications of the ACM* 59, no. 7 (June 24, 2016): 96–104, https://doi.org/10.1145/2818717.

31 Jacob Ratkiewicz et al., "Detecting and Tracking Political Abuse in Social Media," *ICWSM* 11 (2011): 297–304.

32 Ferrara et al., "The Rise of Social Bots."

33 Lorraine Murphy, "The Curious Case of the Jihadist Who Started Out a Hacktivist," *Hive*, accessed August 1, 2017, https://www.vanityfair.com/news/2015/12/isis-hacker-junaid-hussain.

34 Erin E. Buckels, Paul D. Trapnell, and Delroy L. Paulhus, "Trolls Just Want to Have Fun," *Personality and Individual Differences*, The Dark Triad of Personality, 67 (September 2014): 97–102, https://doi.org/10.1016/j.paid.2014.01.016.

35 Whitney Phillips, "LOLing at Tragedy: Facebook Trolls, Memorial Pages and Resistance to Grief Online," *First Monday* 16, no. 12 (November 28, 2011), http://firstmonday.org/ojs/index.php/fm/article/view/3168.

36 Peter V. Marsden and Noah E. Friedkin, "Network Studies of Social Influence," *Sociological Methods & Research* 22, no. 1 (August 1, 1993): 127–51, https://doi.org/10.1177/0049124193022001006.

37 Pew Research Center, "Smartphone Ownership and Internet Usage Continues to Climb in Emerging Economies," February 2016.

38 Ibid.

39 Ibid.

40 Ibid.

41 Several ARIS publications discuss the different organizational strategies adopted by resistance groups and the implications of organization on the movement's subsequent behaviors. See chapter 3 of *Human Factors Considerations of Undergrounds in Insurgencies*, and chapter 4 of *The Science of Resistance*.

42 P. Erdos and A. Rényi, "On Random Graphs. I," *Publicationes Mathematicae* 6 (1959): 290–297.

43 D. J. Watts and S. H. Strogatz, "Collective Dynamics of 'Small-World' Networks," *Nature* 393, no. 6684 (1998): 440–442.

44 A. Barabasi and R. Albert, "Emergence of Scaling in Random Networks," *Science* 286 (1999): 509-512.

45 Robin Dunbar, "The Social Brain Hypothesis and Its Implications for Social Evolution," *Annals of Human Biology* 36, no. 5 (August 2009): 562–72.

46 Lerman Kristina, Xiaoran Yan, and Xin-Zeng Wu, "The 'Majority Illusion' in Social Networks," *PLoS ONE* 11, no. 2 (2016), https://doi.org/10.1371/journal.pone.0147617.

47 Based on numerous polling efforts commissioned by the Department of State, USCENTCOM, SOCCENT, and nongovernment organizations. Methods vary from traditional polls to specialized research methods that avoid explicit bias. These data are not publicly available but can be requested through the Joint Staff J39, USCENTCOM J3-IO, or SOCCENT J5.

48 This research uses implicit measures to avoid fear, dishonesty, and explicit bias in data collection. A review of these methods is beyond the scope of this text.

49 Dale T. Miller and Leif D. Nelson. "Seeing Approach Motivation in the Avoidance Behavior of Others: Implications for an Understanding of Pluralistic Ignorance," *Journal of Personality and Social Psychology* 83, no. 5 (2002): 1066–1075.

50 Solomon E. Asch, "Studies of Independence and Conformity: I. A Minority of One against a Unanimous Majority," *Psychological Monographs: General and Applied* 70, no. 9 (1956): 1–70.

51 Ibid.

52 Ibid.

53 Ian McCulloh, "Social Conformity in Networks," *Connections* 33, no. 1 (2013): 35-42.

54 Ibid.

55 Ibid.

56 The latitude of acceptance is the range of possible views on a subject that someone finds acceptable. For example, someone who believes murderers should be incarcerated for life without the possibility of parole may find capital punishment acceptable, but releasing a convict after only serving ten years of their sentence may fall outside of their latitude of acceptance and be considered unreasonable.

57 Robin Shreeves, "Greenpeace and Nestle in a Kat Fight," *Forbes*, March 19, 2010, https://www.forbes.com/2010/03/18/kitkat-greenpeace-palm-oil-technology-ecotech-nestle.html.

58 M. Benigni, "Detection and Analysis of Online Extremist Communities," (dissertation for Carnegie Mellon University) Technical Report CMU-ISR-17-108, 2017.

59 Hemank Lamba, Momin M. Malik, and Juergen Pfeffer, "A Tempest in a Teacup? Analyzing Firestorms on Twitter," *IEEE/ACM International Conference*, 2015, https://www.researchgate.net/publication/301444817_A_Tempest_in_a_Teacup_Analyzing_Firestorms_on_Twitter.

60 The ARIS publication *"Little Green Men:" A Primer on Modern Russian Unconventional Warfare, Ukraine 2013-2014* includes an in-depth discussion of Russia's information warfare tactics. See pgs. 14-19.

61 Emily Banks, "Egyptian President Steps Down Amidst Groundbreaking Digital Revolution," Mashable, *CNN.com*, February 11, 2012.

62 John D. Sutter, "Will Twitter War Become the New Norm?" *CNN.com*, November 15, 2012.

63 Wael Ghonim is an Egyptian who helped organize the mass mobilization in Egypt. For details, see his entry in Wikipedia.

64 Micah L. Sifry, "Did Facebook Bring Down Mubarak?" *CNN online*, February 11, 2011.

65 Ibid.

66 The multiplicative effect of retweets can help create an exponential growth in the reach of a message through auto-repeaters, such as the previously described ISIS app, Dawn of Glad Tidings. Auto-repeaters automatically retweet messages from a specified source without requiring a human to make the retweet.

67 Sutter, "Will Twitter War Become the New Norm?"

68 Raja Abdulrahim, "Egypt Police Try to Improve Image through Facebook," *Los Angeles Times*, February 19, 2011.

69 "Egypt: Military Junta Launches Facebook Page," *Telegraph*, February 17, 2011.

70 Sarah Cook, "China's Growing Army of Paid Internet Commentators," *freedomhouse.org*, October 11, 2012; Ai Weiwei, "China's Paid Trolls: Meet the 50-Cent Party," *newstatesman.com*, October 17, 2012.

71 "Web Brigades," Wikiepedia, accessed July 9, 2017.

72 Patrick Allen, *Information Operations Planning* (Boston, MA: Artech House, 2007), 123-124.

73 Weiwei, "China's Paid Trolls: Meet the 50-Cent Party."

74 Ibid.

75 Ibid.

76 John Koetsier, "China Bans internet Anonymity," *Venturebeat.com*, December 28, 2012.

77 Pete Middleton, "When Chinese Troops Fired on Two Gloster Meteors at Chongdan, the Australians Made Them Regret It," *Military History*, August 2005; Allen, *Information Operations Planning*, 266.

78 Ignas Kalpokas, "Influence Operations: Challenging the Social Media – Democracy Nexus," *SAIS European Journal of Global Affairs*, 2016.

79 Paul Roderick Gregory, "Inside Putin's Campaign of Social Media Trolling and Faked Ukrainian Crimes," *Forbes*, May 11, 2014.

80 Ibid.

81 Ibid.

82 "Pizzagate Consipiracy Theory," Wikipedia, http://www.wikipedia.org/wiki/pizzagate_conspiracy_theory.

83 Baum, Lazer, and Mele, "Combatting Fake News." References within this quote are further cited in the source.

84 P. W. Singer and Emerson T. Brooking, *LikeWar: The Weaponization of Social Media* (New York: Houghton Mifflin Harcourt Publishing Company, 2018), 124.

85 Kevin Drum, "Social Media is Best Used for Distraction, not Argument," *Mother Jones News*, January 17, 2017, https://www.motherjones.com/kevin-drum/2017/01/social-media-best-used-distraction-not-argument/.

86 Ibid.

87 Weiwei, "China's Paid Trolls: Meet the 50-Cent Party."

88 Stephan Lewandowsky et al., "Misinformation and Its Correction Continued Influence and Successful Debiasing," *Psychological Science in the Public Interest*, December 2012.

89 Gregory, "Inside Putin's Campaign of Social Media Trolling and Faked Ukrainian Crimes."

90 John Amato, "Has The Beltway Created A Twitter Media Playhouse?" *crooksandliars.com*, October 10, 2012.

91 Yazan Boshmaf et al., "The Socialbot Network: When Bots Socialize for Fame and Money," Annual Computer Security Applications Conference (ACSAC), Orlando Florida, 5-9 Dec 2011.

92 Paul Roderick Gregory, "Russian TV Propagandists Caught Red Handed: Same Guy, Three Different People (Spy, Bystander, Heroic Surgeon)," *Forbes*, April 12, 2014.

93 Ibid.

94 Ibid.

95 Ibid.

96 Kalpokas, "Influence Operations."

97 Gregory, "Inside Putin's Campaign of Social Media Trolling and Faked Ukrainian Crimes."

98 Ibid.

99 Alan Travis, "Battle Against al-Qaida Brand Highlighted in Secret Paper," *Guardian*, August 26, 2008.

100 "Astroturfing," Wikipedia, June 2017, https://en.wikipedia.org/wiki/Astroturfing.

101 Boshmaf, "The Socialbot Network."

102 Jim Giles, "Fake Tweets by Socialbot Fool Hundreds of Followers," *New Scientist*, Reed Business Information, UK, March 19, 2011.

103 "Massive Networks of Fake Accounts Found on Twitter," *BBC online*, Technology section, January 24, 2017.

[104] Weiwei, "China's Paid Trolls: Meet the 50-Cent Party."

[105] Anton Troianovski, "A Former Russian Troll Speaks: 'It Was like Being in Orwell's World,'" *Washington Post*, February 17, 2018, https://www.washingtonpost.com/news/worldviews/wp/2018/02/17/a-former-russian-troll-speaks-it-was-like-being-in-orwells-world/.

CHAPTER 6.
IMPLICATIONS OF CYBER-PHYSICAL SYSTEMS

INTRODUCTION

The criticality of realizing the vulnerabilities of cyber-physical systems is emphasized in the 2018 National Cyber Strategy. It states:

> America's prosperity and security depend on how we respond to the opportunities and challenges in cyberspace. Critical infrastructure, national defense and the daily lives of Americans rely on computer-driven and interconnected information technologies. As all facets of American life have become more dependent on a secure cyberspace, new vulnerabilities have been revealed and new threats continue to emerge.[1]

IT networks have historically been the target of hackers, as noted in chapter 3. However, with an increasing number of instances of physical processes under the monitoring and control of various types of computing device, those devices and the processes they control will also be an increasingly common target of attack. As large numbers of these systems that cross the cyber and physical appear within the operational environment, military personnel must be aware of these cyber-physical systems and their implications for military operations.

Cyber-physical systems are a general category overlapping that of two related terms: the Internet of things (IoT) and ICS. The interrelated nature of the terms is clear in the National Institute of Standards and Technology (NIST) definition: "Cyber-physical systems (CPS) are smart systems that include engineered interacting networks of physical and computational components."[2] They also provide further amplification, stating:

> CPS generally involves sensing, computation and actuation. CPS involve traditional information technology (IT) as in the passage of data from sensors to the processing of those data in computation. CPS also involve traditional operational technology (OT)

185

for control aspects and actuation. The combination of these IT and OT worlds along with associated timing constraints is a particularly new feature of CPS.[3]

The definition of IoT is similar. An Internet Society white paper states "The term Internet of Things generally refers to scenarios where network connectivity and computing capability extends to objects, sensors and everyday items not normally considered computers, allowing these devices to generate, exchange and consume data with minimal human intervention."[4] As with cyber-physical systems, NIST states that the "IoT involves sensing, computing, communication, and actuation."[5] Common examples of the IoT include printers, routers, video cameras, thermostats, refrigerators, and televisions.

ICS share many of these same attributes. ICS, often sub-categorized into SCADA, distributed control systems (DCS), or programmable logic controllers (PLCs), are computational systems used for the management and control of physical processes. "ICS are typically used in industries such as electric, water and wastewater, oil and natural gas, transportation, chemical, pharmaceutical, pulp and paper, food and beverage, and discrete manufacturing (e.g., automotive, aerospace, and durable goods.)"[6] A key component within an ICS is a control loop, which "utilizes sensors, actuators, and controllers (e.g., PLCs) to manipulate some controlled process."[7] A typical ICS layout might resemble that depicted in Figure 6-1.

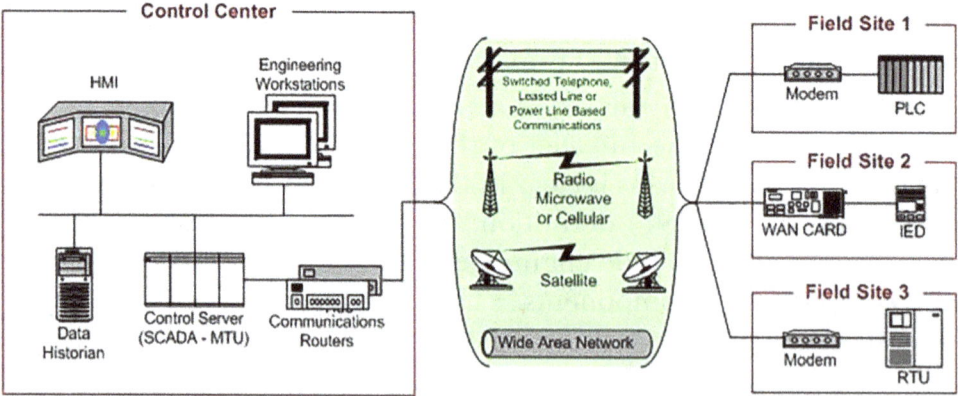

Figure 6-1. ICS system layout.

As noted, these terms have much in common, often distinguished primarily by the use to which they are put (e.g., electric grid operations are typically categorized as ICS, while certain smart grid technologies might utilize the IoT terminology[8]). The two most important shared characteristics are:

- The linking of cyber domain and the physical domains.
- The inclusion of sensing, computing, communication, and actuation functions within one system.

> 6.1 *Cyber-physical systems, to include the IoT and ICS, conduct sensing, computing, communication, and actuation functions.*

SECURITY IMPLICATIONS OF CYBER-PHYSICAL SYSTEMS

Cyber-physical systems pose many security challenges that cannot be easily solved through the use of typical IT or cyberspace defense activities. "The introduction of IT capabilities into physical systems presents emergent behavior that has security implications."[9] First, certain common practices used to defend IT networks, such as automated vulnerability scanning and patching, are not typically used on cyber-physical systems. NIST delineated some common differences between IT systems and ICS (see Table 6-1).

Table 6-1. IT / ICS system differences.

Category	IT System	Cyber-Physical System
System Operation	Systems are designed for use with typical operating systems. Upgrades are straightforward with the availability of automated deployment tools.	Differing and possibly proprietary operating systems, often without security capabilities built in. Software changes must be carefully made, usually by software vendors, because of the specialized control algorithms and perhaps modified hardware and software involved.
Resource Constraints	Systems are specified with enough resources to support the addition of third-party applications such as security solutions.	Systems are designed to support the intended industrial process and may not have enough memory and computing resources to support the addition of security capabilities.
Change Management	Software changes are applied in a timely fashion in the presence of good security policy and procedures. The procedures are often automated.	Software changes must be thoroughly tested and deployed incrementally throughout a system to ensure that the integrity of the control system is maintained. ICS outages often must be planned and scheduled days/weeks in advance. ICS may use operating systems that are no longer supported

Similar considerations apply to IoT devices as well, with "many IoT devices intentionally designed without any ability to be upgraded, or the upgrade process is cumbersome or impractical."[10] However, as the IoT market expanded, some other concerns have become more important. In particular, the sheer numbers of IoT devices, when combined with the typical security concerns, pose a problem themselves:

- "Many IoT deployments will consist of collections of identical or near identical devices. This homogeneity magnifies the potential impact of any single security vulnerability by the sheer number of devices that all have the same characteristics."[11]

- "Many Internet of Things devices, such as sensors and consumer items, are designed to be deployed at a massive scale that is orders of magnitude beyond that of traditional Internet connected devices.[12]

These issues will likely continue to worsen because estimates foresee an increase in the overall size of the population of IoT devices. A list of 2015 projections (from separate sources) provided ranges from twenty-four billion Internet-connected objects by 2019, seventy-five billion networked devices by 2020, and one hundred billion IoT connections by 2025.[13] In sum, it is more difficult to defend cyber-physical systems, and there are increasingly many more of them to defend.

> 6.2 Common cyberspace defense techniques are often ineffective when guarding a cyber-physical system.

IMPACTS OF CYBER-PHYSICAL SYSTEMS IN CYBERSPACE

These security concerns can have impacts in both the cyber and physical spaces. In cyberspace, the number of vulnerable devices represents the inventory that an attacker can control in regard to sensing, computing, communication, and actuation capability. Recent examples include the use of malware to target vulnerable IoT devices and create large-scale botnets.

For example, a type of malware called "Mirai" has been used in high profile DDoS attacks, generating a record bandwidth estimated as high as 1.5 terabits per second in an attack on a French website in September 2016:[14]

> The Mirai malware continuously scans the Internet for vulnerable IoT devices, which are then infected and used in botnet attacks. The Mirai bot uses a short list of 62 common default usernames and passwords to

189

scan for vulnerable devices. Because many IoT devices are unsecured or weakly secured, this short dictionary allows the bot to access hundreds of thousands of devices.[15]

Mirai, which typically targets home routers, network-enabled cameras, and digital video recorders[16] also caused a major website outage in October 2016. The DDoS attack targeted the "Dyn" company, which provides managed Domain Name System (DNS) services to variety of other websites.[17] These attacks caused outages for numerous websites that used Dyn for DNS services:[18]

> But hundreds of thousands, and maybe millions, of those security cameras and other devices have been infected with a fairly simple program that guessed at their factory-set passwords — often "admin" or "12345" or even, yes, "password" — and, once inside, turned them into an army of simple bots. Each one was commanded, at a coordinated time, to bombard a small company in Manchester, N.H., called Dyn DNS with messages that overloaded its circuits.[19]

In the cases of the DDoS attacks, it could be said that an attacker utilized, out of the four key functions of a cyber-physical system, the combined computing and communication capabilities of hundreds of thousands of the devices. However, a member of a resistance organization could subvert these capabilities for covert communications, such as a method of "dead drop" communications. Rather than dropping information in a physical location, the information can be uploaded into a virtual location in cyberspace on an IoT device, limited by the amount of memory available on such a device.

However, the sensing and actuation functions are just as susceptible to misuse. For example, there could be data integrity issues with the authentication of sensors, which may deliberately misidentify themselves.[20] There could also be confidentiality issues with sensing. If a resistance member places a security camera outside a safe house but does not change the default password, then the state security service may readily guess the password and access the camera's feed. Therefore, instead of helping protect members of the resistance, an unsecured IoT device could be used against the resistance movement. Finally, with regarding to the actuation function, if "fed malicious data from other 'things', issues with life-threatening consequences are possible if the actuator operates in a safety-critical environment."[21]

> *6.3 Cyber-physical devices have been used as a means for large-scale cyber attacks, and each of the sensing, computing, communication, and actuation functions has the potential for misuse.*

IMPACTS OF CYBER-PHYSICAL SYSTEMS IN PHYSICAL DOMAINS

In physical domains, the impact of these cyber-physical systems is linked to the criticality of the physical processes involved. Because these processes are often part of critical civilian infrastructure,[22] the impacts can be severe. Additionally, critical infrastructure entities or sectors often have interdependencies, with attacks on one sector chaining into other sectors. An example is the dependency of multiple types of critical infrastructure on electric power:

> Electric power is often thought to be one of the most prevalent sources of disruptions of interdependent critical infrastructures. As an example, a cascading failure can be initiated by a disruption of the microwave communications network used for an electric power transmission SCADA system. The lack of monitoring and control capabilities could cause a large generating unit to be taken offline, an event that would lead to loss of power at a transmission substation. This loss could cause a major imbalance, triggering a cascading failure across the power grid. This could result in large area blackouts that could potentially affect oil and natural gas production, refinery operations, water treatment systems, wastewater collection systems, and

191

pipeline transport systems that rely on the grid for electric power.[23]

These systems and services are tied to the effectiveness of governance and therefore directly affect the success or failure of a resistance organization. Joint doctrine identifies the relationship between governance, legitimacy, and provision of services:

> A state's ability to provide effective governance rests on its political and bureaucratic willingness, capability, and capacity to establish rules and procedures for decision making, as well as its ability to provide public services in a manner that is predictable and acceptable to the local population.[24]

In joint doctrine, these essential services are often identified using sewage, water, electricity, academics, trash, medical, security, and other considerations.[25] "Because essential services are often a clear sign of effective governance, facilities and personnel that provide these services are often perceived as high value targets for insurgents and other adversaries."[26] Most of these services will rely in some way upon ICS or other types of cyber-physical systems, and therefore systems may represent a critical vulnerability for essential services.

> *6.4   Cyber-physical systems will likely be a critical vulnerability for governance, legitimacy, and the provision of essential services.*

General examples of adversarial incidents that might target ICS or cyber-physical systems are listed in Table 6-2.

Table 6-2. Example ICS adversarial incidents.

Threat Event	Description
Denial of Control Action	Control systems operation disrupted by delaying or blocking the flow of information, thereby denying availability of the networks to control system operators or causing information transfer bottlenecks or denial of service by IT-resident services (such as DNS)
Control Devices Reprogrammed	Unauthorized changes made to programmed instructions in PLCs, RTUs, DCS, or SCADA controllers, alarm thresholds changed, or unauthorized commands issued to control equipment, which could potentially result in damage to equipment (if tolerances are exceeded), premature shutdown of processes (such as prematurely shutting down transmission lines), causing an environmental incident, or even disabling control equipment
Spoofed System Status Information	False information sent to control system operators either to disguise unauthorized changes or to initiate inappropriate actions by system operators
Control Logic Manipulation	Control system software or configuration settings modified, producing unpredictable results
Safety Systems Modified	Safety systems operation are manipulated such that they either (1) do not operate when needed or (2) perform incorrect control actions that damage the ICS
Malware on Control Systems	Malicious software (e.g., virus, worm, Trojan horse) introduced into the system.

Beyond a list of notional possibilities, incidents occurred in which an ICS or related critical infrastructure suffered a cyber attack. Some representative events include:

- Stuxnet. Likely the most famous cyber attack targeting ICS or critical infrastructure and discovered in 2010, Stuxnet "included a highly specialized malware payload that was designed to target only specific SCADA systems that were configured to control and monitor specific industrial processes."[27]

- Shamoon. "Saudi Aramco, which is the world's 8th largest oil refiner, experienced a malware attack that targeted their refineries and overwrote the attacked system's Master Boot Records (MBR), partition tables and other random data files."[28] While this attack targeted critical infrastructure, the Shamoon malware does not appear to specifically target ICS.[29]

In comparison with the numerous cyber incidents occurring in typical IT networks, there are relatively few that target ICS specifically. As late as 2015, an overview of cyber aspects in the conflict between Russia and Ukraine included the following statements:

- "However, although an increase in typical cyber skirmishes was reported throughout the crisis, prominent cyber operations with destructive effects have not yet occurred."[30]

- "Neither critical infrastructure nor Ukrainian weapons have been damaged or disrupted."[31]

Soon afterwards, the first known instance of a destructive cyber attack on an electric power system occurred in Ukraine.

This significant escalation in cyber conflict occurred in December 2015 during a cyber attack on several Ukrainian electrical power distribution networks, causing power outages lasting several hours that affected approximately 225,000 people.[32] This attack displayed a variety of tactics, techniques, and procedures (TTPs), including the use of spear-phishing, malware, and virtual private networks (VPNs)[33] to traverse the target networks.[34] The attackers appeared to gain and maintain access to the electrical power networks for at least six months,[35] and the three separate power companies were attacked within thirty minutes of each other,[36] affecting thirty separate electrical substations.[37]

A second attack on Ukrainian electric power systems occurred in December 2016, targeting a single electrical substation but affecting comparable amounts of total power (135 MW in 2015 and 200 MW in 2016).[38] However, the malware utilized in the 2016 attack was more sophisticated. While the 2015 attack required remote attacker interaction with the system, the 2016 malware operated autonomously and was the first instance of a modularized malware targeting electric power.[39]

A final example of a cyber capability developed specifically to target ICS is HatMan malware. This capability focuses specifically on safety systems, which could potentially lead to impacts in multiple infrastructure sectors. "Safety controllers are used in a large number of environments, and the capacity to disable, inhibit, or modify the ability of a process to fail safely could result in physical consequences."[40] This malware, which may have been used in an August 2017 cyber attack on systems in Saudi Arabia,[41] represents an additional increase in sophistication. "HatMan follows Stuxnet and Industroyer/CrashOverride in specifically targeting devices found in industrial control system (ICS) environments, but surpasses both forerunners with the ability to directly interact with, remotely control, and compromise a safety system—a nearly unprecedented feat."[42]

> 6.5 *While still relatively rare, cyber attacks on ICS and supported critical infrastructure are increasing in capability and sophistication.*

KEY TAKEAWAYS

The concept of threat is often described as a function of capability and intent. Cyber-physical systems, to include the IoT and ICS, continue to increase in numbers and in their sensing, computing, communication, and actuation functionality. The demonstrated capability of a cyber attacker to maliciously affect such systems continues to proportionally increase. As these systems become ever more ubiquitous, their capabilities, for use and misuse, grow more important to military personnel. They will become ever more integrated into the provision of essential services and governance, and it can be assumed that resistance organizations will intentionally target such systems.

6.1 Cyber-physical systems, to include the IoT and ICS, conduct sensing, computing, communication, and actuation functions.

6.2 Common cyberspace defense techniques are often ineffective when guarding a cyber-physical system.

6.3 Cyber-physical devices have been used as a means for large-scale cyber attacks, and each of the sensing, computing, communication, and actuation functions has the potential for misuse.

6.4 Cyber-physical systems will likely be a critical vulnerability for governance, legitimacy, and the provision of essential services.

6.5 While still relatively rare, cyber attacks on ICS and supported critical infrastructure are increasing in capability and sophistication.

ENDNOTES

1   Donald J. Trump, "National Cyber Strategy of the United States of America," President of the United States, September 2018.

2   Cyber-Physical Systems Public Working Group, "Framework for Cyber-Physical Systems: Volume 1, Overview," National Institute of Standards and Technology (NIST) Special Publication 1500-201, Version 1.0 (June 2017), vi.

3   Ibid., 2.

4   Karen Rose, Scott Eldridge, and Lyman Chapin, "The Internet of Things: An Overview. Understanding the Issues and Challenges of a More Connected World," The Internet Society, October 2015.

5   Jeffrey Voas, "Networks of 'Things,'" National Institute of Standards and Technology Special Publication 800-183 (July 2016), 1.

6   Keith Stouffer, Joe Falco, and Karen Scarfone, "Guide to Industrial Control Systems (ICS) Security," National Institute of Standards and Technology (NIST) Special Publication 800-82, Revision 2 (May 2015), 1.

7   Ibid., 2–3.

8   Rose et al., "The Internet of Things," 73.

9   Stouffer et al., "ICS Security," 2–1.

10 Rose et al., "The Internet of Things," 34.

11 Ibid.

12 Ibid.

13 Ibid., 8.

14 US Department of Homeland Security, Cybersecurity and Infrastructure Security Agency (CISA) Cyber + Infrastructure, "Heightened DDoS Threat Posed by Mirai and Other Botnets," Alert (TA16-288A), October 14, 2016, last revised October 17, 2017.

15 Ibid.

16 Ibid.

17 Scott Hilton, "Dyn Analysis Summary of Friday October 21 Attack," *Oracle Vantage Point in the News*, October 21, 2016, https://www.pcworld.com/article/3133847/internet/ddos-attack-on-dyn-knocks-spotify-twitter-github-etsy-and-more-offline.html.

18 Ibid.

19 David E. Sanger and Nicole Perlroth, "A New Era of Internet Attacks Powered by Everyday Devices," *New York Times*, October 22, 2016.

20 Voas, "Networks of 'Things,'" 25.

21 Ibid.

22 Stouffer et al., "ICS Security," 2–1.

23 Ibid., 2–3.

24 US Joint Chiefs of Staff, "Counterinsurgency," Joint Publication 3-24 (JP 3-24), April 25, 2018, I-6.

25 US Joint Chiefs of Staff, "Stability," Joint Publication 3-07 (JP 3-07), August 3, 2016, III-59.

26 Ibid.

27 Stouffer et al., "ICS Security," C-12.

28 Ibid.

29 US Department of Homeland Security, Cybersecurity and Infrastructure Security Agency (CISA) Cyber + Infrastructure, "Shamoon/DistTrack malware (Update B)," ICS Joint Security Awareness Report (JSAR-12-241-01B), October 16, 2012, last revised April 18, 2017.

30 Sven Sakkov, "Foreword" in *Cyber War in Perspective: Russian Aggression Against Ukraine*, ed. Kenneth Geers (Tallinn, Estonia: NATO Cooperative Cyber Defence Center of Excellence, 2015), 8.

31 James Andrew Lewis, "'Compelling Opponents to Our Will': The Role of Cyber Warfare in Ukraine," in *Cyber War in Perspective: Russian Aggression Against Ukraine*, ed. Kenneth Geers (Tallinn, Estonia: NATO Cooperative Cyber Defence Center of Excellence, 2015), 41.

32 Robert M. Lee, Michael J. Assante, and Tim Conway, "Analysis of the Cyber Attack on the Ukrainian Power Grid Defense Use Case," SANS and Electricity Information Sharing and Analysis Center (E-ISAC), March 18, 2016, https://www.nerc.com/pa/CI/ESISAC/Documents/E-ISAC_SANS_Ukraine_DUC_18Mar2016.pdf, 2.

33 A VPN uses cryptography to securely emulate a point-to-point link while using a shared or public network.

34 Lee et al., "Cyber Attack on the Ukrainian Power Grid," 5.

35 Ibid., 6.

36 Ibid., 2.

37 Ibid., 1.

38 Ibid., 3.

39 Ibid.

40 US Department of Homeland Security, Cybersecurity and Infrastructure Security Agency (CISA) Cyber + Infrastructure, "HatMan—Safety System Targeted Malware (Update A)," Malware Analysis Report (MAR-17-352-01), April 10, 2018, 2.

41 Nicole Perlroth and Clifford Krauss, "A Cyberattack in Saudi Arabia Had a Deadly Goal. Experts Fear Another Try," *New York Times*, March 15, 2018.

42 US Department of Homeland Security, "HatMan," 2.

CHAPTER 7.
HUMAN FACTORS CONSIDERATIONS OF UNDERGROUNDS IN CYBER RESISTANCE

INTRODUCTION TO THE CYBER UNDERGROUND

Unconventional warfare in cyberspace may be the future of special warfare,[1, 2] which requires a better understanding of not only cyberspace but also its unique ecology.[3] The emergence of computer-based communication technologies has not changed the nature of irregular warfare, but it altered the characteristics, such as speed, reach, and effectiveness of the psychological battle to inform and influence various target audiences. Irregular conflict is no longer geographically constrained or relegated to the grievances of local in-groups but often waged, violently and nonviolently, globally for a local political objective. From peaceful social mobilization[4] to "internet guerrilla warfare,"[5] cyberspace has become an increasingly contested operational environment. SOF will need to include the cyber domain in their planning considerations as it, more so than any ungoverned physical territory, has the potential to enable disproportionate effects on small groups and individuals, and these asymmetries should be exploited to advance US objectives.[6]

The goal of this chapter is to provide a theoretical, empirical, and operational update to previous ARIS research[7] from a cyber-psychological perspective to help lay the intellectual foundations of modern instances of unconventional, political, and psychological warfare. The focus is not on state-sponsored cyber warfare or cyber organizations such as the People's Liberation Army's Unit 61398, the Islamic Revolutionary Guard Corps' Iranian Cyber Army, Russia's "Information Troops," or the Israeli Defense Forces' Unit 8200. The structure, function, personnel, and operations of such state organizations are considerably different than those of non-state actors included here.

This chapter is based largely on previous work published in *Undergrounds in Insurgent, Revolutionary, and Resistance Warfare*[8] and *Human Factors Considerations of Undergrounds in Insurgencies*,[9] with a specific emphasis on the psychological factors associated with the clandestine component of resistance in the cyber domain. As with the latter book, the term "human factors" refers to "the psychological, cultural, behavioral, and other human attributes that influence decision-making, the flow of information, and the interpretation of information by individuals and groups at any level in any state or organization."[10] Despite the limits of this definition,[11] the human factors associated with cyber resistance include how interconnected computer technologies affect organizational design, leadership, social influence, mental health, and the other related topics in the behavioral sciences. The chapter seeks to 1) integrate the technological concepts addressed in greater detail elsewhere in this volume and 2) outline how cyber affects the traditional underground functions of leadership and organization, recruiting, intelligence, financing, logistics, training, communications, security, subversion and sabotage, and psychological operations.

THE UNDERGROUND IN CYBER RESISTANCE

The term "underground," outside the study of insurgencies, typically refers to the relatively inaccessible subculture of any particular sector; the cyber underground encompasses individuals who live part of their lives in the world of information and communication technology (ICT).[12] The cyber underground is a type of playing field for aspiring hackers as neither physical prowess, socioeconomic status, or academic achievement is favored. Rather it is the demonstration of ability that is valued. Cyber resistance in all forms throughout its brief history emerged from this concept of the cyber underground.

Cyber resistance can take at least one of three forms: physical, syntactical, and/or semantic.[13] Cyber resistance in the physical domain includes interfering with the material form and/or function of a system—for example, gaining access to a secure computer laboratory and defacing, sabotaging, or destroying hardware.[14] Syntactical cyber resistance entails manipulating the software of a system for a purpose not intended by the developer(s)—for example, some early computer gangs manipulated the code on video game software so it could be copied and shared without others having to purchase it.[15] Finally, semantic cyber resistance entails engaging and undermining the discursive norms and realities of the system.[16] Semantic resistance is the most sophisticated form of resistance as it entails not only an astute understanding of acceptable standards but also the ability to subtly manipulate (or in some cases, not so subtly destroy) those standards.

The cyber underground, as discussed throughout this chapter, is the clandestine component of a cyber resistance movement; it is established to operate in areas denied to, or conduct operations not suitable for, the armed or public components. Undergrounds initiate recruiting, training, and infiltration, establish escape-and-evasion networks, raise funds, establish safe havens, and develop external support.[17] The establishment of formal organizations and training programs and the coordinated penetration of government entities are historic functions of the underground, as is intelligence and logistics support to these operations. Undergrounds also often coordinate the engagement of diaspora communities for financial, logistic, and/or informational support. Many of these functions require overt movements and/or establishment of relationships outside the insurgent institution.[18]

Insurgencies exist in both overt and clandestine domain. Figure 7-1 depicts some of the overt and covert functions of an underground. Much like undergrounds attempt to both hide and operate in the physical domain, cyber undergrounds are designed to do the same in cyberspace. Much of the early activity involves disseminating information to generate internal and external support, shape perceptions, and set conditions for broader mobilization.[19]

Undergrounds may evolve to conduct subversive, psychological operations to undermine and delegitimize the government and cultivate popular support.[20] In the nascent phase, the underground predominates, but as the movement evolves, either the armed and/or public components increase in preeminence.[21]

> *7.1   Unconventional warfare in cyberspace requires a rich contextual understanding of the sociotechnical aspects of the cyber ecology.*

Figure 7-1. Covert and overt functions of an underground.

The Hacker Ethic

The cyber underground did not necessarily arise with malicious intent or even a political objective; it began with students seeking to push the bounds of academic stricture and explore the technological potential of computing. The roots of cyber resistance lie in the telecommunications "phreakers" of the late 1950s and later the early adopters of Internet technologies who equated hacking with revolutionary behavior.[22] The earliest manifestation of a cyber underground culture is this hacker ethic, a concept that helps describe the behaviors and drivers of the early computer science and engineering communities that emerged in technological-intellectual centers around the Massachusetts Institute of Technology (MIT) in Cambridge, Massachusetts, and Stanford University in Palo Alto, California, in the late 1950s. These were the first generation of scientists, engineers, and mathematicians who rejected the bureaucratic obstacles that prevented them from exploring the technological systems that stimulated their intellectual curiosity.[23] The initial cause or belief system was ambiguous beyond the free access to information; however, from this a "hacker ethic," or a set of aesthetic and ethical imperatives, emerged that include a commitment to access, meritocracy, and a belief that computers are the foundation for not only a contemporary form of performance art but also a utopian ideal for society.[24] This would serve as an example for others to follow.[25] The hacker ethic, identified by Steven Levy, is more of an underground code than an official professional society obligation,[26] but most acknowledge six principles:

- *Hands on Imperative:* Barriers to access technology are inherently wrong, and attempts to avoid or break said obstacles are justified.[27] To truly understand a technological system, one must have access to it, and restricting access inhibits freedom.

- *Information Wants to Be Free:* Information should be freely available to the curious without restriction.[28] Most familiar with the hacker ethic consider this to be the key principle.[29] Breaking barriers through technological skill and cunning and/or illegal methods (physical breaking and entering) are justified.[30] To hackers, acquired knowledge is useless if it cannot be shared.[31] An example of this manifestation of the hacker is Aaron Swartz, who downloaded articles from

207

JSTOR, an online academic database, and posted them to public websites.[32] Swartz was a noted online activist before the case, and his actions, which brought him no personal financial gain, resulted in national attention. He was charged with two counts of wire fraud and eleven violations of the Computer Fraud and Abuse Act. During the plea-bargaining stage of his trial, a counter offer was rejected, and Swartz committed suicide.[33] He and his case remain an example of this and other manifestations of the hacker ethic.

- *Mistrust Authority: All centralized, hierarchal, and bureaucratic is not to be trusted. Large institutions,* corporations, universities, and government agencies seek to control and limit individual autonomy.[34] This generalized distrust of authorities was evident in the early cyber gangs, such as Masters of Deception (MoD), later groups like Anonymous, and even individuals who worked for some of those authorities. Major security compromises such as those by Edward Snowden[35] are viewed as justified by many due to this principle. Snowden himself rationalized his behavior using concepts associated with this aspect of the hacker ethic.

- *No Bogus Criteria:* Hackers view the cyber underground as the ultimate meritocracy where individuals are judged by their technical skill and not by "bogus criteria" such as race, age, sex, position, education, or socioeconomic status.[36] The Internet is viewed by hackers as a great leveler where traditional limitations on upward mobility no longer apply. This component of the ethic was valued greatly by hackers from urban areas, notably New York City, who viewed hacking as a new identity without the constraints of poverty and racism experienced in the physical world. [37]

- *Truth and Beauty Can Be Created on a Computer:* Hacking is considered an aesthetic pursuit by hackers; it is a combination of technical skill, artistry, and creativity.[38] Those who see hacking as techno-art consider it less a set of skills or even academic discipline but a philosophy—a means through which one can conceptualize one's world.[39]

- *Computers Can Provide a Social Good:* Hackers view computers and, by extension, the Internet as positive forces in humanity as they can create things that are good, true, and/or beautiful. [40]

Early hackers were driven by curiosity, but by the 1980s, as hacking become more widespread, the same drivers were not necessarily there. By the 1990s, the intent of many hackers became somewhat malicious with the goal of violating computer systems and exchanging information in the underground to build credibility.[41] Some contemporary hackers criticize those who consider themselves hackers for lacking the technical skills of their predecessors.[42] The more technically inclined resent those who aspire to affiliate for the social status instead of pure intellectual curiosity.[43] Some suggest the majority of contemporary hackers not only do not uphold the hacker ethic but are largely ignorant of the original hackers who created it.[44] In fact, some even suggest the modern cyber underground is a "toxic technoculture" that is misogynistic, homophobic, and racist.[45] Much like groups and crowds, the Internet can afford the individual sufficient anonymity to take risks and/or perform actions they would be unwilling to perform as individuals.[46] When taken to the extreme, crude and even cruel behaviors can emerge.

This is not to say that all those who consider themselves hackers are intolerant, but rather that a recent trend in some online communities suggest a radical decrease in tolerance and an increase in bullying. [47] That said, often the recognition and/or perpetuation of these concepts becomes a shibboleth of sorts; members of Anonymous often relied on conveying this information as a demonstration of insider status, further reinforcing a hacker identity. [48]

CYBER UNDERGROUNDS AS ORGANIZATIONS

The underground is one of four components of a resistance movement. It is defined as "a clandestine organization established to operate in areas denied to the armed or public components of conduct operations not suitable for the armed or public components."[49] The other three components are the armed component, the auxiliary, and the public component. The Internet makes it easy to find communities of similar ideological interest, where grievance can be aired, sympathy generated, and success stories shared, all of which increase an individual's willingness to act for a particular cause. However, it is the ease with which tactics, preparatory information, target lists, and the like can be placed into the public sphere, lowering an individual's fear of failure and/or fear of consequence where cyber's revolutionary power is felt the most. Resisters can experience a sense of social belonging, prepare themselves cognitively to act, and have a blueprint for how the action can be undertaken without any direct person-to-person contact.

> 7.2   *The Internet facilitates locating and contacting communities of similar ideological interest. As such, there is no "local" cyber resistance and building insurgent networks may not require a physical footprint.*

Organizational Structure and Function

This section introduces some of the organizational structural and functional theories, research, and practice associated with the cyber underground. Similar to the *Human Factors Considerations of Undergrounds in Insurgencies*, this section employs examples from a variety of organizations, some of which predate the accessibility of the Internet and subsequently integrated cyber capabilities into their organizations. Anonymous, the network of weakly connected activists, hacktivist, and hackers that has grown prominent since its formation in 2003 has been described as a merger of nihilism and idealism, utopianism and dystopianism, individualism and collectivism, and negative and positive liberty ideals.[50] Other types of resistance movements rose in conjunction with communications technologies and proliferated with the Internet. Some of the organizations in these types of groups have hierarchal structures, while others are flatter, and some, such as Anonymous, claim to have no structure but are rather amorphous collectives that cannot be defined using traditional industrial and organizational psychology constructs.

Advances in information and communications allowed resistance organizations to compete on a level playing field with (and in some cases develop an asymmetric advantage over) state actors.[51] Computer technologies ease the distribution of information required to support a leaderless or limited hierarchy-type resistance movement. Leaderless organizations are essentially flat; there is no single individual in charge, and in some cases, all members have equal authority. The informality that leaderless groups afford is often appreciated by less-experienced personnel who tend to feel more valued. Leaderless groups can function well when there is a clear objective (or rationale for forming a group), and members are relatively psychographically and/or demographically homogenous. However, it can be difficult to sustain a leaderless organization as individual priority can create not only confusion but also discord among members. This was evident with Anonymous as some Anons believed the collective should become more proactively political, while others eschewed such mainstream forms of activism.[52]

Unlike traditional organizational constructs, these networks do not require physical infrastructure, geographic collocation, or even individual notoriety.[53] Political pressure no longer requires the aggregation and assimilation of committed individuals into an organizational structure, thus potentially broadening the appeal to individuals traditionally disinterested in formally affiliating with a group as a result of the risk associated with physical collaboration or, more banally, the time required to attend meetings. The lower barriers to entry and less-demanding temporal requirement can both broaden the appeal of movements yet hinder their growth due to the diffusion of responsibility or individual inaction as a result of presuming another will act.

> *7.3   Leaderless or limited hierarchy-type resistance movements can maintain operational security without the requirement for sophisticated physical security tradecraft.*

With virtual communities, there is not always a vetting and/or acculturation process, and thus there is not the same discipline or unity of effort seen in small, clandestine organization.[54] In some cases, this has made the groups somewhat easier to infiltrate, while in others it deliberately kept the size of the core group rather small. Politically divisive issues often catalyze the formation of virtual networks that organize to discuss, plan, and sometimes act. An example is the Tibet Autonomous Region, which presents challenges with official relationships with the People's Republic of China (PRC). Because of the geographic isolation of Tibet and the relative inability of Tibetan activists to voice dissent from within the PRC, an online "Free Tibet" network emerged to support the independence movement. The conglomeration of websites in English, Chinese, and Tibetan are hosted on servers outside China (often in Europe and the United States), are linked to one another, and have similar pro-independence sentiments; however, they do not have a common style, format, themes, messages, or strategies.[55] While there is conceptual commonality among these activities, there is no explicit C2 apparatus. This affords freedom of action; however, it comes at a cost because loosely affiliated networks can be more easily co-opted, distracted, and/or delegitimized.[56]

There are at least four types of groups that warrant further investigation: cyber activist, hacktivist, hacker, and cyber terrorist. Some of the distinctions in the following section may seem academic as all of these organizations involve technologically sophisticated individuals interacting on and with communication technologies for a specific sociopolitical objective.

Cyber Activist Organizations

Cyber (or virtual) activism refers to normal, non-disruptive use of the Internet in support of an agenda or cause. Also referred to as online organizing, electronic advocacy, e-campaigning, and e-activism, operations in this area include web-based research, website design and publication, transmission of electronic publications and other materials through email, and use of the web to discuss issues, form communities of interest, and plan and coordinate activities.[44] Activist groups can advance their concerns more rapidly through the spontaneous formation of distributed networks of concerned individuals. These ad-hoc, location-independent, and medium-agency networks self-organize around functional, not geographic, concerns.[57] While the rapidity with which these groups can coalesce and act far surpasses that of a traditional insurgent underground movement, an organizer sacrifices control and message discipline for speed.[58] In some social movements, online activism may be the initial venue; for example, the Arab Spring movement in Egypt largely initiated on Facebook.[59] In other cases, social media simply serves as a distributed broadcast channel for activists already demonstrating on the streets. The Syrian uprising in 2011 began with people openly defying the Assad regime in the streets, and afterward individuals shared their experiences on social media.[60]

Those experiences were shared (and liked) by others, and awareness of and support for the movement grew internationally.[61]

Online activism can be categorized into awareness/advocacy, organization/mobilization, and action/reaction based on a continuation of action required on behalf of the group by non-group members:[62]

- Awareness/advocacy groups aim to publicize a cause and/or provide information about a particular issue, be it a political cause or a social issue.[63] These efforts may include a fundraising component or request some low-level of effort such as donating, signing an online petition, and/or disseminating information. Individuals who support the cause at a low level, perhaps donating a small amount, digitally signing a petition, and/or changing their social media avatar but not necessarily taking any risks are often referred to as "slacktivists."[64]

- Organization/mobilization groups seek to not only raise awareness but generate action on behalf of a particular cause.[65] These actions tend to be planned with sufficient time to ensure turnout at a particular event is large, reasonably well-organized, and/or sufficiently focused.

- Action/reaction groups tend to be those with a relatively tight yet focused core, with a larger set of more regular contributors. These groups tend to be more aggressive and/or geared toward more rapid operations, be they flash mobs or protests in response to a particular event.[66]

The Occupy Wall Street (OWS) protest movement transitioned from virtual to physical on September 17, 2011 in New York City. OWS is an example of an organization/mobilization[67] effort that started online in February 2011 by Canadian anti-consumerist and pro-environment group Adbusters[68] and grew into a civil disobedience movement that involved tens of thousands of participants over months.[69] The movement arose in the wake of the Arab Spring (described later in this chapter) to attempt to hold responsible the private organizations who contributed to the Great Recession and the global income disparity between the haves and have nots. On July 13, 2011, Adbusters distributed an email to an approximately ninety-thousand-person listserv with the hashtag #OccupyWallStreet and a date of September 17, 2011.[70] The popularity of the hashtag grew and spread to social media venues, which resulted in the sharing of revolutionary materials from electronic books, to manifestos, to how to guides.[71] By August 2011, activists began meeting in public parks to plan and organize for September 17, 2011.[72] OWS was a leaderless resistance movement that employed an assembly as a decision-making body, whereby participants attempt to reach consensus.[73] In addition to the main assembly, additional working groups and/or committees emerged to discuss and plan for specific contingencies.[74] Much of the organization was bottom-up and actively avoided the emergence of a single authority or leader. In September 2011, the encampment at Zuccotti Park in Lower Manhattan New York City, where the ad-hoc arrangement of tents produced a shantytown visual that harkened back to the Hoovervilles of Central Park during the Great Depression, housed between one hundred to two hundred individuals. As the numbers grew, so did attention, and the New York City government ordered the park vacated on October 13. There was open defiance to this order, but no attempts to forcibly remove the protesters were made until the New York Police Department cleared the park on November 15. Despite the egalitarian approach to group decision-making, the lack of a centralized set of objectives and/or a coherent narrative was attributed to the lack of strategic success of the movement.[75] Nevertheless, OWS exemplifies the evolution of a social movement from a purely online communication of grievances to a large-scale physical operation that gained global notoriety.[76]

Hacktivist Organizations

Hacktivism refers to the amalgamation of hacking and activism; it is the exploitation of computer systems (hacking) for a political purpose that brings methods of civil disobedience to cyberspace.[77] Hacktivist tactics include a litany of constantly evolving techniques, often at the leading edge of information security. Included among them are virtual sit-ins, automated email bombs, web hacks and computer break-ins, and computer viruses and worms. Hacktivist groups can take the form of cyber-centric organizations whose sole modus operandi is the use of hacking to achieve their objectives or a more traditional resistance movement that exploits cyberspace for various operational purposes. Groups include Critical Art Ensemble 1984 Network Liberty Alliance, Cypherpunk, and the Electronic Disturbance Theater (EDT) amongst others. Many of these groups, tend to use illegal techniques for what they perceive to be a social good. For example, EDT's virtual sit-in to stop the Mexican government's crackdown on the Zapatista revolutionary movement is a form of a DDoS) attack.[78]

Early adopters of Internet-based technologies identified this new medium as an ideal electronic space for cultural and political resistance.[79] On October 16, 1989, computer systems at NASA's Goddard Space Flight Center in Greenbelt, Maryland, were infected by the Worms Against Nuclear Killers (WANK) worm. The attack was executed by a loosely affiliated group of anti-nuclear weapons activists.[49] The disruption was not catastrophic; if the user's terminal became infected, it displayed a WANK logo and short message that the system had been "WANKed."[80] This attack was one of the earliest, and oddest, instances of hacktivism, as the perpetrator was the son of Robert Morris, the chief cryptographer of the NSA.

Anonymous is notable for its ability to maintain a stable, collective identity despite the relatively loose ties between members, and the ethic and socioeconomic heterogeneity amongst its members.[81] What were considered heretofore necessities in developing cohesive organizations, Anonymous eschewed yet was able to achieve specific tactical objectives.[82] Anonymous can best be described as a dynamic, low-density network[83] with the number of nodes increasing and decreasing depending on a particular focus.[84] As the Interest in the Anonymous' operation against the Church of Scientology grew, a geographically based cell structure became evident.[85] The connective strength within cells was greater than that between cells, suggesting a cliquish network more than a homogenous organization.[86]

Hacker Organizations

Hacking refers to the intentional manipulation of computer hardware or software for a purpose for which it was not necessarily originally intended. What began as a subculture within the counterculture movement during the 1960s became an international underground and is now considered a community.[87] Hacking includes gaining access to secure databases, defacing websites, or disrupting Internet traffic. Groups in this category include Anonymous Anarchist Action Group--A(A)A, Cult of the Dead Cow--cDc and/or Hactivismo, 1984 Network Liberty Alliance, LulzSec, Syrian Electronic Army (SEA),

CyberCaliphate, Chaos Computer Club, Global kOS, The Level Seven Crew, globalHell, TeaMp0isoN, Network Crack Program Hacker Group, Masters of Deception (MoD), and Milw0rm.MoD, who was noteworthy for a number of hacks, principal among them was compromising the Regional Bell Operating Company (RBOC) system. MoD used the system by falsifying permissions and access to avoid being charged Bell's fees to communicate with one another and play pranks on rival hacker gangs. Ultimately, five members were indicted and pled guilty to federal charges.[88]

Cyber gangs are hacker groups that resemble street gangs in demographics that are both curious about computer/information technology and socially non-conformist.[89] During the 1980s, cyber gangs emerged in many urban areas. Their members were typically adolescent males who possessed an intellectual curiosity about computing but whose families may not necessarily have had the resources to purchase hardware or subscriptions to publications or online services.[90]The locally based gang model is less prevalent in the twenty-first century, largely as a result of the Internet; however, the incorporation of criminal hacking by many terrorist networks seems to have reinvigorated the gang model.

Some hacker organizations tend to eschew hierarchy on principle and tend to function more as a loosely organized collective than an organization. Hacker organizations tend to lack hierarchy and most members are on an equal level. If there is a leader, it is typically the individual who founded the group. Divisions of labor can be identified, roles and tasks are often established on an ad-hoc basis. Rarely is someone ordered to perform a particular task.[91] Over time, as a situation demands or technical specialties emerge along with a requirement, individuals may fall into set roles or responsibilities within a particular group.[92]

While hacker groups often have strict behavioral norms, hacker organizations may lack the cohesion of underground organizations that have greater physical interaction. The interpersonal relationships, while nonetheless important, do not seem to be as important as they are in more traditional insurgent undergrounds.[93] In many cases, members of hacker organizations more readily cooperate with authorities than organized crime organizations, terrorist cells, or gangs.[94]

Cyber Terrorist Organizations

The term cyber terrorism refers to the use of cyberspace to commit terrorist acts and, like the term terrorism, is more of a classification of tactics than a type of group. Cyber terrorism covers sociopolitically motivated hacking operations intended to cause grave harm such as loss of life or severe economic damage.[95] Operations in this domain include penetrating SCADA systems to interfere with water purification plants, air traffic control, or metropolitan traffic management system, as well as hospital electronic recordkeeping databases and investment bank transaction records. Sophisticated non-state organizations such as al Qaeda, Hizbollah, the Islamic State, and various Mexican drug cartels[96] conducted, or expressed interest in conducting, such operations.[97] Hizbollah, an organization that entails a paramilitary force, a terroristic network, and a legitimate political party, was considered the most sophisticated terrorist organization in cyberspace;[98] however, they likely have been surpassed by the Islamic State, among which are the most sophisticated cyber profiles.[99]

Cyber terrorism, according to the North Atlantic Treaty Organization (NATO), entails a cyber attack using or exploiting computer or communication networks to cause sufficient destruction or disruption to generate fear or intimidate a society into an ideological goal.[100] As of this writing, there are few examples of a group existing solely in the virtual space planning and executing an act of cyber terrorism. More typically, there is a cell or an external group loosely affiliated with a group classified as a terrorist organization conducting operations on behalf or in support of that organization. Cells can be organized to accomplish different workflows and maintain security.[101] When possible, cells may be arranged in series, like an assembly line, or arranged in parallel (see Figure 7-2). These can conduct work independently and report up a chain of command. Parallel cells are also sometimes set up to confirm or disconfirm information independently or set up as backups in case one cell is compromised.[102]

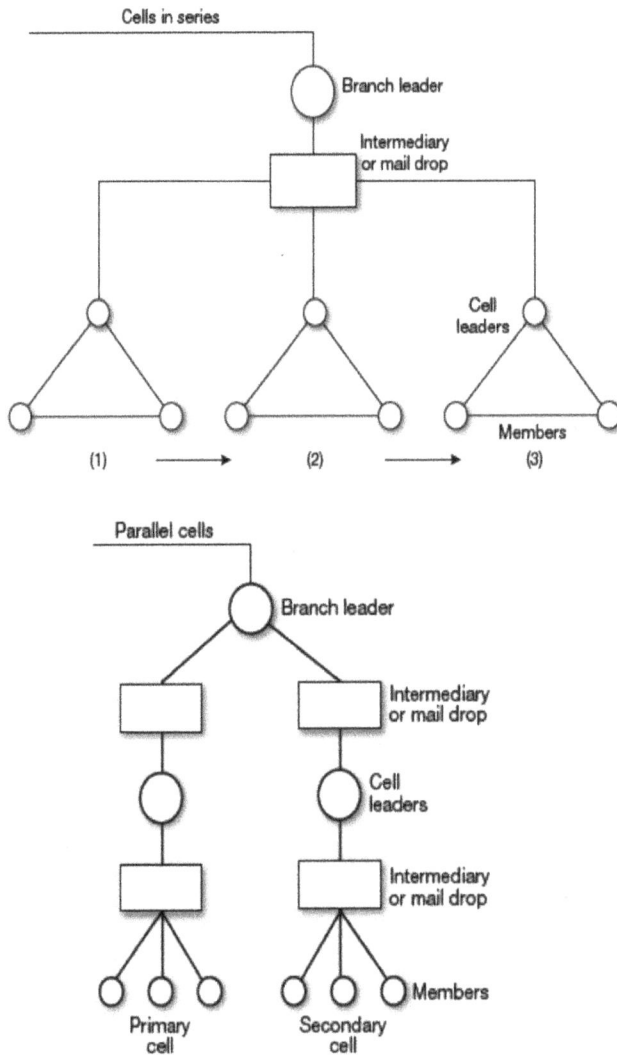

Figure 7-2. Cells in series and parallel.

Anonymizing technologies are increasingly used to maintain operational security, and as a result, physical and/or structural approaches to security are rendered obsolete.[103] The Islamic State advised against using the onion router (TOR), software developed for the US Department of Defense and distributed by the US Department of State that enables anonymous communication by directing Internet traffic through a worldwide network of relays.[104] Instead, they recommend both Tails and Pretty Good Protection (PGP) software on various platforms to communicate within its organization and with promising recruits.[105]

These anonymizing technologies enable a movement to interact with potential supporters and thus grow in personnel and resources, while minimizing the risk of compromising operations security. For more on these technologies, see chapter 4 of this book.

Table 7-1. Cyber organizational taxonomy.

Type of Organization	Description	Example
Cyber gang	Groups that resemble street gangs in demographics (adolescent males in urban areas) that are both curious about computer/information technology and socially non-conformist.	Masters of Deception (MoD)
Cyber Activist	Groups that seek to advance a particular agenda through the normal, nondisruptive use of the Internet in support of an agenda or cause.	MoveOn.org, Act Now to Stop War and End Racism (ANSER)
Hacktivist	Cyber-centric organizations whose sole modus operandi is the use of hacking to achieve their objectives or a more traditional insurgent movement that exploits cyberspace for various operational purposes.	Critical Art Ensemble 1984 Network Liberty Alliance, Cypherpunk, Project Chanology, and the Electronic Disturbance Theater

Type of Organization	Description	Example
Hacker	Groups that exist to perform the intentional manipulation of computer hardware or software for a purpose for which it was not necessarily originally intended.	Anonymous Anarchist Action Group--A(A)A, Cult of the Dead Cow--cDc and/or Hactivismo, 1984 Network Liberty Alliance, LulzSec, SEA, CyberCaliphate, Chaos Computer Club, Global kOS, The Level Seven Crew, globalHell, TeaMp0isoN, Network Crack Program Hacker Group, Milw0rm
Cyber Terrorist	Cyber terrorism covers sociopolitically motivated hacking operations intended to cause grave harm such as loss of life or severe economic damage. These groups are typically cells within organizations classified as "terrorists."	Cells within al Qaeda, Hizbollah, the Islamic State, and various Mexican drug cartels

THE CYBER AUXILIARY

The auxiliary is a component of a resistance movement defined as the support element of the irregular organization whose structure and operations are clandestine in nature and whose members do not openly indicate their sympathy or involvement with the movement.[106] Members of the auxiliary are more likely to be occasional participants of the insurgency with other full-time occupations. The cyber auxiliary is the support element of a cyber organization (or loosely affiliated network) whose structure and operations are clandestine in nature and whose members do not openly indicate their sympathy or involvement with the irregular movement. Given the transient nature of the cyber underground, it is fair to say the vast majority of individuals operating online are more a part of the auxiliary than the underground. While some cyber organizations have a more structured and disciplined underground, many are far less rigid and often allow individuals to come and go. This lower barrier to entry allows for a much larger potential base of support but can also be a security vulnerability as the organization may be more susceptible to penetration by rivals or security forces.

Leadership

An underground must recognize the underlying socioeconomic and/or political grievances and seek to organize that dissatisfaction into a coherent narrative to form a broader base of support. Upstart resistance movements are often headed by leaders who are charismatic. Much of the early activity involves disseminating information to generate internal and external support, shape perceptions, and set conditions for broader mobilization. Several other strong personality traits, some bordering on psychopathologies, may be associated with resistance leadership and may affect, hinder, or sometimes help a would-be underground leader. Charismatic leaders can often co-opt and ultimately redefine the social reality value for the group; this behavior is more prevalent in cults than radical groups, often to a much greater degree. Small groups, however, are vulnerable to co-option by the charismatic individual who can effectively manipulate the perceptions of others through emotional appeals, symbolism, or isolation.[107] The social reality value of the group depends on internalizing group standards of value, including moral standards[108] and can determine whether or not an individual remains in the underground.

Some charismatic hackers either exploit others to gain access and/or manipulate those more skilled to act on their behalf.[109] Chris Goggan, known as Erik Bloodaxe, a founding member of Legion of Doom (LOD) was able to convince others to perform hacks to prove their merit.[110] Bloodaxe's behavior did not seem particularly manipulative, but more of a means of establishing bona fides for potential LOD members.

Some hackers become leaders of a movement as a result of their imprisonment, as a public face of an underground cause. While Kevin

Mitnick did not endure the tribulations or achieve the notoriety of Bobby Sands,[111, 112] his notoriety and capacity to influence others only grew after his arrest and trial. Kevin Mitnick is an American hacker who rose to prominence within the cyber underground through a combination of detailed knowledge of information systems with empathy for employees in particular roles.[113] Mitnick's tactics (described in a later section) spread through the cyber underground via his close associates, particularly those who embodied the hacker ethic, and ultimately via mainstream media after he was imprisoned.[114] Mitnick was viewed as a scapegoat by his supporters, an individual who was punished because of his ability to exploit flaws in software/hardware design as well as organizational cultural ignorance of cybersecurity.[115] Despite never financially profiting from these exploits, Mitnick was charged with the financial value of intellectual property, which was perceived to be inherently unfair amongst the hacker underground.[116] As a result, Mitnick became the temporary face and name of the late twentieth century manifestation of the hacker ethic.

The Islamic State, despite its extremist ideology and near fetishization of violence,[117] is not governed by madmen but rather by individuals with heterogeneous backgrounds, mental models, worldviews, psychological characteristics, and leadership styles.[118] This diversity appears to extend to the CyberCaliphate and other cyber auxiliaries that support the ISIL. Contemporary insurgent movements have seen the emergence of ideological leaders through cyberspace. An example is Anwar al-Awlaki, the American imam and al Qaeda in the Arabian Peninsula (AQAP) leader, who burnished a reputation through his lectures, publishing CDs originally through Islamic bookstores and ultimately online.[119] Awlaki did not rise through prominence through local organizing or battlefield exploits but virtually interacting with individuals who had questions about his lectures and sermons posted online. Over years, his message became increasingly militant, at times directly encouraging travel to Yemen to join AQAP or take up violence closer to home, as was the case with US Army Major Nidal Malik Hasan the psychiatrist who fatally shot thirteen and wounded thirty at Fort Hood, Texas in 2009.[120]

> *7.4 The personalities of key leaders can also have a strong influence on the operations of resistance movements, particularly during their early stages. Charismatic leadership in cyberspace may take on different forms but is typically accompanied by the technological skills often associated with elite hackers.*

Criminal Nexus

Cyber resistance organizations often require contact with career criminals and organized crime elements, particularly those in oppressive regimes, to accomplish tactical objectives. Most armed insurgencies at some point use criminal activities such as theft, smuggling, or extortion as a means of fundraising. Criminal connections are a threat to the groups in more ways than one, threatening their ideological legitimacy and their support from the populace and tempting members to become swept up in a different lifestyle and social network. Criminality may be viewed as distasteful by many of the more ideologically or politically motivated insurgents, particularly those drawn in for political or ideological reasons.[121] Cyber resistance organizations are similar in this case as the nexus between social movements, particularly those in oppressive regimes, and criminal enterprises can be difficult to distinguish.

Computer crimes first became prominent within the corporate world in the 1970s. The majority involved low- to mid-grade employees with computer skills who perpetrated some sort of financial malfeasance or fraud against their employer.[122] A more destructive trend began in the 1980s where the intent was to render information unusable but not necessarily steal it.[123] With the 1990s emerged the dissemination of sophisticated viruses with destructive power that brought notoriety to the perpetrator.[124] The turn of the twentieth century gave rise to the confluence of hacking and organized crime,[125] a convergence with significant operational implications for cyber resistance movements. Evgeniy Mikhailovich Bogavhev, a Russian hacker known as "Slavik" was implicated for association with not only Russian organized crime but also Russian military and intelligence organizations.[126] Slavik developed custom malware and rented his botnets to compromise online banking systems and even doxx Georgian intelligence officers and members of an elite Turkish police unit.[127] A doxx agent maliciously seeks to unearth and expose private information about a person on the Internet. These types of mercenary intelligence operations can be a significant, and possibly necessary, aspect of cyber resistance operational security.

The twenty-first century observed an increasing sophistication in cyber crime; it is increasingly organized, creative, and impactful.[128] These crimes range from bank fraud to hacking casinos,[129] to remotely shutting down industrial equipment and holding the factory[130] or hospital in the United States[131] or United Kingdom[132] for ransom or even commandeer a bank's online operation.[133] Many law enforcement organizations established cyber-crime centers, cells, boards, bureaus, and working groups to assess and interdict these threats. While legislation has not yet caught up to the information age, much support provided to insurgent movements entails material support through the Internet. As cyber law evolves, the set of hacktivist and/or hacking tactics that are considered illegal grows, and thus many cyber activists themselves will commit criminal acts to further a sociotechnical and/or political objective.

MOTIVATIONS UNDERLYING CYBER RESISTANCE

Determining the why of underlying human behavior is one of the most challenging theoretical and methodological aspects of psychological science. Individuals may not truly understand the underlying motivation of their behavior and, even if they do, may be reluctant to share it with either researchers studying them or an organization of security forces interrogating them after an arrest. Nevertheless, a number of inferences can be drawn about why from each what and who within the cyber underground. The following sections discuss some of those inferences.

Underground Cyber Psychology

Cyber Psychology is the study of mental processes and behavior in the context of interaction and communication of both humans and machines.[134] The field of study is broader than those inhabiting the cyber underground, but its principles and research contextualize the individual mentalities and behaviors of those within the cyber underground. The following sections attempt to explain the hacker psyche, as well as some of the demographic and psychographic attributes typically ascribed to hackers either by themselves or researchers studying them.

The contemporary use of "hacker" typically refers to an individual who attempts to penetrate security systems on remote computers.[135] The historical, at least dating to the 1950s, meaning of the term "hacker" referred to an individual capable of creating hacks, or elegant, unusual, and unexpected uses of technology.[136] Hackers tend to be attracted to technical challenges and judge the interest of work or other activities in terms of the degree of difficulty and/or the technological novelty of the tools/systems they explore if successful.[137] Satisfaction derives from the exploration process, the study, and the ability to learn and ultimately overcome the obstacles that restrict access to a computer system.[138] Intellectually curious but not necessarily possessing scholastic aptitude, hackers often consider themselves scientists who use computers as laboratory tools to examine a complex living system that connects humankind.[139] The hacker mindset is independent of the particular medium in which the hacker works;[140] it is not restricted to computer hardware/software. Many apply the convergence of technology and art to a variety of problem solving or exploratory challenges.[141]

As there is no terrorist psychological profile,[142] there is no single hacker profile, although there seems to be more psychological similarities amongst hackers than evidenced amongst terrorists. However, like any composite profile, there are individuals who may not exhibit any of the characteristics described. The most prevalent traits of a typical hacker are a person of above-average intelligence, high intellectual curiosity, and technological self-efficacy.[143] Aspiring hackers tend to demonstrate above-average intelligence (although they tend to under-achieve academically) and often display sound technical and problem-solving abilities when they are interested in the problem.[144] The ability to mentally absorb, retain, and reference large amounts of detailed information is common amongst elite hackers as their analytic intelligence and memory enables them to solve technological challenges intuitively.[145] The stereotype of hackers as intellectually narrow is more a misnomer resulting from their hyper focus on a particular task; they ignore more mundane requirements in pursuit of sating an intellectual curiosity.[146]

> *7.5   There is no single hacker profile, although there are consistently observed behaviors—for example, having an intense focus on a particular technical challenge at the expense of the more mundane.*[147]

Hackers tend to be relatively individualistic and nonconformist,[148] with a general disdain for authority that often manifests in rebellious behavior.[149] Authority figures typically start with their parents and then extend to adults in general, systems administrators, and/or representatives of the legal system.[150] Adolescent hackers are often indiscreet and attention seeking, sometimes fabricating exploits to gain notoriety and respect in the underground.[151] Unstable home lives including parental conflicts, custody changes, and/or financial challenges requiring frequent moves are not uncommon amongst hackers.[152] The resultant instability can manifest in fear, distrust, and/or insecurity from which individuals escape by turning to the Internet for companionship, social support, and/or and intellectual distraction from emotional distress.[153] Frustration can also manifest if hackers, particularly adolescents, do not measure up to peers academically and/or athletically and thus perceive themselves to be lower in the social hierarchy and unable to advance.[154] The solace found in online life can appear as social alienation to their peers,[155] and seeking out and operating within the underground is a form of escapism. [156]

The aforementioned tendency to come from unstable home lives often provides a motive to creating an underground cyber resistance. The combination of escapism, in this case from the discomfort of domestic reality, provides a productive and creative outlet. [157] For many hackers, the satisfaction they find in challenging the authorities, first among them the police and security professionals, feeds their egos.[158] Some hackers who demonstrate high ego strength, a combination of confidence in their ability and a strong sense of their identity, view law enforcement efforts as feeble, bordering on the humorously inept, and thus derive satisfaction from being chased.[159] The prevalent anti-establishment views of the underground world in which many hackers operate are usually aimed at organizations and agencies that in their eyes want to hinder technological development and free circulation of information through a market monopoly.[160] This aim began with the telephone companies but extended to software giants, Internet service providers, and government regulatory bodies. [161]

The in-group bias of hackers suggests they tend to have limited capacity for emotional awareness and often struggle with relationships.[162] It is not clear whether they suffer from specific impairments or whether their behavior is a consequence of their tendency toward self-absorption, intellectual arrogance, and impatience with people and tasks perceived to be wasting their time.[163] Hackers often comply with the stereotype threat[164, 165] of the geek: withdrawn, relationally incompetent, sexually frustrated, and desperately unhappy when not submerged in his or her craft.[166] While a cliché, the archetype resonates with many hackers. Hackers tend to be especially poor at confrontation and negotiation,[167] often succumbing to the projection bias whereby they presume others have the same worldview, skillsets, and objectives.

Hackers' motivations are typically not political or financial. However, as the underground evolved, so have the inhabitants, and contemporary motivations involve illegal activities for either financial gain or community notoriety.[168] Hackers are generally only very weakly motivated by conventional rewards such as social approval or money[169] and tend to value self-efficacy and autonomy over material wealth.[170] Satisfaction derives from the technical conquest over a given system.[171] This pursuit, when taken to the extreme, can manifest in delusions of grandeur and an over-exaggerated sense of self-efficacy regarding their ability to gain and maintain access to a particular system.[172] High self-esteem serves as a counterweight to the frustration caused by lack of recognition from those outside the hacker underground who cannot appreciate the skill or determination required for a particular hack. [173]

Many hackers have an altruistic view of their activities and prefer to be remembered for having changed things for the better, contributing to improved computers and making them more powerful and user friendly.[174] Some consider hacking as a tool with which to face many political and social problems—techniques that can be used to defend oneself from violations of the principles that govern the online world, from the attacks of the physical world they consider morally corrupt, such as the attacks against the civil liberties of both hackers and all other users.[175] Most hackers are eager to share the discoveries, know-how, and information acquired during their raids with other members of the underground.[176] For hackers, the real crime is not hacking but rather hiding the truth.[177] Hackers perceive their role as defending the right to information and making information free and accessible to those who seek it, despite the attempts of those organizations (corporations, government agencies) who seek to control information and profit from its restriction.[178] Those organizations often criticize hackers as anarchistic; however, the objective of most hackers is to abolish all rules, thus replacing unfair, existing rules with new ones that guarantee greater security and equal access to all users.[179]

Hackers tend to be somewhat meticulous, but that fastidiousness is often relegated to their own setup (or space where they conduct their hacking) and/or procedures, often preferring specific tactics or programming languages over others to the point of intolerance.[180] This personality trait can manifest in a kind of technological intolerance or even bigotry that causes interpersonal rifts.[181] Accordingly, many tend to be careful and orderly in their intellectual lives yet chaotic elsewhere, demonstrating a keen mental composition but the outward appearance of absentmindedness.[182] Many addictive and obsessive behaviors can be found in the hacker underground;[183] however, the research to date has not identified whether these individuals actually meet the diagnostic criteria for a mental disorder or the behavior is simply the hyperfocus many believe necessary to accomplish particularly difficult hacks. Hackers tend to dislike tedium, nondeterminism, and/or the banality of everyday life.[184] Often the demonstration of one's skills and abilities can have a therapeutic effect for some,[185] and thus hacking can be a conscious form of self-soothing but occasionally at the expense of other aspects of daily life.

Many hackers have a strong sense of humor,[186] which often manifests in pranking behavior or attempts at "lulz."[187, 188] A portion of the cyber underground also views hacking as a modern manifestation of technological pranking. A number of individuals involved with the Anonymous collective prioritized having fun and viewed themselves as technologically sophisticated tricksters, not malevolent cyber actors.[189] The purpose of an Anonymous 2006 raid on the online game "Haboo Hotel" was to protest what they perceived as a racist game played by racist placers.[190] Anons manipulated the game's avatars to reflect stereotypical symbols of African American culture as a means of brining this perceived racism to light.[191] This latter behavior typifies Anonymous in that the majority of Anons participate to display their wit.[192]

A hacker is a technologically proficient individual with a deep knowledge of and affinity for computing, while a hack is a non-obvious and innovative use of technology.[193] Table 7-2 includes a set of labels and descriptions for different types of hackers. It is important to note that these are labels computer security researchers use to classify individuals to better understand various phenomena. They are not necessarily how an individual hacker may describe him or herself. This categorization is a logical starting point to better understand any potential differences. There is a secondary categorization with the aforementioned categories that were further classified by four motivations: revenge, financial, notoriety, and curiosity.[194] The limitation of this circumplex[195, 196] model to revolutionary and insurgent warfare is that it describes cyber criminals. While many revolutions and insurgencies are required to break the law in some manner, labeling those groups and/or individuals as "criminal" may not be accurate. First, it presupposes the inherent rightness of the relevant legal system, and second, it undermines the moral and/or political goals of those individuals and groups. This typology, or classification based on psychological factors, is unlikely to apply to those in groups that view cyberspace as a medium through which art and civil disobedience should converge for the common good. Furthermore, the labels are not necessarily mutually exclusive, and an individual may meet the criteria for a set of labels over his or her career or even at one time as many hackers tend to have multiple online identities.

Table 7-2. Hacker taxonomy.

Type	Description
White Hat Hackers	White Hat hackers find and exploit computer systems vulnerabilities to make the systems more secure.[197] They tend to adhere to the hacker ethic, with slight deviations as much of their work is to secure proprietary information.[198]

Type	Description
Black Hat Hackers (Crackers)	Black Hats are malicious hackers (or "crackers) who find and exploit system vulnerabilities generally for personal gain, be it profit or notoriety.[199] The term cracker originally referred to those who removed the protection from commercial software programs. The term has evolved to refer to hackers with more malicious intent; those that seek to do damage for the sake of damage and are thus rarely involved in computer security.[200]
Gray Hat Hackers	Gray Hats are those who are not easily classified as white or black and may actually perform functions of both at different times.[201] Some consider the Gray Hat hacker the most ideologically congruent with the early hackers as they tend to be highly skilled yet disdain labels.[202]
Wannabe Lamer	Lamers are aspiring crackers who lack experience, skills, and the wherewithal to acquire the abilities via study and/or trial and error without learning.[203] They are attracted to the underground because of the camaraderie as well as the faddishness of hacking. They tend to lack the intellectual curiosity, diligence, and/or ethics of traditional, or aforementioned hackers.[204]
Script-kiddie	Script-kiddies are aspiring hackers that use existing computer scripts or code to gain access to a system for the purposes of defacement or information exfiltration.[205] These individuals lack the skills and/or experience of full-fledged hackers and, at times, may be manipulated into performing illegal activities.[206]
Military Hackers	These are professional hackers whose work is a component of their service in their nation's armed forces.[207] They are typically highly skilled and well trained, some of who were recruited from or at least had experience in the underground prior to enlisting or recruitment.

Type	Description
Government Agent	These hackers, often recruited from the underground, are typically employed by the intelligence apparatus of a nation. They seek to gain access to systems and/or information for the purpose of espionage or counterespionage.[208]
Phreakers	Also called Phone Phreakers or Blue Boxers, there are few of this type left. These are people who attempt to use technology to explore and/or control the telephone system. Originally, these activities involved the use of "blue boxes" or tone generators, but as the phone company began using digital instead of electromechanical switches, the phreaks became more like hackers.[209] Phreaking is a technique that consists of using computers or electrical circuits to generate special tones with specific frequencies or to modify the voltage of a telephone line.[210]
Virus Writers	Virus writers are people who write code that attempts to reproduce itself on other systems without authorization.[211] These individuals develop code that is often implanted by other hackers who seek to compromise a system to display a message, play a prank, or destroy a computer or computer network.[212]
Pirates	Pirates are modern-day crackers whose primary goal is to overcome measures used to prevent the unauthorized duplication of software.[213] Those who accept the hacker ethic simply gives away the software once it is copied, while others seek to profit from their efforts.[214]
Cypherpunks (cryptoanarchists)	Cypherpunks freely distribute the tools and methods for making use of strong encryption. Some cryptoanarchists advocate strong encryption as a tool to completely evade the state, by preventing any access whatsoever to financial or personal information.[215]

Type	Description
Anarchists	Anarchists are committed to distributing information, particularly illegal information, at any cost. Some take the "Information is Free" dictum of the hacker ethic to the extreme and may facilitate illegal sharing of child sexual abuse depictions or other illicit transactions on the dark web.[216]
Cyberpunk	Cyberpunks are amalgamations of the hacker and punk rock scenes with interests in bio-hacking (particularly their own bodies), science fiction, and non-mainstream applications of technology.[217]
Carders	Carding, credit card number fraud, is a technique that consists of appropriating credit card numbers, usually obtained by violating the systems of banks or financial agencies and using them to make long-distance phone calls or to buy goods without the cardholder's knowledge.[218]
Cyber Mercenary	These individuals are often former military hackers or government agents who venture out on their own. Some are Gray Hats, but many in this category who work for profit are typically considered Black Hats.[219]

A conscious consideration when addressing Part II of *Human Factors Considerations of Undergrounds in Insurgencies* was dispelling the "terrorist/insurgent as psychopath" myth. A similar approach is necessary here to address the "autistic hacker" misconception[220, 221] that seems almost as prevalent.[222, 223] This misconception attributes antisocial cyber behavior to the underdeveloped social skills typically associated with those along the autism spectrum.[224] Much like the notion that because terrorism is violent crime, those participating in it must exhibit higher rates of psychopathology, there is little empirical support for individuals along the autism spectrum being more likely to engage in cyber crime than others.[225] In both misconceptions, there are likely examples that support the myth[226] but far more that do not as more nuanced descriptions of roles beyond the broad categorization of "terrorist" or "hacker."

Hackers are capable of violating not only information systems but social norms.[227] The failure to accept technological limits often extends to the social world and can lead to a variety of abhorrent behaviors as hackers search the bowels of the underground.[228] These abhorrent behaviors can range from drug abuse to illegal gambling to human trafficking. When taken to the extreme, hackers tend to employ complex solutions that often require considerable thought, time, and expertise when less-elegant but perhaps acceptable solutions would have been not only viable but also accomplished at a fraction of the time and effort.[229] This tendency has often been used to explain the failures of sometimes brilliant hackers to function adequately in the corporate world of professional computer hacking.[230]

Much like an insurgent movement has overt and covert components, hackers tend to have at least two identities: a physical identity in the physical world and at least one in the online underground.[231] Hackers tend to increase their power through alternative identities; the handle (or handles) is a nickname or call sign that serves as both a *nom de guerre* as well as operational security, which attempts to simultaneously build a reputation amongst the in-group while remaining ostensibly anonymous to the outgroup.[232] These parallel lives can increase stress on hackers, often resulting in insecurity bordering on the paranoid.[233] These feelings are caused by the constant fear of arrest and the uncertainty caused by never knowing with whom they communicate online.[234] The most frequent complaints of underground hackers tend to be insomnia and its resultant effects.[235] Those not only comfortable with but also thrive as a result of this duel existence make ideal underground and/or auxiliary personnel.

Joining, Staying In, and Leaving the Cyber Underground

The type of individual an insurgency attracts and the nature of their motivations for joining changes as a resistance movement evolves. In the early stages of an underground movement, recruitment is selective and restricted as much as possible to known, trusted associates of current members. Strong ideological sympathies are most important at this point, when joining entails greater risk and thus requires stronger convictions. If they are successful and survive to later stages of expansion and militarization, undergrounds move into mass recruitment. The motivations for joining an underground movement during later phases are more complex, with no single, dominant, motivating reason.[236]

Joining

Many hackers start accidently and some often in their pre-teen years. A minority do not begin until they are in their late teens or early twenties, but the majority tend to start hacking in their early teens.[237] Typically, hacking starts as a peripheral interest and tends to develop into a sole hobby or preoccupation (if not an actual occupation) once sufficient skills are developed and recognized by others.[238] Like many adolescents, aspiring hackers seek both identity and affiliation, and the lure of the cyber underground can provide both. It can provide not only a sense of belonging but also a sense of purpose greater than the superficial interests of many adolescents.[239] In some cases, the allure of the underground is the perceived appeal of an idealized technocracy where skill alone determines whether one is accepted. In some cases, the cyber underground is an alternative to joining a gang, while in other cases, the cyber path is an extension of an existing gang life.[240] Having gang experience and/or criminal records (particularly robbery) seems to be more prevalent amongst American hackers than those in other countries, and those Americans with gang ties tend to be from urban areas.[241]

While many hackers prefer to work alone, either for security purposes or to reduce the burden of dealing with less skilled "lamers," organizations provide a degree of protection and/or safety for the individual.[242] Typically, only the most skilled hackers are permitted entry to a group, with the more skilled and/or experienced often controlling access through invitation-only sites.[243] Providing hacking bona fides is often a requirement; however, there is no standard approach. The demonstration can range from a simple question and answer about specific system parameters, claiming credit for a previous hack, or gaining access to a specific target. Gaining access requires demonstration of skill and not simply braggadocio in open forums.[244] More sophisticated groups only recruit hackers with a wide range of skills and knowledge.[245]

Apart from curiosity, other push-factors driving hackers to join an underground group or movement can be political or ideological motives,[246] such as organized efforts against the Church of Scientology (see the later section on Project Chanology) or support for countering human trafficking.[247] For example, Ethical Hackers against Pedophilia (EHAP) is a nonprofit organization made up of hackers and ordinary citizens who use unconventional and legal tactics to try to combat graphic depictions of child sexual abuse.[248] Some hackers try to recruit others to engage in this form of hacktivism and are occasionally successful in mobilizing a collective to engage in coordinated web defacement for political motives.[249] At other times, these attempts are met with scorn and/or criticism for "*moralfags*"[250] taking an issue too personally.

Revenge is not an uncommon motivation for hacking; anger and frustration are common risk factors for participation in violent extremist movements.[251] These tend to be common motivations for script-kiddies to engage in more insidious hacking behaviors.[252] Junaid Hussain (or "TriCk"), a British national of Pakistani descent who rose to become a prominent propagandist in the CyberCaliphate, cites revenge toward an online gaming opponent as his motivation to begin hacking.[253] Hussain is the rare example of the slippery slope model of radicalization[254] as he progressed from exploring hacking to more aggressive hacking within TeaMp0ison to cyber terrorism.[255]

> **7.6** *Many aspiring hackers seek both identity and affiliation, and the lure of the cyber underground can provide not only a sense of belonging but also a sense of purpose and/or an idealized technocracy where skill alone determines whether one is accepted.*

Anonymous members actively eschew organizational terms or definitions and values cultural familiarity. New Anons, or those seeking legitimacy, are expected to understand the community and interact within a set of established norms.[256] Individuals violating such norms are typically admonished to "lurk more," a phrase which refers to observing without actively participating.[257] The phrase can also be used as a derogatory attempt to assert one's dominance over another. This approach seems to strengthen the relationships amongst established insiders through shared cultural knowledge.[258] Because participation is nameless, affiliation based upon shared personal characteristics is nonexistent so the distinction between expert and novice (or insider and outsider) strengthened the bonds; with each new outsider engaging with the insiders, the group of insiders grew.

Rarely evidenced in criminal organization, street gangs and/or insurgent groups join for the fun of it. Many who affiliated with Anonymous during Project Chanology made clear that, while they supported demonstrating against Scientology, their main purpose for affiliation was enjoyment.[259] Many of these individuals also held strong beliefs against institutions (government or private) and individuals who are perceived to impede the free flow of information on the Internet.[260] Thus, it is unclear whether their primary motivation was political, personal, or their perceived obligation to uphold the hacker ethic.

Staying In

Understanding the hacker mentality and work ethic is useful in determining why they choose to remain underground. Many hackers prefer to maintain their own schedules, working day and night if they are sufficiently motivated, but are not necessarily comfortable with or willing to conform to someone else's timeline.[261] Hacking can become an unconventional way of living, thinking, and viewing reality, as well as a means to solve problems that cannot otherwise be faced. In these cases, hacking is not limited to the computer world but moves into other areas.

Group dynamic theory distinguishes between two sources of attraction to a group: the value of material group goals and the value of the social reality created by the group.[262] Material goals include the obvious rewards of group membership, such as progress toward common goals, congeniality, status, and security. Less obvious is the social reality value of the group where the group is the sole source of certainty for many questions of value.[263] Remaining active in the cyber underground often relies on the social reality value of the group;[264] affiliation/participation provides a degree of certainty and/or insulation against anxiety.[265] The more transient nature of the cyber underground allows for individuals to moderate their participation, such as logging on less often and/or taking a less active role in discussions, in ways not available to a traditional underground of an insurgent movement.

When the motivation is affinity for technology, hacking consists in exploring, which is rarely destructive, at least initially.[266] Hackers who hack out of curiosity or for fun are unlikely to do so for money or for the perverse pleasure of damaging infrastructure and/or harming someone. For them, a clever task, which holds technical difficulties, is fun.[267] This intangible quality to the hacker ethic appears to be among the more prevalent drivers to remaining a component of a hacker collective. During Project Chanology, Anonymous oriented its protests at the Church of Scientology; however, for some, it was more of a demonstration of their own ideals than a directed operation against Scientology. For many insiders, this approach fit within the normative boundaries of the broader collective,[268] but for others, they resisted the politicization of the collective.

The indoctrination of individuals with computer skills to illicit activities from hacktivism to hacking for mercenary reasons to providing communications/infrastructure support to other illicit activities is an important consideration in underground operations. While the majority of individuals affiliate with groups through preexisting contacts (or human bridges), an increasing number of individuals are learning about different ideologies, groups, and/or causes through the Internet. In some cases, these individuals are "self-radicalized" where their affinity toward a particular issue is increased and refined through exposure to various forms of information ranging from staged videos, to blogs, to sermons. Individuals may or may not remain in the cyber underground after being radicalized, depending on the circumstances and/or the degree of radicalization.

Group dynamics and the degree to which one's preferences and/or psyche are subsumed by those of the group may also contribute to an individual remaining in the cyber underground. One of the important psychological phenomena associated with group dynamics of undergrounds is deindividuation. Deindividuation pertains to anti-normative behavior observed in groups in which individuals are not considered as individuals; their immersion in a group is sufficiently intense whereby the individual ceases to be seen as such.[269] Deindividuation through the reinforcement of the social and group/collective identities is typically looked upon as favorable from the group standpoint as it builds cohesion and engenders loyalty. This is done through a variety of tactics: indoctrination (including the exploitation of cultural, religious, and martial symbolism and ritual), training, and/or the use of uniforms to obscure the physique, particularly the face.[270] Numerous laboratory and naturalistic research suggests deindividuation can result in increased rates of aggressive behavior and diminished aversion to risk.[271] Deindividuation does not necessarily entail the loss of personal identity; an individual is still him or herself but acts as a faceless part of a larger whole. The anonymity afforded an individual, be it through a physical mask (such as those worn by many Venezuelan protestors in 2017)[272] or a virtual one, tends to decrease individual accountability and thus increase aggressiveness. Cyber undergrounds do not necessarily employ the traditional methods often associated with the armed components of insurgent movements; however, the use of aliases or nicknames and the lack of physical interaction with others seems to have a suitably deindividuating (loss of self-awareness) effect. Research on Anonymous and LulzSec suggests this phenomenon may be evident in various Internet Relay Chat (IRC) rooms used by such groups.[273]

There are points in time at which an ordinary, psychologically healthy person first crosses the boundary between good and evil to engage in an illegal, immoral, and/or cruel behavior.[274] This significant transformation of human character likely occurs in settings in which social situational forces are sufficiently powerful to overwhelm, or set aside temporally, personal attributes of morality, compassion, or a sense of justice and fair play.[275] In anonymous online forums, deindividuation enables individuals to create and maintain their identities. Using increasingly provocative speech, and/or acts are often a means of gaining and maintaining credibility. The deindividuation phenomenon is often used to explain how adolescents, with no underlying psychopathology or history of violence, can communicate cruel and inhumane threats to others via social media.[276] Be it cyber-bullying or simply trolling, anonymity affords an individual to be more confrontational and/or aggressive than they would typically be in the physical world.

Moral disengagement, the process of convincing oneself that normative values, ethical standards, and/or legal structures do not apply in a particular context, is a consequence of deindividuation[277] and is an important phenomenon to consider when discussing cyber radicalization. Moral disengagement is a necessary step for many individuals with adept computer skills but lesser developed consciences to progress from nuisance acts to cyber crime and/or cyber terrorism.[278] The superego, a component of the human psyche responsible for self-criticism and adherence to standards of learned behavior, is more susceptible to influence during adolescence, and, as such, individuals who may not be fully psychologically mature and are not afforded the opportunity to witness the effect of actions taken against another may be less likely to accurately assess the negative impact of their behavior on others.[279] The anonymity, lack of social-emotional cues, ease of communicating via social networks, and/or media attention afforded by cyberspace also contribute to cyber-bullying,[280] cyber aggression, or deliberate actions taken to intimidate or threaten others.[281]

Psychological experiments indicate perceived anonymity of the aggressor is a predictor for cyberbullying behavior,[282] and some of those findings have been observed in the cyber underground. GamerGate was an organized online harassment campaign against two female game developers.[283] The controversy started after the release of Depression Quest, an interactive fictional game, received critical acclaim. Some claim it was undeserved attention resulting from the lead developer being female, and the online debate grew increasingly vitriolic and ultimately devolved into misogynistic stereotyping and even threats of rape and death.[284] Some critics allege that the outrage was deliberately manufactured to generate buzz about the game, while others suggest the vitriol was a form of satiric trolling. The incident, however, provides examples of how dark some online discourse can become when individuals are allowed to publish anonymously.[285]

The deindividuation process weakens an individual's capacity to resist performing harmful or socially disapproved actions. It also heightens individual responsiveness to external cues resulting from increased implicit suggestibility. In prosocial groups, this tends to be positive, while in radical groups, it facilitates the loss of individual accountability. The result of the latter is uninhibited behavior that may be deliberately harmful to another and/or other departures from normative behaviors. Deindividuation also increases adherence to norms that emerge with the group. The reestablished standard heightens susceptibility to conformity through social influence. The emergence of new group norms often leads to groupthink, a phenomenon characterized by faulty decision-making in a group.[286] Some of these phenomena appear to be at work within early cyber gangs, such as MoD, as well as the Anonymous collective. While individuals developed reputations affiliated with their aliases, many believed they were shielded from consequences (physical, financial, and/or legal) by their anonymity. Often, early success buoys confidence, and increased risk tolerance sometimes results in a false sense of security and/or superiority (relative to others as well as security forces). A common downfall seems to be the increasing tolerance for risk without considering either improved capabilities on behalf of security forces or inside information provided to them by defectors.

Leaving

Some hackers leave the underground as the responsibilities of adulthood require more time and effort; individuals with a job and/or family lack the time to interact with fellow hackers on IRC.[287] Others leave the underground involuntary; these are the individuals who are caught by law enforcement and either imprisoned, compelled to relinquish their underground activities,[288] or have their true identities revealed through doxxing by fellow hackers.[289] Like any group that becomes an operational priority by security forces, cyber underground groups risk penetration by military, intelligence, and/or law enforcement personnel. Sometimes the penetrations can be undercover operations, while others result from defection or "turning" a member based on comprised material. Hector Monsegur, aka Sabu, was an integral member of Anonymous and LulzSec's hacking efforts against both private corporations and governments.[290] Once identified and subsequently arrested by the Federal Bureau of Investigation (FBI), he became an informant. His cooperation was leveraged out of fear of his cousins (over whom he had guardianship) being put into the New York City foster care system.[291] Sabu continued to interact with Anonymous colleagues and hack along with them; however, he simultaneously provided incriminating information on them to the FBI, leading to the arrest and prosecution of others.[292]

Project Chanology attracted a number of activists whose purpose was more against Scientology than for Anonymous, and, as such, most left once the post-operation culture reverted to political agnosticism. Some Anons preferred to use their skills in a more politicized environment and drifted toward opportunities to do so.[293] While some Anons left as a result of politics (Anonymous became either too political for some or not political enough for others), others left simply because they became bored with the receptiveness of the interaction.[294]

Cyber Administrative Operations

Resistance movements developed varied and often sophisticated means of both operational and administrative functions.[295] The latter term, used in the 1963 *Undergrounds in Insurgent, Revolutionary, and Resistance Warfare*,[296] is not to be confused with the military staff function of administration (S/G/J-1). Rather, administrative operations refer to the underground functions that support military and paramilitary operations.[297] As undergrounds adopt certain practices, the security forces invariably develop countermeasures that destroy the effectiveness of these practices, and consequently, both governments and undergrounds constantly change techniques and develop new ones.[298] This section describes some of the necessary administrative functions of an insurgency and the methods with which they are performed solely within or aided by components of the cyber domain.

Recruiting

Recruiting remains the *sin qua non* of resistance warfare,[299] and closely related activities include indoctrination and radicalization. Cyberspace is a medium for social interaction,[300] and thus the related phenomena require investigation from various sociotechnical perspectives. The first interaction an individual will have with a social movement, be it nonviolent or violent, is likely through the Internet.[301] The use of the Internet to recruit potential insurgents/terrorists has long been a research interest in the psychology of terrorism,[302] and the use of the Internet is by now a well-established tactic used by groups such as Hizbollah, al Qaeda and its affiliates, and the Islamic State (or Da'esh).

The cyber underground serves as a venue for individuals who may lack the knowledge about phenomena that affect their lives ranging from concern about disenfranchised populations to subjective experience with mental health challenges.[303] Reddit, an online community that does not require disclosing one's identity to post information, hosts both antisocial and prosocial content.[304] Some Reddit sites, or subreddits, contain material espousing child abuse, spousal abuse, racism, and/or other forms of hate speech,[305] while others serve as essentially peer-support networks for those coping with a variety of life's challenges.[306] In some cases, those challenges may bring together a heretofore anonymous group of individuals whose collective grievances can manifest in violence. An example is the involuntarily celibate or Incels, an online subculture of men who believe they are unrightfully marginalized and thus react with virulent and sometimes violent forms of misogyny.[307] In one such example, Alek Minassian drove a van into a crowded public shopping area in Toronto, Canada, killing ten. Prior to executing the assault, Minassian posted *"The Incel Rebellion has already begun!"* on Facebook.[308] In more pro-social areas of the cyber underground, individuals suffering from depression use Reddit to communicate with one another to establish virtual peer-support networks, and a growing body of scientific literature suggests there is indeed a benefit from participation in these forums.[309]

Social media, and the algorithms that underlie the search mechanisms, suggestions, and content delivery tend to exacerbate biases.[310] Videos of successful terrorist attacks or guerilla missions are popular features of sites focused on recruitment. Technologically sophisticated developers with malicious intent can exploit both the medium and code to not only facilitate but also accelerate the radicalization process. For example, if a user were to view an extremist video on YouTube, the algorithm might suggest others. Those additional videos are often more extreme as the algorithm is designed to keep the viewer interested and engaged with the content.[311] Some sites and/or smartphone applications employ machine learning and other statistical techniques to curate content and/or develop user profiles. Unintentionally, these approaches enable self-radicalization.[312] These approaches, widely used by marketing firms (and enabled by technology companies) gain favor in operational influence contexts.[313] ISIL adopted a hierarchy of needs[314] approach to this process, whereby potential recruit behavior, including their communication habits, is mined to identify their susceptibility to different themes, messages, and/or targeting approaches.[315] Videos often show extreme graphic violence, sometimes accompanied by audio or text commentary. Producing these videos is important enough that a number of violent groups, including Hizbollah, the Chechen resistance, and al Qaeda, routinely include a videographer as an essential part of an operational team.

Videos serve several functions. First, videos attract attention and excite passions of sympathizers, particularly young males who may be recruited to perform these types of actions.[316] Recent trends, documented in a 2017 article by Ariel Lieberman, suggest that a growing number of young women are also avid consumers of this content produced by ISIL in particular.[317] Lieberman notes that in 2014, ISIL recruited three teenage girls from Denver, Colorado via social media who took steps to travel to the Middle East.[318] Second, videos create mental imagery, allowing recruits to imagine themselves as successful operatives, and repetitive video reinforces the message that attacks are likely to be successful. Reading about a successful attack is not as compelling as watching one unfold in real time from the vantage points of someone involved who lived to deliver the video footage. This is not unlike the tactics many militaries, at least those in nations without universal conscription, use to recruit young people to enlist. Third, videos begin the process of desensitizing recruits to violence and the parallel process of dehumanizing opponents.[319] Online recruitment also provides prospective members access to libraries of books, poems, speeches, and/or other art forms to which they may not have access (or of which, they are completely unaware);[320] the exposure to the aesthetic aspects of jihadist insurgent culture may further legitimize the organization in an historical context. Research on the Islamic State's videography skills indicate the group replicates the imagery and techniques used in popular Hollywood movies and first-person shooter video games.[321]

Video games have proven to be a valuable recruiting tool. Desensitization has been demonstrated in many different settings. Most of the relevant research has been done with children. Children who witness adults behaving aggressively, for example, by pummeling a stuffed animal, tend to imitate that aggression. Children who watch violent television or play violent video games demonstrate aggressive thoughts and behaviors, less empathy toward victims, and lower physiological reactions when witnessing violence; children exposed to actual violence show a range of negative stress reactions that persist long after the events.[322] Similar results have been found with adults.[323] Witnessing violence does not universally cause violence, and no amount of watching violent television or playing violent video games will make a child violent if they are not predisposed.[324] Exposure to violence will lead some to commit more acts of violence, through desensitization or simple imitation, and desensitization on a large scale can affect how quickly people intervene or punish incidents and generally weaken cultural mores that prevent violence.[325] Hizbollah created and distributed a video game called "Special Forces," a first-person shooter where players can perform target practice by shooting at an avatar of then-Israeli prime minister and Israeli Defense Force chief of staff.[326] Players can then play in recreated missions against Israeli tanks, helicopters, and fortified positions. Instructions for play are available in Arabic, English, French, and Farsi; the game claims to have sold more than ten thousand copies.[327] Video games, particularly networked multi-player games, have also been fertile recruiting grounds for right-wing extremist organizations in the United States.[328]

The ability to recruit talent and work from distant locations are also key advantages of resistance movements that have relationships, overt or covert, with Internet-based journalism. The Tamilnet.com site began in Sri Lanka but benefitted from the assistance of Tamils living overseas, including a computer programmer from Norway, a systems analyst in London, and "dotcom entrepreneurs" from the United States.[329] Distributing the technology does not obviate the danger for local reporters; a Tamilnet.com reporter was killed by a grenade tossed through his study window in 2000.[330] However, the decentralized nature of the Internet provides more mobility to content providers, so endangered local writers and editors can also more easily move locations while continuing to produce content. Translating, editing, and some writing can also be done extraterritorially. As an example of this, Mark Whitaker describes a time when the editor of Tamilnet.com filed a story from Canada.[331] The story, leaked by a member the Sri Lankan government's own Human Rights Commission, was about an instance of police brutality against a Tamil detainee and included accounts of prisoners tortured with boiling water, forced to eat cow dung, and inflicted with a string of similar abuses. The editor crafting the story was sitting in his nephew's bedroom in a suburb of Toronto trying to tune out six noisy nieces and nephews.[332]

Targeted recruitment takes on a much different form for cyber undergrounds and may involve considerably different personalities. Some suggested psychological differences amongst Black Hats and White Hats;[333] limited primary psychological research provides empirical support, but it remains an ongoing research interest. Recent research identifies that the combination of psychopathy, Machiavellianism, and narcissism (known as the Dark Triad) are evident in those who engage in cyber bullying.[334, 335] This combination of psychological characteristics is typically associated with maladaptive behavior and seems to be evident in online trolling.[336] The Islamic State is particularly adept at exploiting existing psychological vulnerabilities and in-group biases by appealing to culturally resonant imagery, be it through a verbal reference or, in many cases, iconic imagery to influence their target audiences.[337] In many cases, both Abu Bakr al Baghdadi[338] (ISIL's leader) and Abu Mohammed Adnani[339] (ISIL's spokesman and director of foreign operations) appealed to multiple target audiences in the same speech.[340] ISIL's thematic content suggests conscious attempts to exploit multiple risk factors for radicalization.[341] The continuity of messages across multiple demographics (males aged twelve to eighteen, nineteen to thirty-nine, and forty to sixty-five, respectively), and psychological vulnerabilities create a coherent master narrative by appealing to the concomitant crises that need to be resolved at the corresponding psychosocial stages: adolescents (identity versus role confusion), young adults (intimacy versus isolation), and middle adulthood (generativity versus stagnation).[342]

The Internet enabled resistance organizations to compete on a level playing field with state actors,[343] perhaps nowhere as transformative as in social media. Some implicate social media as an enabling factor in the rise of right-wing extremism in the United States since 2016.[344] At some point in the recruitment and radicalization process, recruits usually make personal contact with mentors or organizational liaisons to further training or coordinated activity. Junaid Hussain of the Cyber-Caliphate, drawing on his own experience as a disaffected youth who sought solace through online gaming, convinces similar individuals to heed the call to support ISIL, either physically or virtually.[345] Technology facilitates this process by allowing users to move from viewing a website to sending an email, posting on a discussion board, or joining a real-time chat room. The transition from passively viewing to interacting involves increased risk for both parties. Revelation of personal details or concrete biographical facts increases the chances of identification and geolocation. The Internet's anonymity provides partial protection, but both recruiters and recruits can be "spoofed" by opponents or law enforcement playing a role.[346] Such spoofing attempts can be used as satire, as it seems to be in Keyonstone, United Kingdom,[347] or as more deliberate attempts to undermine an ideology, such as efforts in Minnesota.[348]

> 7.7 *Recruiting remains arguably the most essential activity of resistance warfare,[349] and the use of the Internet includes online forums, social media, and even video games to advertise, solicit, and recruit potential members/supporters.*

Intelligence

Intelligence and counterintelligence are among the most important functions of the underground. Drawing on the strength of clandestine networks and relationships, undergrounds often gain access to intelligence by virtue of confederates operating throughout the theater of conflict. Underground intelligence networks most often extend beyond the borders of the movement's native country, and it is not uncommon for undergrounds to distribute cells throughout the world among populations sympathetic to the cause.[350] Cyberspace enabled these networks to not only grow but also form extraterritorially, often globalizing a social movement and thus individuals willing to provide intelligence to the underground within hours or even minutes. In some cases, underground organizations develop an organic capability, while in others, they rely on the broader cyber underground for mercenaries willing and able to conduct intelligence operations for hire.[351] The mercenaries, in turn, established their own marketplace where corporations can clandestinely engage in industrial espionage, and/or an individual can hire someone to serve as a virtual private investigator to collect information on a target of interest, be it a spouse or prospective employee.[352]

Perhaps no cyber intelligence tactic has become more dangerous to clandestine organizations than doxxing, the research and online publication of private or identifying information about an individual, typically with malicious intent.[353] Doxxing can be employed as not only a defensive counterintelligence tactic but also a means of leverage to recruit a potential intelligence agent.[354] Anonymity online affords individuals a degree of security; it insulates them from the consequences of provocative or malicious action. The loss of this protection could undermine their reputation and/or willingness to participate in organized activities. As such, the ability to identify these individuals by name and/or address is a technique used by both hackers themselves as well as security forces. Anons often attempted to doxx new members who tried to shift the conversation or focus of effort without establishing the requisite bona fides of the group. Sometimes the threat of doxxing brought members back in line, while other times, outing them was necessary to silence them.[355] After a series of rallies and counter rallies organized by White Nationalists in Charlottesville, Virginia in November 2017 turned violent, remote supporters of the counter-rallies employed doxxing to discover and publicize the names, faces, and addresses on many White Nationalist protestors.[356] In an attempt to both hold those committing violence accountable as well as publicly reveal the identities of those affiliated with the more extremist fringes of the white supremacist movements,[357] the debate over doxing as a form of social justice or vigilante targeting was made national. In some cases, once individuals were identified, Twitter was used to publish the relevant information and, in some cases, publicly shame those supporters.[358]

The ability to collect and correlate information is an essential skill for those in the cyber underground. The human factor has long been acknowledged as the most significant vulnerability in cybersecurity (also see Key Takeaway 4.2). In the cybersecurity literature, social engineering refers to the psychological manipulation of an unwitting individual to gain information.[359] More specifically, the use of social engineering (the use of deception to manipulate individuals into divulging confidential or personal information that may be used for fraudulent purposes) is well documented as the entry point for many cyber-security compromises.[360] Social engineering employs several techniques: pretexting, phishing, online social engineering, shoulder surfing, and dumpster diving.[361] Social engineering attacks human nature on a social psychological level. The three aspects it addresses include: "alternative routes to persuasion (i.e., central route and peripheral route), attitudes and beliefs that affect human interactions, and techniques for persuasion and influence."[362] These three aspects serve as useful tools to social engineers.

Kevin Mitnick, a notorious American hacker, provides detailed examples of methods used by hackers to gain access to various information systems. One of Mitnick's most common approaches was social engineering, a set of tactics used to deceive others in gaining information.[363] Social engineering is commonly used by hackers; however, it is not simply exploiting a vulnerability or tricking an unknowledgeable individual into providing information. Social engineering requires either a specific understanding of the target or a generalized ability to empathize to exploit.[364] The approach can be used in person, over the phone, or via email. Mitnick combined a detailed knowledge of information systems, particularly the telephone system in southern California with empathy for employees in particular roles.[365]Mitnick conducted detailed reconnaissance of facilities and technologies; often reading service manuals and employee handbooks. In one incident in 1981, Mitnick and two colleagues socially engineered their way into a Pacific Bell facility in Los Angeles to gain access to the computer system for mainframe operations. After stealing a system manual, Mitnick later used the knowledge gained from careful study to develop more effective pretexting tactics when socially engineering Pacific Bell employees.[366] Through trial and error, he attempted to remotely gain access to different systems, sometimes making phone calls to specific offices and assuming the identity of technicians, managers, and/ or security personnel.[367] He used these approaches to add services to his account without paying and determine whether his and his families' phones were tapped.[368] Mitnick's tactics, which ultimately landed him in prison, were a combination of psychological and technological manipulation, which were often emulated by other hackers, including some associated with Anonymous.

While some Anons relied entirely on technological means to exploit system vulnerabilities, others preferred social engineering. In some cases, there was a deliberate differentiation of labor with assigned tasks, while in others, it was more a competition of who could gain the desired information first, with greater credit going to those who used technological means.[369] A particular form of remote social engineering is phishing, the use of deceptive emails to acquire personal information, such as user names and passwords, bank account numbers, and/or information that can be exploited by hackers for a variety of purposes.[370] A more precisely targeted form of phishing, called spearphishing, targets a specific individual or organization and often appears as though it originates from a trusted source.[371] Typically, when the email is opened or an embedded link clicked, malicious code is downloaded to the unsuspecting user's system. Spearphishing requires careful collection of information, analysis, and technical skill to simulate the appearance of authenticity. This combination of social and technical approaches has been widely replicated and is responsible for a considerable amount of identity theft. ISIL employed spearphishing against an indigenous Syrian movement called Raqqah as being Slaughtered Silently (RSS). When RSS began publishing accounts of ISIL atrocities, its members were targeted with phishing emails and other attempts to doxx them.[372] Individuals contacted RSS members under the guise of publicizing their activism internationally, and a number fell victim and wound up providing their identities and/or locations.[373] In some cases, malware was downloaded by RSS members trying to establish secure communications with would-be journalists.[374] While the malware used in the attempts was not explicitly traced to ISIL, ISIL demonstrated the willingness to attack (both physically and virtually) an organization critical of their policies and actions, and the code itself was sufficiently different from techniques and code used by the Syrian regime.[375] While it is unclear whether the spearphising attempts directly led to the deaths of RSS members, there are RSS members who were murdered for their work.[376]

> *7.8 Counterintelligence is among the most important functions of the underground. Social engineering (the psychological manipulation of an unwitting individual to gain information[377]) and doxing (the practice of online researching and broadcasting private or personally identifiable information about an individual or organization[378]) are two signature intelligence collection tactics used by cyber undergrounds.*

Financing

An underground organization requires financial resources to function: agents must be compensated, psychological operations require funds for products and media resources, and headquarters and administrative sections require office supplies. Regardless of how they manage their finances, resistance movements need money to survive,[379] and those operating solely or partly in the cyber domain continue to develop and/or exploit relevant technologies to transfer and maintain access to those funds. Some organizations, such as the Islamic State, dedicated financial cells and/or committees, while others employed more ad-hoc arrangements.[380] Many virtual organizations are increasingly reliant on online services; for example, one of Anonymous' higher profile attacks was on PayPal after it prohibited WikiLeaks, another virtual organization, from accepting electronic donations.[381] Finding new ways of soliciting funds over the Internet is another fast-changing cat-and-mouse game between a resistance movement and its opponents.[382] Financial transactions are easier to track than other kinds of information that flow over the Internet, making fundraising more difficult for insurgents than recruitment. Publicity commerce is an important aspect of private Internet usage, although it is not as often a part of insurgent fundraising.

The first generation of insurgent websites included explicit appeals for online donations; however, subsequent legislation preventing fundraising for terrorist organizations forced this activity underground. Aboveground websites sometimes make money by selling souvenirs and may imply that the money supports the insurgent cause.[383] The 32 County Sovereignty Movement, a group associated with the Real Irish Republican Army (RIRA) at one time joined Amazon.com's "Associates" program and received a cut from book sales when it redirected visitors to buy those books at Amazon; however, the company removed the RIRA from the Associates program when it was informed of the group's insurgent ties.[384]

The advent of cryptocurrencies, such as Bitcoin, and online service providers and/or black marketplaces, such as Megaupload, Silk Road, and Pirate Bay, have made online fundraising and money laundering more accessible to resistance movements, many of which, the Islamic State in particular have done with success.[385] The cryptocurrency Bitcoin is built around a blockchain, a shared distributed database built on cryptographically secured transactions,[386] and offers the ability to conduct financial transactions with enhanced anonymity and security. Cryptocurrencies are commonly used on cyber-criminal marketplaces, such as Silk Road,[387] and their rapid development and volatile valuation make cryptocurrencies risky foreign exchange investments. Although, their adoption by technologically sophisticated illicit organizations continues to present challenges for law enforcement.[388] For more on cryptocurrencies and their underlying technology, see chapter 8.

Logistics

The principle changes to underground logistical functions are the safe and secure operation of information systems to organize supply systems,[389] as well as the requirement for the accompanying hardware and software to support informationized warfare.[390] Collectives and limited-hierarchy organizations may not require the same functions as a traditional insurgency, and thus individuals who volunteer often bring their own equipment. Not unlike many local militia organizations, there is no centralized supply system (although an individual may purchase certain goods in bulk and then distribute to save money), and cyber collectives often rely upon auxiliary support to organize, train, and equip themselves. The increasing capability of web-based services and marketplaces enables insurgent (or any other organization or individual) movements to buy vice build their own capability. In some cases, this can include armies of trolls or botnets that can be essentially rented for short-duration attacks with only minimal interaction with the underground.[391] The appropriate use of sophisticated tradecraft from cutouts to encryption render these interactions almost untraceable in many cases.

Training

After recruiting, training is arguably one of the most important functions of the underground. Developing capability and building capacity are necessary to achieve larger effects (be they physical or virtual) on a populace or an oppressive government. A variety of resistance movements experimented with use of the Internet as a channel for training in operational techniques.[392] There are numerous training manuals, videos, and even instructional software available to anyone with access. However, it is unclear to what extent it is realistic to expect that an individual interested in supporting a resistance movement can trail him or herself to a level of competency to make a serious contribution to a physical resistance effort. Bomb-making techniques, for example, require considerable practice and expertise. In other complex domains, successful e-learning usually requires some personal interaction, contact, and feedback from experts. Self-training is probably more effective for disseminating new techniques or countermeasures to already-trained operatives than for training novices. Self-training manuals such as *The Terrorist's Handbook, The Anarchist's Cookbook,* and *The Mujahedeen Poisons Handbook* are available online. *The New York Times* interviewed a Palestinian called Abu Omar, who was employed as a trainer in Iraq and taught foreign fighters how to make bombs and stage roadside attacks; at the time, he worked with two cameramen, who videotaped his bomb-making classes, to produce Internet instructional videos.[393] The Islamic State took this a step further and ha created both synchronous and asynchronous virtual training models along the lines of university distance learning programs.[394]

The training, indoctrination, and radicalization processes within hacktivist and/or hacking organizations are very much intertwined. Often individuals seeking to acquire additional knowledge and develop skills become influenced by more experienced and potentially manipulative hackers from whom the novice learns.[395] Because the rate of technology advances so quickly, formal training of hackers (particularly Black Hats) is rare, with much of the skill acquired through open-source information, facilitated learning, and trial and error. There are few formal means to teach hacking, and many require one to serve in a government's security apparatus. For an individual interested in the "black" side of hacking, one must seek out others, attempt to gain their trust, and learn from them.

Among other functions, social media also affords the opportunity for informal learning. Twitter, in particular, served as an educational starting point for many interest in the OWS movement.[396] OWS was a protest movement that started on September 17, 2011 in New York City by Canadian anti-consumerist and pro-environment group Adbusters.[397] While one hundred and forty characters is often insufficient space to convey a complex argument or explain how to perform a particular task, tweets often include hyperlinks to websites, videos, chatrooms, and/or maps and calendars—all locations where an individual can learn more.[398]

When attempting to create mayhem or widespread panic, training individuals may not be necessary. For example, Anonymous employed the Low Orbit Ion Cannon (LOIC), a penetration-tasking tool used in cybersecurity that can also be employed as a DOS application, to attack websites associated with the Church of Scientology in 2008.[399] Anonymous also employed the LOIC in 2010 against groups that opposed WikiLeaks, an international nonprofit organization that espouses freedom of information. Many Anons relied on both witting and unwitting computer uses to install (overtly or surreptitiously) the necessary software to use a computer terminal as a LOIC node.[400]

Training is particularly important when positive, not just negative, actions are desired. Noncooperation and civil disobedience are positive acts that involve training, organization, and solidarity on the part of the resisters, whether they operate in the open or clandestinely.[401] More coordinated efforts require skilled operatives to plan and execute operations, many of which require operatives to train others. Project Chanology saw experienced Anons willingly provide information to novices sincerely interested in exposing legal, financial, and moral flaws in the Church of Scientology. Some of this information included the basics of information security: how to properly protect a chat room from infiltration so it can be used to plan, how to use proper operations security when soliciting volunteers for an operation, and/or where to publicize information to attract likeminded people.[402] In other chat rooms, individuals less concerned with peaceful demonstration were provided information on how to deface websites, coopt servers, design botnets, and/or conduct DOS attacks against private and government organizations.[403] In some cases, individuals required tutoring, while in others, helpful Anons simply posted instructions to publicly accessible chat rooms or websites. In the former cases, the process often entailed initiation rituals and/or demonstration of bona fides, for example, demonstrating skill by conducting a low level yet illegal hack.[404] This form of facilitating learning is more labor intensive but can serve the dual purpose of training and radicalizing prospective members. The rate of technology change renders formal training in hacking somewhat rare, with much of the skill acquired through open-source information, facilitated learning, and trial and error. Nevertheless, for larger-scale, more coordinated activities, such as nonviolent resistance, undergrounds rely on facilitated learning provided by more experienced organizers.[405]

Command, Control, and Communications

Communications technologies facilitate clandestine organization and planning; however, government capabilities (although not necessarily government agility) may exceed those of the organizers, and thus redundant systems and plans are necessary. Resistance organizations often develop and employ clandestine methods of communication resembling those used by sophisticated espionage organizations.[406] The emergence of the Internet as a globally accessible communications network changed, and will continue to change, both the internal and external communications of resistance movements.

In the leaderless and limited-hierarchy structures, groups may employ completely decentralized C2, for example, by openly suggesting targets and tactics and hoping that self-managed groups enact them. Private and governmental organizations also spend considerable resources on internal knowledge management. These initiatives may involve creating information systems to ease sharing of information across divisions and creating communities of practice that bring together organizationally separated specialists to share technical information and maintain awareness of new developments. Resistance groups operate under a different set of constraints, however, because of the need for secrecy and compartmentalization. Open, free flow of social contact and information is not a reasonable goal; cross-divisional communication must be more carefully managed or involve more cells.[407]

Throughout the 1990s, al Qaeda used satellite phones and computers to organize and maintain plans and faxed copies of religious rulings issued by bin Laden throughout the Muslim world and Europe, where they were publicized by Arabic-language media outlets.[408] Their communications infrastructure and operations made extensive use of electronic media for mobilization, communication, fundraising, and planning attacks. One of the boldest uses of Internet for C2 is Hizbollah's dedicated fiber optic network. This network was emplaced parallel to Lebanon's legitimate cable television and Internet lines. It was a sign of Hizbollah's political influence that when the Lebanese government threatened to dismantle the network, the organization was able to pressure the government into leaving the network in place.[409]

Cell-phone-based text messaging has been successfully used to coordinate anti-government protests. The WTO protests in Seattle pioneered the use of social media for this purpose. Strategic movement of crowds as a protest tactic has been used for many years but previously relied on pre-planned sequences and formations similar to how military movements were once limited. Cell phones, text messaging, and Internet-based communications allowed much greater flexibility in the movement of demonstrators and diversionary forces in response to police movements. One of the groups involved with the WTO protest, Direct Action Network, utilized communications channels—from cellular phones, to portable computers with an Internet connection, to pagers, police scanners, and two-way radios—to command and control certain nodes and maintain a degree of tactical cohesion. In addition to the organizers' all-points network, individual protesters leveraged protest communications using cell phones, direct transmissions from roving independent media feeding directly onto the Internet, personal computers with wireless modems broadcasting live video, and a variety of other networked communications.[410]

The Internet can also facilitate C2 by serving as a medium through which direction from higher to lower and feedback from lower to higher may be communicated. Internet and cell phone technology are thought by some to play an important mobilizing role in the Iranian protests after the disputed 2009 elections, although this claim is controversial.[411] The use of Twitter during these protests received a great deal of attention.[412] Twitter is a very flexible text messaging service that can be used either to broadcast to a large audience or send personalized messages among friends, and the messages can be broadcast using either the Internet or SMS (cell phone based). It is clear that Twitter, along with other services such as the YouTube video-sharing service, were closely monitored by people outside the country who wanted to follow events. Some authors questioned whether Twitter played an important role in mobilizing and organizing the protests themselves.[413] On June 16, during these protests, the US Department of State contacted Twitter to ask it to delay a scheduled server upgrade that might disrupt Twitter traffic. Later, the Iranian government intentionally disrupted Twitter traffic by shutting down or intentionally degrading both Internet and cell phone services in sections of Tehran.[414]

Project Chanology serves as an example of a social movement initiating in cyberspace but manifesting in physical space. While the initial planning and execution were entirely online, Anonymous' actions against the Church of Scientology eventually transitioned into physical actions, such as demonstrations and confrontations.[415] Anonymous considered this tactical evolution an extension of its unique culture and not necessarily a fundamental shift in strategy or identity.[416] Unlike Egyptians in Tahrir Square, the Anons retained the ability to exert C2 (to the degree that it actually existed) through the Internet but chose to engage in the physical realm to increase the effectiveness of their operations. The technologies utilized by Anonymous, coupled with tradecraft (using handles vice names, not disclosing personally identifiable information, etc.) afforded collaboration amongst a large group.[417] Despite the large volume, the group was reasonably well coordinated for most of the operation;[418] however, during less focused periods, there was evidence of less direction and/or more disjointed collaboration.[419]

Security

Maintenance of security remains a vital underground function, but as undergrounds become increasingly reliant on the Internet, security grows concomitantly complex. Not only is it far more difficult for clandestine members to remain hidden due to aggressive counterinsurgent cyber intelligence, the ongoing training requirements for members require an ever-growing cadre of qualified instructors/security managers. Virtual organizations do not have the same physical security requirements but still require counter surveillance tradecraft. Dark networks are specific portions of the Internet that require specific software, configurations, and/or authorization to gain access and are not indexed by common search engines; thus, individuals who do not have the specific site address could not happen upon it by accident.[420] Dark networks are often components of contemporary illicit trafficking; however, the networks themselves are simply modern manifestations of criminal undergrounds and black markets that existed for centuries. Safe havens are any space, whether physical, legal, financial, or virtual (e.g., invitation to non-public, non-indexable chat room or IRC[421] channel), that enable resistance groups to plan, organize, train, conduct operations, or rest with limited interference from enemy or counterinsurgent forces.

In cyberspace, the clandestine aspects of underground operations are enabled by the technology itself, and thus the analog of the painstaking tradecraft required in the physical domain manifests in the technological skill and discipline when interacting with interconnected technologies. The use of anonymizing tools are necessary to maintain operational security.[422] Understanding the technological means by which these networks establish and maintain, as well as the psychological effect of their existence, are important aspects of the cyber underground and thus should be a component of subsequent research.

Cyber Undergrounds Psychological Operations

To insurgent movements, influencing opinion and attitudes is not an end in itself but a means to communicate their ideology and/or efforts among broad elements of society.[423] Underground psychological operations are conducted in a variety of forms: mass media and face-to-face persuasion; leaflets and theatrical performances; programs for local civic improvement; and threats, coercion, and terror. In the contemporary operational environment, more often than not, the preferred medium for such communication is through social media. Historically, the substantive content of psychological operations during any phase of a campaign is likely to be determined at the highest echelon of the organization;[424] however, the rapidity with which insurgents currently interact with various target audiences resulted in a more decentralized approach to influence. Successful insurgent influence relies upon the ingenuity of the operators at the local level,[425] and nowhere is that more prevalent today than in revolutionary movements with significant cyber components.

Cyber is the logical medium for political and psychological warfare as insurgent groups must effectively engage in battles of persuasion and influence to achieve their sociopolitical objectives.[426] The information domain's impact on the radicalization process in modern insurgency cannot be overstated because the seemingly ubiquitous availability of information, including ideological narratives, success stories, and even the presence of tactics, techniques, and procedures, can have a profound influence on cognitive processes.[427]

> 7.9 *Cyberspace is proving to be the decisive battleground for political and psychological warfare.*

Mass Communication

The basic functions of underground mass communication remain the same regardless of medium; however, the rise of global broadcast media, particularly satellite-linked television and the Internet extended the reach for every important insurgent communications activity, including publicity, recruitment, training, fundraising, and C2. Cyberspace is an ideal venue for influence as a political theater. The Zapatista rebellion, from the perspective of the Mexican government, was no different from many small-scale uprisings of indigenous groups. However, the intellectual and highly stylized aspects of online social influence not only expanded the awareness of the cause to demographics outside Mexico but also transformed the image of the movement itself to one with far more gravitas and/or international significance than a simple local uprising.[428]

The Serbian youth movement Otpor! used public theater and satire through various forms of media in concert with more traditional approaches to nonviolent resistance, such as demonstrations, concerts, electoral politics, general strikes, and even the occupation of government buildings and disruption of traffic. Otpor! ("resistance" in Serbian) was an influential youth movement in Serbia from 1998 to 2003 that engaged in a two-year-long, successful nonviolent struggle against Slobodon Milosevic.[429] Otpor! formed in Belgrade in response to repressive university and media laws introduced earlier that year. The group primarily consisted of members of the Democratic Party Serbia youth wing, members of various nongovernmental organizations that operated in Serbia, and university students (many of whom were veterans of anti-Milosevic demonstrations). The organization quickly gained prominence as anti-regime media outlets started featuring the clenched fist symbol in open defiance of Serbia's information law. In the aftermath of the 1999 NATO bombing, Otpor! demonstrations resulted in nationwide police repression, resulting in the arrests of over two thousand activists, some of whom claimed to have been beaten in custody. After Milosevic's 2000 resignation, the organization became an international resistance cause célèbre and eventually (in 2003) transformed into a political party.[430] Otpor! created publicity in Serbia by spreading handbills, posters, and graffiti showing its symbol (a clenched fist) throughout the country and by having political cartoonists incorporate incongruity and absurdity into its products.[431] While the Otpor! movements predate social media, many of its tactics are evident in more recent resistance demonstrations.

Project Chanology provides an example of the combination of cyber underground culture, memetic information, humor, and online and offline activism that redefined contemporary social movement activism.[432] Anonymous maintained adherence to its cultural norms without alienating many, while taking on a powerful and public target. These actions required the use of the Internet, not only for internal communications but also as a link to various media outlets who then broadcast the messages and actions to a wider audience.[433] Similar to many insurgent groups taking on better resourced adversaries, Anonymous weighed personal security against publicity as small-scale pranks did not have a demonstrable effect on the Church of Scientology. Anonymous adapted to the demands of an operation of this scale by acknowledging the need to communicate with those outside the cyber underground.[434] This interaction required preparations of statements/explanations for traditional broadcast media, which in turn resulted in a further division of labor amongst the hackers with some focusing on "public relations."[435] Few organization demonstrated a more sophisticated combination of technical skills sets and effective uses of a variety of media forms for what many of their collective considered an elaborate prank.[436] Anonymous set a precedent for the sociopolitical effect of blended pranking in the digital age.

More traditional media also benefit from Internet distribution. A twenty-six-page pamphlet with instructions for protestors played an important role in the Egyptian movement and was distributed in either print or PDF format from person to person. Instructions on the front of the pamphlet urged that it not be posted on publicly accessible Internet sites, however, saying: *"Please distribute through e-mail, printing and photocopies ONLY! Twitter and Facebook are being monitored."* This allowed the movement to swell numbers by recruiting local residents, and, more importantly, it aggregated people in the square more quickly and in a manner more difficult to prevent or disperse.[437]

Subversion and Sabotage

The Internet dramatically increased the opportunities for both subversion and sabotage—integral aspects of comprehensive underground psychological operations. Subversion refers to actions designed to undermine the military, economic, psychological, or political strength or morale of a governing authority.[438] Sabotage comprises actions to withhold resources from the government's counterinsurgency effort by acts of destruction. This includes acts with intent to injure, interfere with, or obstruct the national defense of a country by willfully injuring or destroying, or attempting to injure or destroy, any national defense or war materiel, premises, or utilities, to include human and natural resources.[439] The increasing reliance of critical infrastructure in the developed world (and in many places in the developing world) on the Internet creates innumerable opportunities for sabotage. Highly sophisticated and/or well-protected SCADA systems require both a high degree of skill and persistence (the combination being rare in many cyber collectives) but make for logical targets of more politically committed organizations. For more detailed information on SCADA systems, see chapter 6 of this book.

There is often overlap between self-described hackers and Internet trolls. Internet trolls are individuals who deliberately create discord online by instigating quarrels, insulting people (or groups), posting inflammatory, extraneous, or off-topic messages, and/or intentionally violating established normative behaviors.[440] In some cases, trolling may be simply to make others laugh, but more severe manifestations can constitute cyber bullying.[441] Trolling may also be employed as a deliberate means of subversion.

The manipulation of existing or complete fabrication of social movements (or astroturfing) can have profound economic, political, and psychological effects, even at the societal level.[442] Organizations can leverage large-scale subversive capabilities to achieve direct effects or to use as a deception operation to obfuscate true intent. In 2009, the "Faces of Coal" website used stock photos of individuals and groups under headlines that claimed people from all segments of society supported the use of coal.[443] Allegedly, Russian troll farms were behind astroturfing networks during the 2014 incursion into Ukraine,[444] and some researchers suggest the tactic was used by both sides to garner international support for various patriotic social movements.[445] The intentional spread of disinformation through the use of botnets has become a common tactic by state and non-state actors.[446] Forcing security forces and/or political opponents to respond to false information places them at an information disadvantage, preventing them from "getting in front" of any story. In some cases, botnets promulgate disinformation about government atrocities by fabricated humanitarian organizations.[447] A particularly effective approach is deliberately targeting "key influencers," accounts with numerous followers likely to interact with bots, to forward the information and establishing a false sense of credibility via social proof that contributes to the proliferation of misinformation.[448]

Nonviolent Resistance

Nonviolent resistance continues to play a prominent role in many underground and revolutionary activities, and the twenty-first century has witnessed a synthesis of the global technological networks that link computers on the Internet and social networks to result in innovative forms of protest.[449] This trend is likely to continue, as the opportunities for the voiceless to find their voice are numerous and growing, as is discontent with the political status quo and the concomitant passivity. For cyber-enabled nonviolent resistance to be an effective instrument of US statecraft, the "disruptive" thinkers within the United States must recognize their own in other nations and implement the appropriate tactics to enable them to accomplish their sociotechnical and/or political objectives.[450]

Among the most prominent theorists in nonviolent revolution is Dr. Gene Sharp, whose compilation of nonviolent resistance tactics is included in Table 10-1 of *Human Factors Considerations of Undergrounds in Insurgencies*.[451] Sharp's key theme is that political power is not derived from the intrinsic qualities of those in positions of authority but from the consent of the governed, and thus the latter possesses the moral and political authority to take it back. Essentially, despite the government possessing a physical monopoly on the use of force, the people have the moral authority to impose their collective will.[452] Sharp assumes the set of universal human rights published in Article 21 of the United Nations' 1948 *Universal Declaration of Human Rights*[453] and postulated that nonviolent resistance and its' concomitant tactics are a tactically effective means to challenge government authority without sacrificing moral authority. For a more in-depth discussion of Sharp's work as it pertains to insurgent and revolutionary warfare, see chapter 10 of *Human Factors Considerations of Undergrounds in Insurgencies.*[454]

The exhaustive set of tactics Sharp compiled in his research typically focus on the physical domain. However, there are numerous corollaries to cyberspace, and thus, nonviolent cyber resistance manifested in innovative forms of protest emerged in the post-Cold War era. Although few managed to mobilize a sufficient number to displace a regime, they provided a forum for a youthful demographic to engage in creative, often social-media-directed alternatives to the picketing and chants employed by their elders.[455] Participating in nonviolent cyber resistance can take many forms, from changing one's avatar[456] to espousing a form of resistance clothing.[457] Successful social movements tend to 1) directly confront and reframe perceptions about a particular sociopolitical issue, 2) exploit existing social networks and simultaneity to achieve the greatest effect, and 3) connect the ideologues to the mainstream population.[458] Both cyber activism and hacktivism made each of these attributes more accessible to the average individual and/or accentuate the effects of them.

Cyber Activism

Cyber (or virtual or online) activist organizations utilize the Internet in a legal and typically non-disruptive manner in support of an agenda or cause.[459] Internet-enabled activism is particularly evident with groups focused on organization/mobilization[460] as the ability to rapidly inform and mobilize a large, heretofore unrelated, group of individuals to behave in a manner supportive of a particular objective (be it in the virtual or physical domain) considerably greater than that of the underground pamphlet operations on the 1960s.[461] Those fundamental changes also created a vulnerability as security forces also monitor social media as an intelligence source and exploit it to violent quash protest and/or imprison individual nodes who served as sources. In some cases, security forces identified the location of a gathering and the personnel involved through facial recognition software.[462] During an OWS march that turned violent, the New York Police Department used social media posting to document a protester hurling a bottle and accosting an officer.[463]

Obtaining legitimacy is integral to the success of any revolutionary movement, regardless of the means by which the organization presents itself to the general population, its opposition, and external actors;

a movement must be taken seriously as a legitimate actor within the political realm.[464] Those methods include conditions of normative and mystical factors and consensual validation. One method by which leaders of nonviolent resistance movements secure widespread compliance is by cloaking their movement and techniques in the beliefs, values, and norms of society—those which people accept without question.[465] Both OWS and Project Chanology saw the use of memetic appreciation extend from an online community to physical activity. The technique of consensual validation—in which the simultaneous occurrence of events creates a sense of their validity—is often used to coalesce public opinion.[466]

A common approach used to induce tacit withdrawal of popular support of the government can be described as persuasion through suffering.[467] One of the persistent misconceptions of nonviolent resistance is that persuasion through suffering aims only to persuade the opponent and the supporting populace by forcing a guilty change of heart and/or inducing a sense of remorse that will ultimately inhibit the aggressor.[468] This presumes that only two actors are involved in the process of nonviolent resistance—the suffering resister and the opponent—when, nonviolent resistance operates within a framework involving three actors: the suffering nonviolent resister, the opponent (the government and/or security forces), and the larger audience (the population). Social media is often used as a sociotechnical means of increasing awareness/advocacy for a particular cause;[469] in nonviolent resistance, the cause is often to publicize the case of an individual or alienated groups' suffering.[470] The interconnectedness of social media users ensures communication of victims can occur despite deliberate censorship attempts.[471] This communication does not necessarily translate into physical action on behalf of a particular cause;[472] however, "slacktivism" may also increase the perception of public support[473] and, in turn, consensual validation.

Social media also provides access to younger populations, the mobilization of whom can present both challenges and opportunities for established political organizations. The mobilization of civic youth organizations under the well-trained Otpor! youth movement leaders from Serbia and Georgia provided a level of maturity and nonviolence that was critical for the effective presentation of a united front against the sitting administration.[474]

> *7.10 The convergence of accessible technology, a social trend toward increased online sharing, and the ability to organize virtually and share experiences in real time via social media fundamentally changed nonviolent resistance in the twenty-first century by enabling an unrelated group to rapidly spread information, mobilize, and behave in a manner supportive of a particular sociopolitical objective.*

Hacktivism

Hacktivism, the exploitation of computer systems (hacking) for a political purpose, brings methods of civil disobedience to cyberspace.[45] Hacktivist tactics include a litany of constantly evolving techniques, often at the leading edge of information security. Included among them are virtual sit-ins, automated email bombs, web hacks and computer break-ins, and computer viruses and worms. A virtual sit-in is the cyberspace equivalent of a blockade where the objective is to disrupt normal operations, thus calling attention to the perpetrator. In 1998, the EDT organized a series of web sit-ins against a series of Mexican and US government websites, as well as the Frankfurt Stock Exchange, to demonstrate solidarity with the Mexican Zapatistas.[475] This variation of a DOS attack encourages supporters to visit the specified sites to overwhelm the servers and limit accessibility. An email bomb is another form of virtual blockade in which a particular email address is (or group of addresses are) inundated with messages, preventing the effective use of a particular account or server. In 1998, a Tamil group sympathetic to the Liberation Tigers of Tamil Eelam (LTTE) swamped Sri Lankan embassies with thousands of email messages from the Internet Black Tigers. Website defacement is a form of hacking that does not necessarily seek to exfiltrate information or corrupt a network but rather to replace existing public content with a political message. Also in 1998, a group of Portuguese hackers modified the sites of forty Indonesian servers to add a "Free East Timor" banner.[48] Hacktivists use computer viruses, worms, and other malicious code to disseminate propaganda and damage target computer systems.[476] The WANK worm and other hacktivist tactics are migrations (with some mutation) of techniques used in the physical domain to the cyber domain.

Hacking

Hacking can be a means of political resistance—a deliberate attempt to either expose perceived government injustice, advance a particular sociopolitical agenda, and/or provide a voice to disenfranchised individuals.[477] Hacking can be used tactically as a means of gaining information or strategically such that the threat of being hacked raises the operating costs for organizations or governments that a resistance movement seeks to affect. The main targets of computer attacks, especially for web defacing, are government systems or websites, particularly military sites and those belonging to large corporations (mainly financial); those that perform critical functions for security or for the economy; telecommunications companies, Internet providers, and hardware producers; and schools and universities. Educational institutions, however, are typically used only as launch pads or staging points for attacks against other targets due to their relatively permissive access polices.[478]

A common tactic in nonviolent resistance is the use of humor as a means of mobilization, facilitating a culture of resistance, and an inversion of oppression.[479] Humor attracts members through its engendered energy, creativity, and enjoyment, thus increasing in-group cohesion by boosting morale by comically exacerbating in-group/out-group differences.[480] Humor can also be used to provoke an enemy and demonstrate contempt for[481] or logical fallacies within[482] extremist messaging.

Humor as a component of resistance in cyberspace often manifests as trolling, the act of making intentionally offensive or provocative online posts with the aim of upsetting someone or instigating an angry response. Trolling is typically executed by trolls, often anonymous individuals who use targeted ridicule to delegitimize a target, be it an organization, individual, or political movement. In some cases, trolling can take the form of comments posted in chat rooms, while in other cases, more elaborate trolling entails hacking into Twitter accounts, manipulating content, and sending messages wholly different from the user's prior behavior. For example, hackers took control of self-identified Islamic State soldiers' accounts and replaced the jihadist messaging and imagery with pornography,[483] while others attempt to technologically disrupt the extremist infrastructure.[484] Some sought to openly engage with supporters and challenge the legitimacy of their belief system.[485] Not all trolls act as part of a larger sociopolitical movement; many simply offend for the sake of offending or for fun.

Anonymous' employed trolling regularly with a variety of targets, ranging from corporations to terrorist organizations. An example of the latter occurred on December 11, 2015, or "trolling day," in a concerted effort to undermine the propaganda of ISIL through mockery, direct antagonism, an/or humiliation.[486] Project Chanology, however, raised trolling to a contemporary form of social protest.[487] Project Chanology was an operation conceptualized by Anonymous to inform the public of Scientology's questionable practices. The operation is an exemplar of social movements in the digital age, demonstrating how the social role of the internet troll, and the culture that gave rise to it, resulted in novel forms of social protest.[488] Anonymous sought to inhibit the church's ability to function normally both online and offline, ultimately changing the societal perception of the religion.[489] The operation was largely structured around the idea of "offline trolling," or extending the concept of provocative pranks to the physical domain.[490] Many early supports of Project Chanology often associated with online communities hosted at 4chan, which gave rise to a sub-community of participants who extolled trolling, intimidating, and even bullying others.[491] The skills developed in 4chan, the ability to rapidly respond to comments with witty counters, rapidly uncovering the identity of a poster (or "doxxing"), and/or gaining consensus to defeat an idea or demean the ideologue, proved useful in defending Anons who participated.[492] Often, these caustic arguments served as a distraction for hackers to reorganize in a protected chat room to plan and execute more sophisticated operations against the Church of Scientology.[493] These arguments also reinforced solidarity by elevating the culture of humor and memetic knowledge, the proverbial "inside joke," to a distinction between the in-group from the out-group.[494] This group distinction resulted in a tactical distinction, making the effort novel and thus drawing both recruits and media attention.[495]

The contemporary use of memes tends to not only rapidly proliferate but also readily adapt to changes in circumstances; images may be manipulated or a concept deliberately misapplied to further extend the relevance of an inside joke and/or insult those not in on said joke. As a result, Anonymous adapted to the unfolding circumstances in the physical domain and updated the memes to reflect said changes, mutating both the culture and the participants. This bidirectional flow of culturally unique information afforded Anonymous a degree of psychological nimbleness supported by crowdsourced participants.[496] As such, Project

Chanology exemplifies the contemporary social movement enabled by and conducted with and through the cyber underground.

Cyber Terrorism

Terrorism, as conceptualized in *Human Factors Considerations of Undergrounds in Insurgencies*,[497] is a particularly violent form of psychological warfare. Chapter 11 of that book discusses not only the individual and social psychological effects of terrorism but also the planning and justification processes behind the decision to use this tactic. NATO considers cyber terrorism a cyber attack using or exploiting computer or communication networks to cause sufficient destruction or disruption to generate fear or to intimidate a society into an ideological goal.[498] This is particularly evident when considering the psychological effects of a physical attack on a trusted information infrastructure from transportation networks, to banks, to hospitals, to voting machines. More often than not, however, organizations use cyberspace to spread terror without necessarily damaging infrastructure.

Terrorism is a deliberate, and often highly effective, manifestation of psychological warfare that exists "along the edge of a nightmare."[499] The Islamic State appears to consider terror among its principal psychological objectives.[500] The resultant anxiety and dysphoria associated with acts of terror create not only an increased fear but also awareness of death. This leads individuals to affiliate with those of similar worldviews and to be more willing to sacrifice their civil liberties to charismatic (and authoritarian) leaders.[501]

The Islamic State, an insurgent organization that drew its lineage from al Qaeda in Iraq, successfully overran Iraqi and Syrian forces to govern large swaths of territory in Iraq and Syria. In June 2014, the group's leader, Abu Bakr al-Baghdadi, declared that territory to be the modern incarnation of the Islamic caliphate and thus himself the caliph.[502] ISIL was particularly brutal in its use of violence against both military and civilian targets. The latter was particularly terroristic in its intimacy, often employing a combination of creativity and cruelty to intimidate those who might resist.[503] A group of hackers claiming affiliation with ISIL, identified as the "CyberCaliphate" not only helped publicize ISIL video depictions of its brutality but also conducted a series of hacks in support of the organization.[504] Most noteworthy was hacking into the official Twitter and YouTube accounts of US Central Command. The hackers defaced the sites and included pictures that hinted ISIL infiltrated US military formations.[505] While the attacks were notable for their brazenness, the damage caused was not physical, and it is unclear whether there was a significant psychological impact on either US forces or civilians. ISIL advanced the role of cyber in blended attacks or the combination of virtual and physical actions to accomplish a specific objective. [506]

Cyber terrorism, particularly incitement to others to act on behalf of an organization or movement via the Internet, is a low-cost means of expanding the reach of a terrorist group. Al Qaeda, AQAP, and ISIL have all incited others to commit acts on their behalf without requiring individuals to travel to the Middle East or Central Asia, formally affiliate with a group, and/or receive any training or material support to conduct the attacks.[507] Anwar al-Awlaki and AQAP's English language magazine, *Inspire*, also advocated "Open Source Jihad," where readers could take action close to home using readily accessible tools such as knives, vehicles, and hunting rifles to spread terror.[508]

Terrorist organizations seek to manipulate two principal audiences: the organization's constituency (in-group) and the enemy (out-group).[509] The principal objective of the former is to demonstrate strength, while the goal of the latter is to intimidate and/or paralyze the citizenry and provoke the enemy. [510] In-group messages stress the necessity of violent resistance to accomplish the desired end state, that negation is acquiescent to tyrannical authority, and that the adversary is vulnerable. ISIL not only uses language describing in gruesome detail the brutality of its actions against soldiers on the battlefield as well as prisoners[511] but also uses professionally shot and edited footage in its video releases.[512] ISIL's use of violence-related themes describing prisoner executions and the subsequent humiliation of the groups those victims represent is unapologetic and direct.[513] ISIL's use of ritualistic decapitations is staged to maximize their terrifying effect through shock.[514] While decapitation may indeed be homage to medieval Islam, the intimacy of the act and the human revulsion to perpetrating it may contribute to the voyeuristic nature of the videos.[515] Though ISIL's beheadings are a terroristic act of psychological warfare, they still hold religious significance. ISIL's rationalization of prisoner beheadings is a selective interpretation of *Surah* 47:4, deliberately taken out of context.[516] ISIL not only rationalizes but also seems to take pride in its use of terrorism by closely binding violent imagery with Quranic references, providing a degree of legitimacy and constituted authority for its selected target audiences. [517]

While the emergence of a cyber terrorist organization that exists solely online has not yet come to fruition,[518] individuals have been charged with cyber terrorism. In October 2015, the US Department of Justice charged Ardit Ferizi, a citizen of Kosovo living in Malaysia, with stealing the data belonging to the US service members and passing it to the members of the Islamic State with the intent to use the information in terrorist operations against the individuals themselves.[519] The trove contained email address, passwords, and other contact information from 1,351 US service members.[520] This type of doxxing at scale presents a significant threat to members of security forces and their families. While the CyberCaliphate's and Ferizi's actions demonstrated skill and raised concerns amongst security forces and cybersecurity scholars, the physical (and certainly, existential) threat of cyber terrorism has not yet materialized.[521] Often, the threat is exaggerated by cybersecurity proponents seeking to harden infrastructure and who argue that waiting for the demonstration of such capability incurs too much risk.[522] Nevertheless, a cyber warfare component is typically present in most terrorist organizations, and use of such tactics does not show signs of abating. That said, contemporary counterterrorism forces prioritize the disruption or destruction of adversary organizations' ability to plan and execute operations online.[523]

KEY TAKEAWAYS

7.1	Unconventional warfare in cyberspace requires a rich contextual understanding of the sociotechnical aspects of the cyber ecology.
7.2	The Internet facilitates locating and contacting communities of similar ideological interest. As such, "local" cyber resistance may not require a physical footprint.

7.3   Leaderless or limited hierarchy-type resistance movements can maintain operational security without the requirement for sophisticated physical security tradecraft.

7.4   The personalities of key leaders can also have a strong influence on the operations of resistance movements, particularly during their early stages. Charismatic leadership in cyberspace may take on different forms but is typically accompanied by the technological skills often associated with elite hackers.

7.5   There is no single hacker profile, although there are consistently observed behaviors—for example, having an intense focus on a particular technical challenge at the expense of the more mundane.

7.6   Many aspiring hackers seek both identity and affiliation, and the lure of the cyber underground can provide not only a sense of belonging but also a sense of purpose and/ or an idealized technocracy where skill alone determines whether one is accepted.

7.7   Recruiting remains arguably the most essential activity of resistance warfare, and the use of the Internet includes online forums, social media, and even video games to advertise, solicit, and recruit potential members/ supporters.

7.8   Counterintelligence is among the most important functions of the underground. Social engineering and doxing are two signature intelligence collection tactics used by cyber undergrounds.

7.9   Cyberspace is proving to be the decisive battleground for political and psychological warfare.

7.10 The convergence of accessible technology, a social trend toward increased online sharing, and the ability to organize virtually and share experiences in real time via social media fundamentally changed nonviolent resistance in the twenty-first century by enabling an unrelated group to rapidly spread information, mobilize, and behave in a manner supportive of a particular sociopolitical objective.

ENDNOTES

1   C. R. Eidman and G.S. Green, "Unconventional Cyber Warfare: Cyber Opportunities in Unconventional Warfare" (unpublished master's thesis, Naval Postgraduate School, 2014).

2   E. D. Knapp, "Unconventional Warfare in Cyberspace" (unpublished master's thesis, U.S. Army War College, 2012).

3   David S. Maxwell, "The Cyber Underground-Resistance to Active Measures and Propaganda" 'The Disruptors Motto' – 'Think for Yourself'," *Small Wars Journal*, (2017): http://smallwarsjournal.com/jrnl/art/the-cyber-underground---resistance-to-active-measures-and-propaganda-"the-disruptors"-mot-0. Accessed 11/25/2018.

4   Ellen Barry, "Sound of Post-Soviet Protest: Claps and Beeps," *New York Times Online*, July 14, 2011: http://www.nytimes.com/2011/07/15/world/europe/15belarus.html?pagewanted=all.

5   J. P. Carlin, *Dawn of the Code War: America's Battle Against Russia, China, and the Rising Global Cyber Threat* (New York: Public Affairs, 2018).

6   US Army Special Operations Command, "SOF Support to Political Warfare," USASOC white paper, March 10, 2015, http://maxoki161.blogspot.com/2015/03/sof-support-to-political-warfare-white.html.

7   "Assessing Revolutionary and Insurgent Strategies (ARIS) Studies," US Army Special Operations Command, https://www.soc.mil/ARIS/ARIS.html.

8   Robert Leonhard et al., *Undergrounds in Insurgent, Revolutionary, and Resistance Warfare*, 2nd ed. (Alexandria, VA: US Army Publications Directorate, 2013), http://www.soc.mil/ARIS/ARIS.html.

9   Nathan D. Bos et al., *Human Factors Considerations of Undergrounds in Insurgencies*, 2nd ed. (Alexandria, VA: US Army Publications Directorate, 2013), http://www.soc.mil/ARIS/ARIS.html.

10   Defined by the Director of Central Intelligence Directive (DCID) in Ibid., 3.

11   Ibid.

12   Raoul Chiesa, Stefania Ducci, and Silvio Ciappi. *Profiling Hackers: The Science of Criminal Profiling as Applied to the World of Hacking* (Boca Raton, FL: Taylor and Francis, 2012),

13   J. Lane, "Digital Zapatistas" *TDR/The Drama Review* 47, no. 2 (2003): 129-144.

14   Ibid.

15   Ibid.

16   Ibid.

17   Leonhard et al., *Undergrounds in Insurgent, Revolutionary, and Resistance Warfare*.

18   Bos et al., *Human Factors Considerations of Undergrounds in Insurgencies*.

19   Ibid.

20   Ibid.

21   Ibid.

22   Critical Art Ensemble, *Electronic Civil Disobedience* (Brooklyn, NY: Autonomedia, 1996), http://critical-art.net/electronic-civil-disobedience-1996/.

23   Steven Levy, *Hackers: Heroes of the Computer Revolution* vol. 4 (New York: Penguin Books, 2001).

24 Gabriella Coleman and A. Golub, "The Anthropology of Hackers," *Atlantic*, September 21, 2010. https://www.theatlantic.com/technology/archive/2010/09/the-anthropology-of-hackers/63308/.

25 Julian Assange, "The Curious Origins of Political Hacktivism," *Counterpunch*, November 25, 2006. https://www.counterpunch.org/2006/11/25/the-curious-origins-of-political-hacktivism/.

26 Levy, *Hackers*.

27 Ibid.

28 Ibid.

29 Ibid.

30 Michele Slatella and Joshua Quittner, *Masters of Deception: The Gang That Ruled Cyberspace* (New York: Harper Collins, 1995).

31 Chiesa, Ducci, and Ciappi, *Profiling Hackers*.

32 Marcella Bombardieri, "The Inside Story of MIT and Aaron Swartz," *Boston Globe*, March 29, 2014, https://www3.bostonglobe.com/metro/2014/03/29/the-inside-story-mit-and-aaron-swartz/YvJZ5P6VHaPJusReuaN7SI/story.html?arc404=true.

33 Ibid.

34 Levy, *Hackers*.

35 Edward Snowden was a National Security Agency (NSA) contractor who illegally downloaded and exfiltrated highly classified information and provided it to members of the press. As of 2017, he has been living in Russia, which denied US requests to extradite him to prosecute him for espionage.

36 Levy, *Hackers*.

37 Slatella and Quittner, *Masters of Deception*.

38 Levy, *Hackers*.

39 Gabriella Coleman, *Hacker, Hoaxer, Whistleblower, Spy: The Many Faces of Anonymous* (Brooklyn, NY: Verso Books, 2014).

40 Levy, *Hackers*.

41 Ibid.

42 Chiesa, Ducci, and Ciappi, *Profiling Hackers*.

43 Ibid.

44 Coleman, *Hacker, Hoaxer, Whistleblower, Spy*.

45 Adrienne Massanari, "# Gamergate and The Fappening: How Reddit's Algorithm, Governance, and Culture Support Toxic Technocultures," *New Media & Society* (2015): 1461444815608807.

46 Chuck Crossett and Jason Spitaletta, *Radicalization: Relevant Psychological and Sociological Concepts* (Fort Meade, MD: Asymmetric Warfare Group, 2010).

47 Massanari, "# Gamergate and The Fappening."

48 Patrick Underwood and Howard T. Welser, "'The Internet is Here': Emergent Coordination and Innovation of Protest Forms in Digital Culture," *Proceedings of the 2011 iConference* (ACM, 2011): 304-311.

49 Bos et al., *Human Factors Considerations of Undergrounds in Insurgencies*.

50 Luke Goode, "Anonymous and the Political Ethos of Hacktivism," *Popular Communication* 13, no. 1 (2015): 74-86.

51 Murtaza Hussain, "The New Information War," *Intercept,* November 25, 2017, https://theintercept.com/2017/11/25/information-warfare-social-media-book-review-gaza/.

52 Coleman, *Hacker, Hoaxer, Whistleblower, Spy.*

53 Bos et al., *Human Factors Considerations of Undergrounds in Insurgencies.*

54 Ibid.

55 Ibid.

56 Ibid.

57 Ibid.

58 Ibid.

59 Paolo Gerbaudo, *Tweets and the Streets: Social Media and Contemporary Activism* (London: Pluto Press, 2018).

60 Jina Moore, "Social Media: Did Twitter and Facebook Really Build a Global Revolution?" *Christian Science Monitor,* June 30, 2011, https://www.csmonitor.com/World/Global-Issues/2011/0630/Social-media-Did-Twitter-and-Facebook-really-build-a-global-revolution.

61 Ibid.

62 Sandor Vegh, "Classifying Forms of Online Activism: The Case of Cyberprotests Against the World Bank" in *Cyberactivism: Online Activism in Theory and Practice,* eds. Martha McCaughey and Michael D. Ayers (New York: Routledge Press, 2013), 81-106.

63 Ibid.

64 Dana Rotman, Sarah Vieweg, Sarita Yardi, Ed Chi, Jenny Preece, Ben Shneiderman, Peter Pirolli, and Tom Glaisyer, "From Slacktivism to Activism: Participatory Culture in the Age of Social Media" in *CHI'11 Extended Abstracts on Human Factors in Computing Systems* (Vancouver, Canada: ACM, 2011), 819-822.

65 Vegh, "Classifying Forms of Online Activism."

66 Ibid.

67 Ibid.

68 Kevin M. DeLuca, Sean Lawson, and Ye Sun, "Occupy Wall Street on the Public Screens of Social Media: The Many Framings of the Birth of a Protest Movement," *Communication, Culture & Critique* 5, no. 4 (2012): 483-509.

69 Benjamin Gleason, "# Occupy Wall Street: Exploring Informal Learning About a Social Movement on Twitter," *American Behavioral Scientist* 57, no. 7 (2013): 966-982.

70 Nathan Schneider, "From Occupy Wall Street to Occupy Everywhere," *Nation,* October 12, 2011, https://www.thenation.com/article/occupy-wall-street-occupy-everywhere/.

71 Ibid.

72 Ibid.

73 Manissa McCleave Maharawal, "Occupy Wall Street and a Radical Politics of Inclusion," *Sociological Quarterly* 54, no. 2 (2013): 177-181.

74 Ibid.

75 Craig Calhoun, "Occupy Wall Street in Perspective," *British Journal of Sociology* 64, no. 1 (2013): 26-38.

76 Michael D. Conover, Emilio Ferrara, Filippo Menczer, and Alessandro Flammini, "The Digital Evolution of Occupy Wall Street" *PloS one* 8, no. 5 (2013): e64679.

77 D. E. Denning, "Activism, Hacktivism, And Cyberterrorism: The Internet as a Tool for Influencing Foreign Policy" in eds. J. Arquilla and D. Ronfeldt, *Networks and Netwars: The Future of Terror, Crime, and Militancy* (Santa Monica, CA: RAND Corporation, 2001).

78 Colin Lecher, "Massive Attack: How a Weapon of War Became a Weapon Against the Web," *Verge*, April 14, 2017, https://www.theverge.com/2017/4/14/15293538/electronic-disturbance-theater-zapatista-tactical-floodnet-sit-in.

79 Critical Art Ensemble, *Electronic Civil Disobedience.*

80 Assange, "The Curious Origins of Political Hacktivism."

81 Underwood and Welser, "'The Internet is Here."

82 Ibid.

83 Ibid.

84 Coleman, *Hacker, Hoaxer, Whistleblower, Spy.*

85 Underwood and Welser, "'The Internet is Here."

86 Ibid.

87 Coleman and Golub, "The Anthropology of Hackers."

88 Slatella and Quittner, *Masters of Deception.*

89 Ibid.

90 Ibid.

91 Coleman, *Hacker, Hoaxer, Whistleblower, Spy.*

92 Ibid.

93 Scott Atran, Hammad Sheikh, and Angel Gomez, "Devoted Actors Sacrifice for Close Comrades and Sacred Cause," *Proceedings of the National Academy of Sciences* 111, no. 50 (2014): 17702-17703.

94 Coleman, *Hacker, Hoaxer, Whistleblower, Spy.*

95 Denning, "Activism, Hacktivism, and Cyberterrorism."

96 J. C. Castillo, "The Mexican Cartels' Employment of Inform and Influence Activities (IIA) as Tools of Asymmetrical Warfare" in *Information Security for South Africa (ISSA),* (IEEE, 2014), 1-8.

97 Eidman, "Unconventional Cyber Warfare."

98 Bos et al., *Human Factors Considerations of Undergrounds in Insurgencies.*

99 Douglas C. Derrick, Karyn Sporer, Sam Church, and Gina Scott Ligon, "Ideological Rationality and Violence: An Exploratory Study of ISIL's Cyber Profile," *Dynamics of Asymmetric Conflict* 9, no. 1-3 (2016): 57-81.

100 Yonah Alexander and Michael S. Swetnam, *Cyberterrorism and Information Warfare: Threats and Responses* (London: Transnational Publishers, Incorporated, 2001).

101 Bos et al., *Human Factors Considerations of Undergrounds in Insurgencies.*

102 Ibid.

103 Zlatogor Minchev and Mitko Bogdanoski, "Hybrid Challenges to Human Factor in Cyberspace," *Countering Terrorist Activities in Cyberspace* 139 (2018): 32.

104 Abdel Bari Atwan, *Islamic State: The Digital Caliphate* (Oakland, CA: University of California Press, 2015).

105 Ibid.

106 Bos et al., *Human Factors Considerations of Undergrounds in Insurgencies.*

107 Crossett and Spitaletta, *Radicalization*.

108 Ibid.

109 Chiesa, Ducci, and Ciappi, *Profiling Hackers*.

110 Slatella and Quittner, *Masters of Deception*.

111 Bobby Sands was the Officer Commanding of the Irish Republican Army (IRA) members in Long Kesh prison in Northern Ireland. During 1981, Sands led hunger strikes, which ultimately led to his and nine other IRA volunteers' deaths. Bobby Sands' hunger strike lasted long enough for him to be elected to the British parliament during the strike. The results of the hunger strikes were a partial success for each side; the IRA received some concessions on prison conditions and, more importantly, galvanized support around the martyred Bobby Sands. Margaret Thatcher's government demonstrated resolve by allowing ten strikers to die first and in the end made only partial concessions.

112 Bos et al., *Human Factors Considerations of Undergrounds in Insurgencies*.

113 Kevin Mitnick, *Ghost in the Wires: My Adventures as the World's Most Wanted Hacker* (Paris: Hachette UK, 2011).

114 Ibid.

115 Diana S. Dolliver and Kevin Poorman, "Understanding Cybercrime," in P. Reichel and R. Randa, *Transnational Crime and Global Security* (2018): 139-160.

116 Ibid.

117 Jason A. Spitaletta, "Terror as a Psychological Warfare Objective: ISIL's Use of Ritualistic Decapitation," in eds. J. Giordano and D. DiEuliis, *White Paper on Social and Cognitive Neuroscience Underpinnings of ISIL Behavior and Implications for Strategic Communication, Messaging, and Influence* (Washington, DC: Strategic Multilayer Assessment Office, Office of the Secretary of Defense, 2015), http://nsiteam.com/sma-publications/.

118 Derrick, Sporer, Church, and Ligon, "Ideological Rationality and Violence," 57-81.

119 Carlin, *Dawn of the Code War*.

120 Jeremy Scahill, *Dirty Wars: The World is a Battlefield* (New York: Nation Books, 2013).

121 Bos et al., *Human Factors Considerations of Undergrounds in Insurgencies*.

122 Chiesa, Ducci, and Ciappi, *Profiling Hackers*.

123 Ibid.

124 Ibid.

125 Ibid.

126 Garret M. Graff, "Inside the Hunt for Russia's Most Notorious Hacker," *Wired*, March 21, 2017, https://www.wired.com/2017/03/russian-hacker-spy-botnet/.

127 Ibid.

128 Chiesa, Ducci, and Ciappi, *Profiling Hackers*.

129 John McMullan and Aunshul Rege, "Cyberextortion at Online Gambling Sites: Criminal Organization and Legal Challenges," *Gaming Law Review* 11, no. 6 (2007): 648-665.

130 Masarah Paquet-Clouston, David Décary-Hétu, and Olivier Bilodeau, "Cybercrime is Whose Responsibility? A Case Study of an Online Behaviour System in Crime," *Global Crime* 19, no. 1 (2018): 1-21.

131 I. Glenn Cohen, Sharona Hoffman, and Eli Y. Adashi, "Your Money or Your Patient's Life? Ransomware and Electronic Health Records," *Annals of Internal Medicine* 167, no. 8 (2017): 587-588.

132 Lily Hay Newman, "The Ransomware Meltdown Experts Warned Us About," *Wired*, May 12, 2017, https://www.wired.com/2017/05/ransomware-meltdown-experts-warned/.

133 Andy Greenberg, "How Hackers Hijacked a Bank's Entire Online Operation," *Wired*, April 4, 2017, https://www.wired.com/2017/04/hackers-hijacked-banks-entire-online-operation/.

134 Kent L. Norman, *Cyberpsychology: An Introduction to Human-Computer Interaction* (Cambridge, UK: Cambridge University Press, 2017).

135 Chiesa, Ducci, and Ciappi, *Profiling Hackers*.

136 Steven Mizrach, "Is There a Hacker Ethic for 90s Hackers," in eds. R. Molander and S. Siang, *The Legitimization of Strategic Information Warfare: Ethical Consideration* (Professional Ethics Report, XI (4), 1997).

137 Eric S. Raymond, *The Jargon File* version 4.4.8, accessed April 22, 2017, http://www.catb.org/~esr/jargon/.

138 Chiesa, Ducci, and Ciappi, *Profiling Hackers*.

139 Ibid.

140 Ibid.

141 Coleman, *Hacker, Hoaxer, Whistleblower, Spy.*

142 Bos et al., *Human Factors Considerations of Undergrounds in Insurgencies.*

143 Raymond, *The Jargon File.*

144 Chiesa, Ducci, and Ciappi, *Profiling Hackers.*

145 Raymond, *The Jargon File.*

146 Ibid.

147 Ibid.

148 Ibid.

149 Chiesa, Ducci, and Ciappi, *Profiling Hackers.*

150 Ibid.

151 Ibid.

152 Ibid.

153 Ibid.

154 Ibid.

155 Raymond, *The Jargon File.*

156 Chiesa, Ducci, and Ciappi, *Profiling Hackers.*

157 Ibid.

158 Ibid.

159 Ibid.

160 Ibid.

161 Coleman and Golub, "The Anthropology of Hackers."

162 Raymond, *The Jargon File.*

163 Ibid.

164 Stereotype threat is a situational predicament where individuals are (or perceive themselves to be) at risk of conforming to preexisting beliefs about their social group. This social psychological phenomenon has been shown to reduce academic performance in groups who are negatively stereotyped. In some cases, for example, hackers, individuals

may be aware of preexisting beliefs about a desired social group and alter their behavior to conform to the social norms.

[165] Claude M. Steele and Joshua Aronson, "Stereotype Threat and the Intellectual Test Performance of African Americans," *Journal of Personality and Social Psychology* 69, no. 5 (1995): 797.

[166] Raymond, *The Jargon File.*

[167] Ibid.

[168] Denning, "Activism, Hacktivism, and Cyberterrorism."

[169] Raymond, *The Jargon File.*

[170] Chiesa, Ducci, and Ciappi, *Profiling Hackers.*

[171] Ibid.

[172] Coleman, *Hacker, Hoaxer, Whistleblower, Spy.*

[173] Chiesa, Ducci, and Ciappi, *Profiling Hackers.*

[174] Ibid.

[175] Ibid.

[176] Ibid.

[177] Ibid.

[178] Ibid.

[179] Ibid.

[180] Raymond, *The Jargon File.*

[181] Ibid.

[182] Ibid.

[183] Chiesa, Ducci, and Ciappi, *Profiling Hackers.*

[184] Raymond, *The Jargon File.*

[185] Chiesa, Ducci, and Ciappi, *Profiling Hackers.*

[186] Ibid.

[187] "Lulz" is a derivation of LOL, or "laugh out loud," a colloquialism used originally in text messaging but that spread to common parlance. Many hackers tend to appropriate and edit colloquial language for their own use, thus creating a system of jargon not unlike other specialized communities. Gabriella Coleman, "Anonymous: From the Lulz to Collective Action," *The New Everyday: A Media Commons Project* 6 (2011).

[188] Coleman and Golub, "The Anthropology of Hackers."

[189] Underwood and Welser, "'The Internet is Here'," 304-311.

[190] Coleman, *Hacker, Hoaxer, Whistleblower, Spy.*

[191] Coleman and Golub, "The Anthropology of Hackers."

[192] Coleman, "Anonymous."

[193] Chiesa, Ducci, and Ciappi, *Profiling Hackers,* 217-235.

[194] M. K. Rogers, "The Psyche of Cybercriminals: A Psycho-Social Perspective," in *Cybercrimes: A Multidisciplinary Analysis* (Berlin: Springer, 2011), 217-235.

[195] The interpersonal circle or interpersonal circumplex is a model for conceptualizing, organizing, and assessing interpersonal behavior, traits, and motives. James A Russell, "A Circumplex Model of Affect," *Journal of Personality and Social Psychology* 39, no. 6 (1980): 1161.

[196] Bos et al., *Human Factors Considerations of Undergrounds in Insurgencies.*

[197] Chiesa, Ducci, and Ciappi, *Profiling Hackers.*

[198] Coleman, *Hacker, Hoaxer, Whistleblower, Spy.*

[199] Chiesa, Ducci, and Ciappi, *Profiling Hackers.*

[200] Ibid.

[201] Ibid.

[202] Coleman, *Hacker, Hoaxer, Whistleblower, Spy.*

[203] Mizrach, "Is There a Hacker Ethic for 90s Hackers."

[204] Chiesa, Ducci, and Ciappi, *Profiling Hackers.*

[205] Ibid.

[206] Coleman, *Hacker, Hoaxer, Whistleblower, Spy.*

[207] Chiesa, Ducci, and Ciappi, *Profiling Hackers.*

[208] Ibid.

[209] Mizrach, "Is There a Hacker Ethic for 90s Hackers."

[210] Chiesa, Ducci, and Ciappi, *Profiling Hackers.*

[211] Mizrach, "Is There a Hacker Ethic for 90s Hackers."

[212] Ibid.

[213] Ibid.

[214] Ibid.

[215] Ibid.

[216] Ibid.

[217] Ibid.

[218] Chiesa, Ducci, and Ciappi, *Profiling Hackers.*

[219] Mizrach, "Is There a Hacker Ethic for 90s Hackers."

[220] Autism Spectrum Disorders (ASDs) are a range of neurological disorders characterized by difficulties in interacting with and empathizing for others along with limited language acquisition and a restricted and repetitive repertoire of behaviors and interests; individuals along the ASD spectrum may also experience marked disturbances in sensory processing and other neurological and psychological disorders. The "hacker as autistic" myth is essentially that poor (or limited) social skills are the proximal reason individuals turn to the cyber underground to sate social needs. This, like many myths, trivializes ASD and tends to oversimplify why some individuals prefer online to physical social interaction.

[221] American Psychiatric Association, *Diagnostic and Statistical Manual of Mental Disorders: DSM-5* (Arlington, VA: American Psychiatric Association, 2013); David Kushner, "The Autistic Hacker," *IEEE Spectrum*, 2011, http://spectrum.IEEE.org/telecom/internet/the-autistic-hacker.

[222] K. C. Seigfried-Spellar, C. L. O'Quinn, and K.N. Treadway, "Assessing the Relationship Between Autistic Traits and Cyber Deviancy in a Sample of College Students," *Behaviour & Information Technology* 34, no. 5 (2015): 533-542.

[223] R. Ledingham and R. Mills, "A Preliminary Study of Autism and Cybercrime in the Context of International Law Enforcement," *Advances in Autism* 1, no. 1 (2015): 2-11.

[224] Ibid.

[225] Ibid.

[226] Kushner, "The Autistic Hacker."

[227] Chiesa, Ducci, and Ciappi, *Profiling Hackers.*

[228] Ibid.

[229] 229   Raymond, Eric S. The Jargon File, version 4.4.8. 2004 http://www.catb.org/~esr/jargon/. Accessed 4/22/2017.

[230] 230   Levy, Steven. Hackers: Heroes of the computer revolution. Vol. 4. New York: Penguin Books, 2001.

[231] Chiesa, Ducci, and Ciappi, *Profiling Hackers.*

[232] Ibid.

[233] Ibid.

[234] Ibid.

[235] Ibid.

[236] Bos et al., *Human Factors Considerations of Undergrounds in Insurgencies.*

[237] Chiesa, Ducci, and Ciappi, *Profiling Hackers.*

[238] Ibid.

[239] Ibid.

[240] Slatella and Quittner, *Masters of Deception.*

[241] Chiesa, Ducci, and Ciappi, *Profiling Hackers.*

[242] Ibid.

[243] Ibid.

[244] Coleman and Golub, "The Anthropology of Hackers."

[245] Coleman, *Hacker, Hoaxer, Whistleblower, Spy.*

[246] Chiesa, Ducci, and Ciappi, *Profiling Hackers.*

[247] Coleman and Golub, "The Anthropology of Hackers."

[248] Chiesa, Ducci, and Ciappi, *Profiling Hackers.*

[249] Ibid.

[250] Internet slang for an individual who expresses moral opinions in chatrooms that are ostensibly amoral.

[251] Crossett and Spitaletta, *Radicalization.*

[252] Chiesa, Ducci, and Ciappi, *Profiling Hackers.*

[253] Carlin, *Dawn of the Code War.*

[254] Crossett and Spitaletta, *Radicalization.*

[255] Chiesa, Ducci, and Ciappi, *Profiling Hackers.*

[256] Underwood and Welser, "'The Internet is Here'," 304-311.

[257] Ibid.

[258] Ibid.

[259] Ibid.

[260] Ibid.

[261] Chiesa, Ducci, and Ciappi, *Profiling Hackers.*

[262] Crossett and Spitaletta, *Radicalization.*

[263] Ibid.

[264] Ibid.

[265] Ibid.

[266] Chiesa, Ducci, and Ciappi, *Profiling Hackers*.

[267] Ibid.

[268] Underwood and Welser, "'The Internet is Here'," 304-311.

[269] Crossett and Spitaletta, *Radicalization*.

[270] Ibid.

[271] Ibid.

[272] Vinson Cunningham, "The Masks in Venezuela and the Pathos of Protest Art," *New Yorker*, May 13, 2017, https://www.newyorker.com/culture/photo-booth/ the-masks-in-venezuela-and-the-pathos-of-protest-art?mbid=rss.

[273] Parmy Olson, *We are Anonymous: Inside the Hacker World of LulzSec, Anonymous, and the Global Cyber Insurgency* (New York: Little, Brown and Company, 2012).

[274] P. G. Zimbardo, *The Lucifer Effect: Understanding How Good People Turn Evil* (New York: Random House Publishing Group, 2008).

[275] Ibid.

[276] A. K. Goodboy and M. M. Martin, "The Personality Profile of a Cyberbully: Examining the Dark Triad," *Computers in Human Behavior* 49 (2015): 1-4.

[277] A. Bandura, C. Barbaranelli, G.V. Caprara, and C. Pastorelli, "Mechanisms of Moral Disengagement in the Exercise of Moral Agency," *Journal of Personality and Social Psychology* 71, no. 2 (1996): 364-374.

[278] R. Young, L. Zhang, and V.R. Prybutok, "Hacking into the Minds of Hackers," *Information Systems Management* 24, no. 4 (2007): 281-287.

[279] K. C. Runions and M. Bak, "Online Moral Disengagement, Cyberbullying, and Cyber-aggression," *Cyberpsychology, Behavior, and Social Networking* 18, no. 7 (2015): 400-405.

[280] The American Psychological Association (APA) considers cyberbullying the sending of hurtful or threatening emails or instant messages, spreading rumors, or posting embarrassing photos of others.

[281] Runions and Bak, "Online Moral Disengagement, Cyberbullying, and Cyber-aggression," 400-405.

[282] Christopher P. Barlett, Douglas A. Gentile, and Chelsea Chew, "Predicting Cyberbullying from Anonymity," *Psychology of Popular Media Culture* 5, no. 2 (2016): 171-180.

[283] Massanari, "# Gamergate and The Fappening."

[284] Ibid.

[285] Ibid.

[286] Crossett and Spitaletta, *Radicalization*.

[287] Chiesa, Ducci, and Ciappi, *Profiling Hackers*.

[288] Ibid.

[289] Nellie Bowles, "How 'Doxxing' Became a Mainstream Tool in the Culture Wars," *New York Times*, August 30, 2017, https://www.nytimes.com/2017/08/30/technology/doxxing-protests.html.

[290] Aki Ito, "A Former Anonymous Hacker's Search for Redemption," *Bloomberg Technology*, March 6, 2018. https://www.bloomberg.com/news/features/2018-03-06/a-former-anonymous-hacker-s-search-for-redemption.

[291] Ibid.

[292] Coleman, *Hacker, Hoaxer, Whistleblower, Spy.*

[293] Ibid.

[294] Ibid.

[295] Andrew R. Molnar, William A. Lybrand, Lorna Hahn, James L. Kirkman, and Peter B. Riddleberger, *Undergrounds in Insurgent, Revolutionary, and Resistance Warfare* (Washington, DC, 1963).

[296] Ibid.

[297] Ibid.

[298] Ibid.

[299] Leonhard et al., *Undergrounds in Insurgent, Revolutionary, and Resistance Warfare.*

[300] C. Fuchs, "The Self-organization of Cyberprotest," *Advances in Education, Commerce & Governance,* The Internet Society II, (2006): 275-295.

[301] Jason A. Spitaletta, "Use of Cyber to affect neuroS/T based Deterrence and Influence," in eds. D. DiEuliis, W. Casebeer, J. Giordano, N. Wright, and H. Cabayan, *White Paper on Leveraging Neuroscientific and Neurotechnological (NeuroS&T) Developments with Focus on Influence and Deterrence in a Networked World* (Washington, DC: Strategic Multilayer Assessment Office, Office of the Secretary of Defense, 2014), http://nsiteam.com/sma-publications/.

[302] Crossett and Spitaletta, *Radicalization.*

[303] Albert Park and Mike Conway, "Harnessing Reddit to Understand the Written-Communication Challenges Experienced by Individuals with Mental Health Disorders: Analysis of Texts from Mental Health Communities," *Journal of Medical Internet Research* 20, no. 4 (2018): e121.

[304] Martin Shelton, Katherine Lo, and Bonnie Nardi, "Online Media Forums as Separate Social Lives: A Qualitative Study of Disclosure Within and Beyond Reddit," *iConference 2015 Proceedings* (2015).

[305] Crossett and Spitaletta, *Radicalization.*

[306] Park and Conway, "Harnessing Reddit to Understand the Written-Communication Challenges Experienced by Individuals with Mental Health Disorders," e121.

[307] Lulu Garcia-Navarro, "What's An 'Incel'? The Online Community Behind the Toronto Van Attack," National Public Radio, April 29, 2018, accessed April 30, 2018, https://www.npr.org/2018/04/29/606773813/whats-an-incel-the-online-community-behind-the-toronto-van-attack\.

[308] Sigal Samuel, "Canada's 'Incel Attack' and Its Gender-Based Violence Problem" *Atlantic,* April 28, 2018, https://www.theatlantic.com/international/archive/2018/04/toronto-incel-van-attack/558977/.

[309] Park and Conway, "Harnessing Reddit to Understand the Written-Communication Challenges Experienced by Individuals with Mental Health Disorders," e121.

[310] Editorial Board, "The New Radicalization of the Internet," *New York Times,* November 25, 2018, https://www.nytimes.com/2018/11/24/opinion/sunday/facebook-twitter-terrorism-extremism.html. Accessed 11/25/2018.

[311] Mark Alfano, J. Adam Carter, and Marc Cheong, "Technological seduction and self-radicalization," *Journal of the American Philosophical Association* 3 (2018): 298-322.

[312] Ibid.

[313] Jason A. Spitaletta, ed., *Bio-Psycho-Social Applications to Cognitive Engagement* (Washington, DC: Strategic Multilayer Assessment Office, Office of the Secretary of Defense, 2016).

[314] Abraham Maslow, "A Theory of Human Motivation," *Psychological Review* 50 (1948): 370–396.

[315] David H. McElreath, Daniel Adrian Doss, Leisa McElreath, Ashley Lindsley, Glenna Lusk, Joseph Skinner, and Ashley Wellman, "The Communicating and Marketing of Radicalism: A Case Study of ISIL and Cyber Recruitment," *International Journal of Cyber Warfare and Terrorism (IJCWT)* 8, no. 3 (2018): 26-45.

[316] Derrick, Sporer, Church, and Ligon, "Ideological Rationality and Violence," 57-81.

[317] Ariel Victoria Lieberman, "Terrorism, the Internet, and Propaganda: A Deadly Combination," *Journal of National Security Law and Policy* 9 (2017): 95-124.

[318] Ibid.

[319] Bos et al., *Human Factors Considerations of Undergrounds in Insurgencies.*

[320] Thomas Hegghammer, ed., *Jihadi Culture* (Cambridge, UK: Cambridge University Press, 2017.)

[321] Cori Dauber and Mark Robinson, "ISIL and the Hollywood Visual Style," *Jihadology,* July 6, 2015, http://jihadology.net/2015/07/06/guest-post-ISIL-and-the-hollywood-visual-style/.

[322] Bos et al., *Human Factors Considerations of Undergrounds in Insurgencies.*

[323] Ibid.

[324] Anna T. Prescott, James D. Sargent, and Jay G. Hull, "Metaanalysis of the Relationship Between Violent Video Game Play and Physical Aggression Over Time," *Proceedings of the National Academy of Sciences* 115, no. 40 (2018): 9882–9888.

[325] Ibid.

[326] Timothy L. Thomas, *Hezballah, Israel, and Cyber Psyop* (Fort Leavenworth, KS: Foreign Military Studies Office (Army), 2007).

[327] Bos et al., *Human Factors Considerations of Undergrounds in Insurgencies.*

[328] Jessica Mueller, and Ronn Johnson, "11 Emerging Trends in Technology and Forensic Psychological Roots of Radicalization and Lone Wolf Terrorists," *Emerging and Advanced Technologies in Diverse Forensic Sciences* (2018).

[329] Bos et al., *Human Factors Considerations of Undergrounds in Insurgencies.*

[330] Ibid.

[331] Mark P. Whitaker, "Tamilnet.Com: Some Reflections on Popular Anthropology, Nationalism, and the Internet," *Anthropological Quarterly* 77, no. 3 (Summer 2004): 469–498.

[332] Ibid.

[333] Steve Gold, "Get Your Head Around Hacker Psychology," *Engineering & Technology* 9, no. 1 (2014): 76–80.

[334] Cyber bullying is a form of harassment using electronic media. Instances increased with the advent of social media and its adoption by younger individuals. In some cases, individuals were cyber harassed severely enough to take their own lives. In the United States, awareness of and protections against cyber bullying increased since 2010.

[335] Alan K. Goodboy and Matthew M. Martin, "The Personality Profile of a Cyberbully: Examining the Dark Triad," *Computers in Human Behavior* 49 (2015): 1–4.

[336] Erin E. Buckels., Paul D. Trapnell, and Delroy L. Paulhus, "Trolls Just Want to Have Fun," *Personality and Individual Differences* 67 (2014): 97–102.

[337] Spitaletta, "Terror as a Psychological Warfare Objective."

[338] Baghdadi's given name is Ibrahim Awad Ibrahim al Badri al Samarrai.

[339] Adnani's given name is Taha Subhi Falaha.

[340] Jason Spitaletta, "Comparative Psychological Profiles: Baghdadi & Zawahiri," in eds. H. Cabayan and S. Canna, *Multi-Method Assessment of ISIL* (Washington, DC: Strategic Multi-layer Assessment Office, Office of the Secretary of Defense, 2014), http://nsiteam.com/sma-publications/.

[341] Ibid.

[342] Ibid.

[343] Hussain Murtaza, "The New Information War," *Intercept*, November 25, 2017, https://theintercept.com/2017/11/25/information-warfare-social-media-book-review-gaza/.

[344] Editorial Board, "The New Radicalization of the Internet."

[345] Carlin, *Dawn of the Code War.*

[346] Bos et al., *Human Factors Considerations of Undergrounds in Insurgencies.*

[347] Jon Locket, "'See the World, Join ISIL' Spoof Recruitment Posters for ISIL Appear Across East London," *Sun*, July 29, 2016, https://www.thesun.co.uk/news/1520886/spoof-recruitment-posters-for-ISIL-appear-across-east-london/.

[348] Leonie Schmidt, "Cyberwarriors and Counterstars: Contesting Religious Radicalism and Violence on Indonesian Social Media," *Asiascape: Digital Asia* 5, no. 1-2 (2018): 32–67.

[349] Leonhard et al., *Undergrounds in Insurgent, Revolutionary, and Resistance Warfare.*

[350] Jason A. Spitaletta, Summer D. Newton, Nathan D. Bos, Charles W. Crossett, and Robert R. Leonhard, "Historical Lessons on Intelligence Support to Countering Undergrounds in Insurgencies," *Inteligencia y Seguridad: Revista de AnálISIL y Prospectiva*, no. 13 (2013), 101–128.

[351] Ishan Pandya, Hitanshu Joshi, Biren Patel, and Harshil Joshi, "Threats that Deep Web Possess to Modern World," JIRST: National Conference on Latest Trends in Networking and Cyber Security, March 2017, http://www.ijirst.org/articles/SALLTNCSP033.pdf.

[352] Ibid.

[353] Bowles, "How 'Doxxing' Became a Mainstream Tool in the Culture Wars."

[354] Coleman and Golub, "The Anthropology of Hackers."

[355] Coleman, *Hacker, Hoaxer, Whistleblower, Spy.*

[356] Emma Grey Ellis, "Whatever Your Side, Doxing is a Perilous Form of Justice," *Wired*, August 17, 2017, https://www.wired.com/story/doxing-charlottesville?mbid=nl_81717_p2&CNDID=13902615.

[357] Bowles, "How 'Doxxing' Became a Mainstream Tool in the Culture Wars."

[358] Avi Selk, "A Twitter Campaign is Outing People Who Marched with White Nationalists in Charlottesville," *Washington Post*, August 14, 2017, https://www.Washingtonpost.com/news/the-intersect/wp/2017/08/14/a-twitter-campaign-is-outing-people-who-marched-with-white-nationalists-in-charlottesville/?noredirect=on&utm_term=.c5427dd74a72.

[359] Christopher Hadnagy, *Unmasking the Social Engineer: The Human Element of Security* (Indianapolis, IN: John Wiley & Sons, 2014).

[360] Ibid.

[361] Xin (Robert) Luo, Richard Brody, Alessandro Seazzu, and Stephen Burd, "Social Engineering: The Neglected Human Factor for Information Security Management," *Information Resources Management Journal* 24, no. 3, (July-September 2011): 4.

[362] Ibid.

[363] Michael Workman, "Gaining Access with Social Engineering: An Empirical Study of the Threat," *Information Systems Security* 16, no. 6 (2007): 317.

[364] Hadnagy, *Unmasking the Social Engineer.*

[365] Mitnick, *Ghost in the Wires.*

[366] Ibid.

[367] Ibid.

[368] Ibid.

[369] Coleman, *Hacker, Hoaxer, Whistleblower, Spy.*

[370] Jildau Borwell, Jurjen Jansen, and Wouter Stol. "Human Factors Leading to Online Fraud Victimization: Literature Review and Exploring the Role of Personality Traits," in *Psychological and Behavioral Examinations in Cyber Security*, IGI Global, 2018.

[371] Ibid.

[372] Arun Vishwanath, "Spear Phishing: The Tip of the Spear Used by Cyber Terrorists," in *Combating Violent Extremism and Radicalization in the Digital Era*, IGI Global, 2016.

[373] Arthur Deegan, Yasir Khalid, Michelle Kingue, and Aldo Taboada, "Cyber-ia: The Ethical Considerations Behind Syria's Cyber-War," *Small Wars Journal* (2017), https://smallwarsjournal.com/index.php/jrnl/art/cyber-ia-the-ethical-considerations-behind-syria%E2%80%99s-cyber-war.

[374] Ibid.

[375] John Scott-Railton, and Seth Hardy, "Malware Attack Targeting Syrian ISIS Critics," Citizen Lab, December 18, 2014, https://citizenlab.ca/2014/12/malware-attack-targeting-syrian-isis-critics/.

[376] Hugh Naylor and Eric Curringham, "Anti-Islamic State activist and his friend beheaded in Turkey," *Washington Post*, October 30, 2015, https://www.washingtonpost.com/world/anti-islamic-state-activist-and-his-friend-beheaded-in-turkey/2015/10/30/c3340038-7f05-11e5-bfb6-65300a5ff562_story.html?utm_term=.5f6082863ea3.

[377] Hadnagy, *Unmasking the Social Engineer.*

[378] Coleman and Golub, "The Anthropology of Hackers."

[379] Bos et al., *Human Factors Considerations of Undergrounds in Insurgencies.*

[380] Gina S. Ligon, Mackenzie Harms, John Crowe, Leif Lundmark, and Pete Simi, "An Organizational Profile of the Islamic State: Leadership, Cyber Expertise, and Firm Legitimacy," in *Cabayan, Hriar and Canna, Sarah. A Multi-Method Assessment of ISIL* (Office of the Secretary of Defense, Strategic Multilayer Assessment, 2014).

[381] Andrea Lynch Pyon, "Disrupting Terrorist Financing: Interagency Collaboration, Data Analysis, and Predictive Tools," *Forensics Journal* 6 (2015): 42–51.

[382] Bos et al., *Human Factors Considerations of Undergrounds in Insurgencies.*

[383] Douglas C. Derrick, Gina Ligon, Mackenzie Harms, and William R. Mahoney, "Cyber-Sophistication Assessment Methodology for Public-Facing Terrorist Web Sites," *Journal of Information Warfare* 16, no. 1 (2017): 13–30.

[384] Bos et al., *Human Factors Considerations of Undergrounds in Insurgencies.*

[385] Carla E. Humud, Robert Pirog, and Liana Rosen, "Islamic State Financing and U.S. Policy Approaches," Congressional Research Service Report R43980, April 10, 2015.

386 Neil B. Barnas, "Blockchains in National Defense: Trustworthy Systems in a Trustless World," Research Report in Partial Fulfillment of Graduation Requirements, Air University, June 2016, 19.

387 Lillian Ablon, Martin C. Libicki, and Andrea A. Golay, Markets for Cybercrime Tools and Stolen Data Hacker's Bazaar (Santa Monica, CA: RAND, 2014), PDF e-book, 11–12.

388 Kim-Kwang Raymond Choo, "Cryptocurrency and Virtual Currency: Corruption and Money Laundering/Terrorism Financing Risks?" in *Handbook of Digital Currency* (Academic Press, 2015).

389 M. Y. Lai, C. F. Jin, K. Nie, and J. H. Zhao, "Cyber Physical Logistics System: The Implementation and Challenges of Next-generation Logistics System," *Systems Engineering* 4, no. 008 (2011).

390 "Informationized warfare" is a PRC concept of twenty-first century conflict that prioritizes IT and population control through strategic social influence. For more, see: https://jamestown.org/program/chinas-new-military-strategy-winning-informationized-local-wars/.

391 Dan Goodin, "New IoT Botnet Offers DDoSes of Once-unimaginable Sizes for $20," *Ars Technica*, February 1, 2018, https://arstechnica.com/information-technology/2018/02/for-sale-ddoses-guaranteed-to-take-down-gaming-servers-just-20/.

392 Bos et al., *Human Factors Considerations of Undergrounds in Insurgencies.*

393 Ibid.

394 Derrick, Sporer, Church, and Ligon, "Ideological Rationality and Violence," 57-81.

395 Olson, *We Are Anonymous.*

396 Benjamin Gleason, "#Occupy Wall Street: Exploring Informal Learning About a Social Movement on Twitter," *American Behavioral Scientist* 57, no. 7 (2013): 966–982.

397 DeLuca, Lawson, and Sun, "Occupy Wall Street on the Public Screens of Social Media," 483-509.

398 Gleason, "#Occupy Wall Street."

399 Coleman, *Hacker, Hoaxer, Whistleblower, Spy.*

400 Ibid.

401 Bos et al., *Human Factors Considerations of Undergrounds in Insurgencies.*

402 Olson, *We Are Anonymous.*

403 Ibid.

404 Ibid.

405 Bos et al., *Human Factors Considerations of Undergrounds in Insurgencies.*

406 Ibid.

407 Ibid.

408 Ibid.

409 Ibid.

410 Ibid.

411 Mehdi Yahyanejad, "The Effectiveness of Internet for Informing and Mobilizing in the Events After the Iranian Presidential Election," MIT CSAIL, Fall 2010, http://groups.csail.mit.edu/mac/classes/6.805/admin/admin-fall-2010/weeks/week12-Yahyenejad.pdf.

412 Bos et al., *Human Factors Considerations of Undergrounds in Insurgencies.*

413 Ibid.

[414] Ibid.

[415] Underwood and Welser, "'The Internet is Here'," 304-311.

[416] Ibid.

[417] Ibid.

[418] Ibid.

[419] Coleman, *Hacker, Hoaxer, Whistleblower, Spy.*

[420] Jorg Raab and H. Brinton Milward, "Dark Networks as Problems," *Journal of Public Administration Research and Theory* 13, no. 4 (2003): 413–439.

[421] Internet Relay Chat (IRC) is an application layer protocol that facilitates text communications on a client/server networking model. IRC clients are computer programs installed on a computer designed for group discussion forums (or channels as well as private encrypted messaging and/or file sharing).

[422] Zlatogor Minchev, "Hybrid Challenges to Human Factor in Cyberspace," in *Countering Terrorist Activities in Cyberspace* 139 (2018): 32.

[423] Bos et al., *Human Factors Considerations of Undergrounds in Insurgencies.*

[424] Ibid.

[425] Ibid.

[426] Ibid.

[427] Ibid.

[428] Jill Lane, "Digital Zapatistas," *Drama Review* 47, no. 2 (2003): 129–144.

[429] Bos et al., *Human Factors Considerations of Undergrounds in Insurgencies.*

[430] Ibid.

[431] Ibid.

[432] Underwood and Welser, "'The Internet is Here'," 304-311.

[433] Ibid.

[434] Max Houlka, "What Anonymous Can Tell Us About the Relationship Between Virtual Community Structure and Participatory Form," *Policy Studies* 38, no. 2 (2017): 168–184.

[435] Coleman, *Hacker, Hoaxer, Whistleblower, Spy.*

[436] Underwood and Welser, "'The Internet is Here'," 304-311.

[437] Bos et al., *Human Factors Considerations of Undergrounds in Insurgencies.*

[438] Ibid.

[439] Ibid.

[440] Jonathan Bishop, "The Psychology of Trolling and Lurking: The Role of Defriending and Gamification for Increasing Participation in Online Communities Using Seductive Narratives," in *Virtual Community Participation and Motivation: Cross-Disciplinary Theories,* IGI Global, 2012.

[441] Robert Slonje, Peter K. Smith, and Ann Frisén, "The Nature of Cyberbullying, and Strategies for Prevention," *Computers in Human Behavior* 29, no. 1 (2013): 26–32.

[442] Romy Kraemer, Gail Whiteman, and Bobby Banerjee, "Conflict and Astroturfing in Niyamgiri: The Importance of National Advocacy Networks in Anti-Corporate Social Movements," *Organization Studies* 34, no. 5-6 (2013): 823–852.

[443] Caroline W. Lee, "The Roots of Astroturfing," *Contexts* 9, no. 1 (2010): 73–75.

444 Patrick M. Duggan, "Strategic Development of Special Warfare in Cyberspace," *Joint Force Quarterly* 79, no. 4 (2015): 46–53.

445 Tetyana Lokot, "Public Networked Discourses in Ukraine-Russia Conflict: 'Patriotic Hackers' and Digital Populism," *Irish Studies in International Affairs* 28 (2017): 99–116.

446 Chengcheng Shao, Giovanni Luca Ciampaglia, Onur Varol, Kai-Cheng Yang, Alessandro Flammini, and Filippo Menczer, "The Spread of Low-Credibility Content by Social Bots," *Nature Communications* 9, no. 1 (2018): 4787.

447 William Reno and Jahara Matisek. "A New Era of Insurgent Recruitment: Have 'New' Civil Wars Changed the Dynamic?" *Civil Wars* (2018): 1–21.

448 Massimo Stella, Emilio Ferrara, and Manlio De Domenico, "Bots Increase Exposure to Negative and Inflammatory Content in Online Social Systems," *Proceedings of the National Academy of Sciences* (2018): 201803470.

449 Jon Kleinberg, "The Convergence of Social and Technological Networks," *Communications of the ACM* 51, no. 11 (November 2008): 66–72.

450 Maxwell, "The Cyber Underground-Resistance to Active Measures and Propaganda."

451 Bos et al., *Human Factors Considerations of Undergrounds in Insurgencies.*

452 Ibid.

453 UN General Assembly, "Universal Declaration of Human Rights," *UN General Assembly* (1948).

454 Bos et al., *Human Factors Considerations of Undergrounds in Insurgencies.*

455 Barry, "Sound of Post-Soviet Protest."

456 Greg Satell, "How Social Movements Change Minds," *Harvard Business Review*, July 28, 2015, https://hbr.org/2015/07/how-social-movements-change-minds.

457 Cody Delistraty, "When Wearing a Graphic T-Shirt is a Revolutionary Act," *Garage*, March 30, 2018, https://garage.vice.com/en_us/article/zmgmd4/graphic-t-shirt-revolutions.

458 Satell, "How Social Movements Change Minds."

459 Bos et al., *Human Factors Considerations of Undergrounds in Insurgencies.*

460 Vegh, "Classifying Forms of Online Activism."

461 Tufekci, "Twitter and Tear Gas."

462 Danielle Lottridge, Frank Bentley, Matt Wheeler, Jason Lee, Janet Cheung, Katherine Ong, and Cristy Rowley, "Third-wave livestreaming: teens' long form selfie," in *Proceedings of the 19th International Conference on Human-Computer Interaction with Mobile Devices and Services*, ACM, 2017.

463 Luke Winslow, "'Not Exactly a Model of Good Hygiene': Theorizing an Aesthetic of Disgust in the Occupy Wall Street Movement," *Critical Studies in Media Communication* 34, no. 3 (2017): 278–292.

464 Bos et al., *Human Factors Considerations of Undergrounds in Insurgencies.*

465 Ibid.

466 Ibid.

467 Ibid.

468 Ibid.

469 Vegh, "Classifying Forms of Online Activism."

470 Gerbaudo, *Tweets and the Streets.*

471 Richard Hanna, Andrew Rohm, and Victoria L. Crittenden, "We're All Connected: The Power of the Social Media Ecosystem," *Business Horizons* 54, no. 3 (2011): 265–273.

472 Sebastian Valenzuela, "Unpacking the Use of Social Media for Protest Behavior: The Roles of Information, Opinion Expression, and Activism," *American Behavioral Scientist* 57, no. 7 (2013): 920–942.

473 Dana Rotman, Sarah Vieweg, Sarita Yardi, Ed Chi, Jenny Preece, Ben Shneiderman, Peter Pirolli, and Tom Glaisyer, "From Slacktivism to Activism: Participatory Culture in the Age of Social Media," in *CHI'11 Extended Abstracts on Human Factors in Computing Systems* (Vancouver, Canada: ACM, 2011), 819–822.

474 Bos et al., *Human Factors Considerations of Undergrounds in Insurgencies.*

475 Ibid.

476 Amy Harmon, "'Hacktivists' of All Persuasions Take Their Struggle to the Web," *New York Times*, October 31, 1998, http://www.nytimes.com/1998/10/31/world/hacktivists-of-all-persuasions-take-their-struggle-to-the-web.html.

477 Stefan Wray, "Electronic Civil Disobedience and the World Wide Web of Hacktivism: A Mapping of Extraparliamentarian Direct Action Net Politics," *Nova Iorque* (1998).

478 Chiesa, Ducci, and Chiappi, *Profiling Hackers.*

479 Bos et al., *Human Factors Considerations of Undergrounds in Insurgencies.*

480 Ibid.

481 Ibid.

482 Charlotte Heath-Kelly and Lee Jarvis, "Affecting Terrorism: Laughter, Lamentation, and Detestation as Drives to Terrorism Knowledge," *International Political Sociology* 11, no. 3 (2017): 239–256.

483 Miron Lakomy, "Cracks in the Online 'Caliphate': How the Islamic State is Losing Ground in the Battle for Cyberspace," *Perspectives on Terrorism* 11, no. 3 (2017).

484 Ralph Martins, "'Anonymous' Cyberwar Against ISIL and the Asymmetrical Nature of Cyber Conflicts," *Cyber Defense Review* 2, no. 3 (2017): 95–106.

485 Anne Speckhard and Mubin Shaikh, *Undercover Jihadi: Inside the Toronto 18, Al Qaeda Inspired, Homegrown, Terrorism in the West* (Advances Press, 2014).

486 Andrew Griffin, "Anonymous 'Trolling Day' Against ISIL Begins, with Group's 'Day of Rage' Mostly Consisting of Posting Mocking Memes,'" *Independent*, December 15, 2011, https://www.independent.co.uk/life-style/gadgets-and-tech/news/anonymous-trolling-day-against-ISIL-begins-with-group-s-day-of-rage-mostly-consisting-of-posting-a6769261.html.

487 Houlka, "What Anonymous Can Tell Us About the Relationship Between Virtual Community Structure and Participatory Form," 168-184.

488 Ibid.

489 Ibid.

490 Underwood and Welser, "'The Internet is Here'," 304-311.

491 Ibid.

492 Ibid.

493 Coleman, *Hacker, Hoaxer, Whistleblower, Spy.*

494 Houlka, "What Anonymous Can Tell Us About the Relationship Between Virtual Community Structure and Participatory Form," 168-184.

495 Underwood and Welser, "'The Internet is Here'," 304-311.

[496] Ibid.

[497] Bos et al., *Human Factors Considerations of Undergrounds in Insurgencies*.

[498] Alexander and Swetnam. *Cyberterrorism and Information Warfare*.

[499] Paul M. A. Linebarger, *Psychological Warfare* 2nd Edition (Washington, DC: Combat Forces Press, 1954).

[500] Spitaletta, "Terror as a Psychological Warfare Objective."

[501] Ibid.

[502] Ibid.

[503] Ibid.

[504] Michael Martinez, "Cyberwar: CyberCaliphate Targets US Military Spouses; Anonymous Hits ISIS," *CNN*, February 11, 2015, https://www.cnn.com/2015/02/10/us/isis-cybercaliphate-attacks-cyber-battles/index.html.

[505] Emma Graham-Harrison, "Could ISIL's 'Cyber Caliphate' Unleash a Deadly Attack on Key Targets?" *Guardian*, April 12, 2015.

[506] Carlin, *Dawn of the Code War*.

[507] Ibid.

[508] Ibid.

[509] Bos et al., *Human Factors Considerations of Undergrounds in Insurgencies*.

[510] Ibid.

[511] Lawrence A. Kuznar and William H. Moon, "Thematic Analysis of ISIL Messaging," in H. Cabayan and S. Canna, eds., *Multi-Method Assessment of ISIL* (Washington, DC: Strategic Multilayer Assessment Office, Office of the Secretary of Defense, 2014), http://nsiteam.com/sma-publications/.

[512] Anne Barnard, "Children, Caged for Effect, to Mimic Imagery of ISIS," *New York Times*, February 20, 2015, https://www.nytimes.com/2015/02/21/world/middleeast/activists-trying-to-draw-attention-to-killings-in-syria-turn-to-isis-tactic-shock-value.html.

[513] Kuznar and Moon, Thematic Analysis of ISIL Messaging."

[514] Spitaletta, "Terror as a Psychological Warfare Objective."

[515] Ibid.

[516] Peter Lentini and Muhmmad Bakashmar, "Jihadist Beheading: A Convergence of Technology, Theology, and Teleology?" *Studies in Conflict & Terrorism* 30, no. 4 (2007): 303–325.

[517] Kuznar and Moon, Thematic Analysis of ISIL Messaging."

[518] Paulo Shakarian, Jana Shakarian, and Andrew Ruef, *Introduction to Cyber-Warfare: A Multidisciplinary Approach* (Syngress, 2013).

[519] Tal Pavel, "Physical Threats in Online Worlds–Technology, Internet and Cyber under Terror Organization Services; a Test Case of 'The Islamic State,'" *ICT Information and Communications Technologies* 6, no. 1 (2017): 73–82.

[520] Ibid.

[521] Shakarian et al., *Introduction to Cyber-Warfare*.

[522] Ibid.

[523] R. Kim Cragin and Ari Weil, "'Virtual Planners' in the Arsenal of Islamic State External Operations," *Orbis* 62, no. 2 (2018): 294–312.

CHAPTER 8.
ATTRIBUTION AND CYBER
RESISTANCE MOVEMENTS

THE NEED FOR CYBER ANONYMITY

Resistance movements should avoid attribution when they want to protect themselves. While they often seek attribution to their cause when they want credit for an action, they still want to maintain anonymity of the individual members to avoid reprisals. Achieving and sustaining anonymity is essential to avoid attribution as a member of the resistance.

Cybersecurity is as important as physical security in terms of avoiding compromise. Every person and organization in the modern world has some sort of cyber footprint. Under certain circumstances, that footprint can lead to the attribution of the person or the organization. In the physical world, a resistance member uses a *nom de guerre* to keep his or her real identity unknown to the state security service. In cyberspace, one resistance member may use multiple cyber user names or personae to maintain anonymity while undertaking cyber activities. Maintaining multiple user names and personae is similar to having a set of passports that can be used in different circumstances.[1] Without personae in cyberspace, the anonymity of a resistance member could be readily breached. For example, it is believed that there is no longer any true cyber anonymity in China due to the degree of control the state security service has over the Internet in that country.[2]

In a similar manner, the state security service often needs to provide some level of anonymity to its members, especially those who need special protection. That anonymity should also extend into cyberspace for the same reason—to separate the cyber footprint from the physical footprint of the member. State security service members in certain key positions often need as much anonymity in cyberspace as key resistance members. This chapter, however, focuses on resistance movement attempts to avoid attribution. For those seeking cyber anonymity, the current complexity of the Internet provides many opportunities for nonattributable or misattributable activities.

To better understand cyber anonymity, we introduce the differences between the web, the deep web, and the dark web. This chapter then addresses attributed entities and the types of attribution and methodologies that enable attribution. Technical details describe the analysis required to attribute an entity. To conclude, this chapter addresses methods to minimize attribution of the resistance movement to the state security services.

THE WEB, THE DEEP WEB, AND THE DARK WEB

The common language of the Internet is called the Internet protocol (IP), with nearly all of Internet traffic consisting of the latest two versions: IPv4 and IPv6. On top of the IP lies a variety of other transmission protocols, such as user datagram protocol (UDP) and transmission control protocol (TCP). These protocols enable data to be logically deconstructed and sent across IP networks in small snippets called "packets" that allow for reconstruction by the end device.

Perhaps the most familiar Internet technology is the worldwide web (WWW), or simply "the web," which is built on top of TCP and IP. The web is what most people experience as the "Internet"—consisting of web pages they view in a web browser. The web comprises a server and browser communicating over the hypertext transport protocol (HTTP), which is a standard of communication built on top of TCP/IP.[3]

While the Internet is in general thought of as a free and open collection of information, it is rather more technically a communications transport for sending data between computers. This transport can be used to openly share information, or it can perform more selective sharing of information between only properly credentialed individuals or organizations. The media has taken to calling the parts of the Internet that are not freely accessible the "deep web." Examples of this include bank account information, document repositories, medical records, etc.

Another deep web component is traffic where both users and servers are nonattributed. Through the use of the onion router (Tor),[4] information can be shared and accessed anonymously. This is referred to as the "dark web." Much like the rest of the Internet, the dark web consists of all types of content. However, because the dark web is anonymous, it serves as a haven of bad content, such as stolen information and child pornography. The dark web also includes anonymizing networks similar to Tor, such as Freenet and I2P.

Resistance movements employ the web, the deep web, and the dark web, depending on their purpose and need for anonymity. When the resistance movement wants to reach an international audience to recruit members or raise finances, they use the web, with the understanding that their presence on that web server will likely be short, as described in chapter 4. When they want to communicate more securely, they may use the deep web, leveraging some degree of encryption. However, most resistance movements rely on the dark web due to its strong encryption and counter-attribution capabilities. This is one reason the US State Department provides financial support to Tor to support resistance movements fighting repression, in spite of the use of Tor by criminal elements as well.[5]

> 8.1 Resistance movements employ the web, the deep web, and the dark web, depending on their purpose and anonymity needs. Most resistance movements rely on the dark web for internal communications due to its strong encryption and counter-attribution capabilities.

ATTRIBUTION CONSIDERATIONS AND ENTITY TYPES

Attribution analysis attempts to use available information to attribute the entity that caused a cyberspace attack or specific communication.

An attributed entity can be a human or a network device, depending on the type and amount of evidence discovered. Attributed entities can fall into four categories: nation-state, a network or specific device, an organization, or an individual. Achieving full attribution to a specific individual is very difficult, often impossible.

Attributing to a Nation-State

The most commonly known type of attribution is that to the nation-state. In the early years of the Internet, network exploitation and attack were often tied to a nation-state because only nation-states possessed advanced capabilities and skilled personnel necessary to execute an attack.

A cyber attack assessed to have nation-state involvement was executed in Estonia in 2007. In this event, financial institutions were subjected to an attack that disabled client access to the targeted institution's websites. This attack was initiated in response to the Estonian government moving a World War II memorial statute of a Soviet soldier from Tallinn's (Estonia's capital) central cemetery to another cemetery. The Estonian government announced its intent to move the statue and received objections from the Russian government. In April, financial institutions resident in Estonia were unable to execute transactions. The attack was traced back to Russian IP space, and the Estonian government stated that the attack was state initiated.[6]

Unfortunately, nation-states have since found ways to deny any involvement in cyber attacks by hiring cyber criminals or hackers as contractors. By claiming no knowledge of individual citizen behaviors, nation-states have been able to choose their targets and deny involvement even when the IP address of the attack traces back to their country.

Attributing to Computer Networks, Devices, and Botnets

Given the tools and capabilities available to conceal the identity of the actor launching an attack, attribution may only extend to a computer network or specific device from which the attack launched. One type of attack that is difficult to attribute is a botnet, which is a type of malware distributed on many geographically distributed machines but which acts in concert to achieve a specified goal. Each bot in the network is malware installed on a machine, almost always without the owner of the machine aware of its presence. The fact that each bot lies undetected even when activated is one of the reasons for the longevity and power of these botnets (see Figure 8-1).

Several botnets have been discovered, each consisting of thousands of infected devices. For example, in 2008, Microsoft discovered a computer worm called Conficker designed to interface with the Window's operating system, which replicated itself and spread via a computer network. Each time the Conficker worm spread, it made the infected computer part of a botnet. By the time the Conficker worm was discovered, it spread to over twelve million devices.[7]

The increasing sophistication of modern botnets make them exceedingly difficult to address. Because it is difficult to identify bot-infected machines, some early success was achieved by identifying the C2 centers of the botnets. Even just a few years ago, Botnets were often controlled by a single C2 machine with an unencrypted communications channel. By analyzing an infected machine, the controller could be identified, and law enforcement could effectively take the botnet offline. However, modern botnets employ a highly distributed control structure and encrypted communications channels. This increase in anonymity and longevity make botnets appealing to resistance movements, nation–states, and criminal organizations.

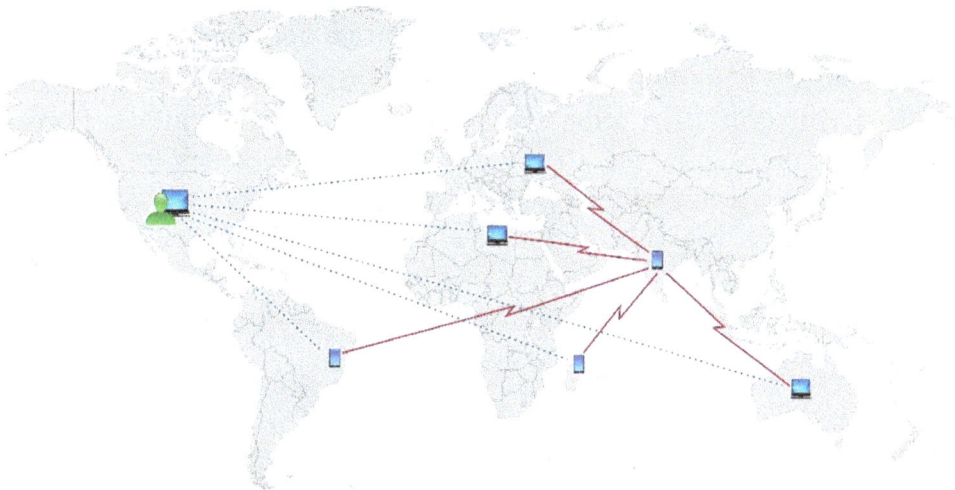

Figure 8-1. Botnet components diagram.

Additionally, with the proliferation of Internet-connected devices such as smart phones, security cameras, game consoles, and appliances—all with various types of vulnerable software and often going years without software patches—botnet operators have many new potential types of targets to add to their networks.

Botnets are also popular tools because of their flexibility in execution.[8] They provide a flexible and inexpensive platform for creating a wide range of effects, such as generating very large DDoS attacks. A DDoS attack happens when a web site is flooded with more requests to access the site than the site can process, and therefore the site shuts down. A botnet aimed at a particular website can effectively shut it down, at little operating cost and almost no risk of attribution.

Botnets are increasingly employed "for hire" to execute cyber attacks on behalf of another party/employer.[9] An actor with very little cyber knowledge or capability can be a formidable adversary against another actor by employing a third party that makes its botnet available for hire. The third party uses the botnet to conduct an attack and minimize the chance of attribution by only using a percentage of available bots and concealing the location of the botnet controller.

The Spamhaus Botnet Controller Advisory is a non-profit organization that tracks spam-email-producing websites, malware, and botnets. When Spamhaus identifies a botnet and its controller, it publishes a list of controller locations by IP address for use by Internet service providers, cybersecurity organizations, and network operators to deny controller access to an infected device.[10]

Unfortunately, disrupting the ability for entities to build a botnet made Spamhaus a target. In 2013, a large DDoS attack to date was executed against Spamhaus. During the DDoS attack, the Spamhaus site and services were inaccessible to clients. This DDoS attack involved geographically distributed bots sending enormous amounts of traffic, dramatically affecting not only Spamhaus, but also multiple Internet service proivders[11] with infected customers.

> *8.2   Resistance movements continue to use botnets for offensive cyber operations, as will cyber criminals and nation-states.*[12]

Attributing to an Organization

Organizational attribution expands beyond the nation-state to include non-nation-state actors, such as extremist groups and profit-motivated criminal organizations.[13] Nation-states and individuals also can employ a criminal organization to act on their behalf, thereby making attribution back to the nation-state or individual very difficult.

Organizational attribution can fall into two subsets: the entity employing a cyber attack and the entity executing the cyber attack. In this first example, a non-state entity known as the Lazarus Group most likely acted on behalf of a nation-state (North Korea), thereby allowing the nation-state to deny involvement in the action.[14] In 2014, Sony Pictures was hacked in what was assessed to be a response to the release of a comedy film called *The Interview*, in which two reporters are hired by the Central Intelligence Agency to assassinate the North Korean Supreme Leader Kim Jong Un. Multiple cybersecurity organizations analyzed the malicious code and concluded that the attack was executed by North Korean actors known as Lazarus Group.[15] North Korea denied any involvement and stated that the attack may have been executed by North Korean supporters and sympathizers.[16] Although the North Korean government denied involvement, it praised the attack as a "righteous deed" because the film was seen as an "act of terrorism."[17]

As a second example, in May 2017, networks in Europe were subjected to a ransomware attack, referred to as WannaCry. A ransomware attack involves a malicious code that hijacks a computer and encrypts its contents. The malicious code operator only unlocks the computer once a ransom is paid or never unlocks the encrypted contents, depending on the purpose of the attack. For example, three days after the detection of the WannaCry attack, the ransomware spread to over two hundred thousand organizations across one hundred and fifty countries.[18]

In 2017, the malware NotPetya spread from the servers of an unassuming Ukrainian software firm to some of the largest businesses worldwide, paralyzing their operations. The cost to major companies included, for example, $870 million to the Pharmaceutical Company Merck, $400 million to Fedex, and $300 million to the shipping company Maersk. Some analysts concluded that, rather than trying to gain ransom money, the malware was designed to send a political message: "If you do business in Ukraine, bad things will happen to you."[19]

Attributing to an Individual

Attribution to a specific individual can be very difficult when the individual has multiple means to disguise his or her identity. Although an attack may be attributed to an IP address, the IP address can be spoofed (made to appear as though it is coming from a different IP address). Even when the source IP address can be identified, the person and motive behind an attack is less definitive and uncertain.[20] An actor can reside in one country and initiate an attack from a device the actor controls in another country (see Figure 8-2). Attribution to an individual may need the cooperation of law enforcement in the country hosting the IP space or attributed actor of the attack. An example of attribution through law enforcement coordination and international cooperation is the FBI's crackdown on the Zeus malware coordinators.

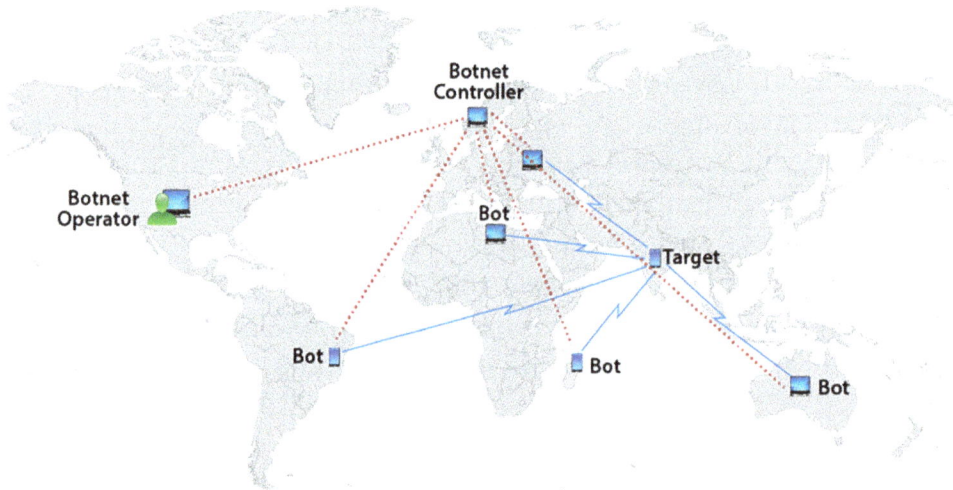

Figure 8-2. Botnet operator and controller separation.

The Zeus malware accesses a computer via a website pop-up window that tells a user that his or her computer is infected with a virus. The pop-up directs a user to a technical support site to remove the malware after paying a "support fee." By interfacing with the website, the user uploads malicious code onto his or her computer, after which the code records financial transactions and associated data (e.g., credit card numbers).

The Zeus malware infected millions of computers and attacked US financial institutions and individual bankers. The botnet consisted of globally distributed devices, but the control of the botnet resided with individual operators. To take legal action, the FBI coordinated with law enforcement agencies within the host countries of the suspected botnet operators. Through cooperation with the United Kingdom, Netherlands, and Ukrainian law enforcement agencies, the Zeus malware ring was disrupted.[21]

Attribution Techniques and Types

In his article "A Survey of Challenges in Attribution," Earl Boebert describes two types of attribution: technical attribution and human attribution. Technical attribution is generally seen within the cybersecurity community as the easier of the two techniques.[22] Technical attribution includes the forensic analysis of malicious code, data analysis, and Internet pathways a code traversed to access a target device. The challenge with human attribution is both discovering the true identity of the actor and understanding the motive behind the action. To achieve conclusive attribution of a cyberspace action to a specific actor requires both technical and human attribution.[23]

Technical Attribution

Another paper[24] by Shamsi et al. further divides technical attribution into two levels, leaving human attribution as the third level. Shamsi considers level one the easiest because the attack evidence is *within* the possession of the target owner, while the evidence is *outside* of the target environment for levels two and three. Table 8-1 captures the three levels of attribution analysis and the dimension that the level investigates.

Table 8-1. Levels and dimensions of attribution.

Technical Attribution		
Level One	Source code analysis	Forensic analysis of events, effects, and methods
Level Two	Path analysis	Investigation to identify the source of the attack
Human Attribution		
Level Three	Actor attribution	Identifying the actual actor (individual or organization) that initiated and controlled the attack

Level one analysis is synonymous with conducting a forensic analysis of the code to discover what happened. Level one analysis assesses, for example, the type of malware used, the actions of the malicious code to the target (e.g., exfiltrated data), the effects to the target (e.g., system crash/non-operable), and the method with which the code penetrated the target (e.g., anti-virus software not updated).

Direct analysis of malware code compares the code to a library of known malware and authorship. Identifying variations in the captured malware can indicate patterns of reuse, possible co-authorship, and intent of the attack. CrowdStrike, a commercial cybersecurity consulting firm, used this method to assign attribution for the 2016 Democratic National Committee breech to Russia.[25]

An example of source code variation is the modification of the Stuxnet code. In 2010, Stuxnet was used to disable Iranian uranium centrifuges supporting the enrichment of uranium in Iran's nuclear capabilities development program. Stuxnet was designed to specifically target the controlling programs used in Iranian centrifuges by disabling their safeguard programs, causing the centrifuges to spin at rapidly varying rates, which disabled them. Since the release of Stuxnet, multiple code variations were discovered in the energy facilities of eight countries. These variants were not related to a specific attack on the newly infected systems, but the code was capable of causing damage if or when activated.[26]

Some tools are custom developed, while other tools are available for public download, such as MetaSploit. MetaSploit is designed for legitimate use by cybersecurity professions to test the security procedures of networks. However, an actor can take a tool like MetaSploit and use it for malicious purposes. Using publically available tools further hides an actor's identity because publically accessible tools can be used by anyone.

> *8.3* *Resistance movements need to continue to use tools to avoid*
> *technical attribution to avoid compromise in the physical world of*
> *the individual or the resistance movement.*

Note that state security services in repressive regimes may not require the "smoking gun" necessary for legal action in open societies. Repressive regimes can identify, arrest, and interrogate likely suspects without complete technical or human attribution. Because of this tendency, resistance movements should follow Key Takeaway 8.4.

> *8.4 Members of resistance movements need to ensure that any attribution obfuscation or misattribution capabilities are not readily accessible on their persons, property, or communications devices.*

Level two analysis involves path analysis, attempting to identify the origin of the attack. When not obfuscated, the IP packets carry substantial, accurate information about the source IP address and the path taken to reach the destination IP address (see Figure 8-3). The IP packet content starts with the packet's logical destination and origin (which can be spoofed, as previously described), followed by the "payload," which contains the protocol and data contained within the packet. The final element of the packet is a cyclic redundancy check (CRC) to determine if the packet was corrupted in transit. If so, error-correcting code is applied, or the packet is sent again.

Destination MAC Address	Source MAC Address	Destination IP Address	Source IP Address	Payload	CRC

Figure 8-3. IP-data packet composition.

Often a data packet traverses multiple hop points to reach its destination. Hop points are general terms for routers that "route" Internet traffic. Similar to a traveler making connecting flights, hop points are the connecting points needed for a packet to traverse the Internet and reach its destination. Detailed pathway analysis is needed because an actor can disguise the pathway taken and code a packet that presents an image that a particular pathway took when in reality it was different.[27]

Russian entities conducted cyberspace attacks on Georgian government networks in August 2008. The attacks consisted of DDoS attacks on Georgian websites executed through botnets. As Russian forces entered the provinces of Abkhazia and South Ossetia, botnets attacked and defaced Georgian websites. Forensic analysis after the attack revealed evidence that these botnets were linked to Russian organized crime groups, including the group known as the Russian Business Network (RBN), which leased botnets for the attack.[28] These attacks were partially executed concurrently with a land invasion of South Ossetia, which was the first time cyberspace operations were fully integrated into major conventional combat operations as part of a larger multi-domain military campaign.[29]

The case of the cyber attack on Georgia highlights that technical attribution is useful in describing the specific execution of a cyber attack, but this is the limit. Who executed the attack and why crosses over into human attribution. Although it can be argued that Russian patriotism answers the why, the answer to whom can responsibility be personally held is less certain.

Human Attribution

According to Boebert, human attribution is often more challenging than technical attribution. Referenced in Table 8-1, there is forensic evidence (e.g., an affected target, altered data, pathway analysis to access a target) left after an attack to conduct technical attribution, but there may not be sufficient evidence to definitively connect the forensics evidence to conclude and assign human attribution. An actor desiring to attack a target can employ a third party to execute the attack on his or her behalf; this is what is suspected in the Georgian government cyber attacks. In some cases, the cyber attacks were traced back to IP space in Russia, but the Russian government denied any involvement.[30] A more concerning aspect of these attacks is nationals who act independently of government sponsorship that attack another party. In such cases, an aggressor government can still deny involvement but favor the actions of the "rogue" cyber operative. Alternatively, a government can discretely approach a third party to act on its behalf while denying any involvement and empirical evidence.[31]

Shamsi's level three attribution focuses on human attribution. It involves identifying the actual actor (individual or organization) that initiated and controlled the attack. Many tools and techniques are available to an actor to conceal his or her identity, though a resistance movement's member's use or possession of such tools may be sufficient for state security services to identify them, as mentioned earlier.

Malicious actors also can create fake identities disguising the actual actor. For example, in 2012, Facebook reported that it identified eighty-three million fake user accounts, although only a small subset were malicious.[32] In May 2014, Reuters reported that Iranian actors created false social networking accounts to spy on high-valued military and political leaders in the United States, Israel, and other countries.[33]

The actors established personas on Facebook and other social media sites and targeted friends and family of the high-valued targets first to build credibility. After befriending friends and family, the actors contacted the high-valued target. Contact was first non-malicious and involved sending news links, such as NewsOnAir.org. After trust was established, additional links with embedded malicious code were sent to the target. FireEye Inc.'s intelligence analysis subsidiary ISight investigated the activity. ISight's level one and two analyses revealed that the malicious code included the security password "Parastoo" to protect the code from modification. ISight previously associated this password with actors operating in Tehran. Additionally, NewsOnAir.org was registered in Tehran.[34] ISight could not conclude if the actors were directly linked to the Iranian government but suspect so because of the complexity of the operation.[35]

Table 8-2 from Shamsi provides examples of cyber attacks and associated level attribution from 2007 to 2015.[36] From left to right, the columns include the attack name, the details of the technical attribution findings, the details of the assessed motivation behind the attack, the details of any conclusions and the actual/greatest degree of attribution. The final column depicts the cumulative result of technical and human attribution from which a level one through three is assigned. Only one of these cases resulted in an attribution at level three, highlighting the challenges of achieving human attribution for non-repressive regimes.

Table 8-2. Cyberspace attribution examples.

Attribution level achieved for cyberattacks.

Cyberattacks	Attack technique	Motivation of the attack	Attribution achieved	Attribution level
Estonia (2007)	DDoS, phishing, spaming, Botnets	Referred to geo-political as it came soon after the removal of a Russian monument Russia–	IP addresses were found from varied locations, mainly Russia.	2
Georgia (2008)	DDoS, structured query language injection, HTTP-based attacks, Botnets	Georgia military conflict	Russian IP addresses (associated with StopGeorgia.ru)	2
Stuxnet (2009)	A malware worm was injected using a USB device to one of the controller machines within the nuclear plant.	Sabotage Iran's nuclear power plant	No definite attribution was achieved. Only the type of malware and its functionality was identified.	1
Operation Aurora (2009)	Trojan. Hydraq—malware (malicious payload)	Referred as counter-espionage of Chinese government by some analysts	Symantec named the group of attackers as Elderwood for a parameter used in attack source code and use of vulnerabilities, but no definite attributes were found).	2
Target Store (2013)	Malware, penetrating to access cash register systems	Financial gain/hacktivism	Based on a Russian malicious tool BlackPOS	2
Spamhaus (2013)	DDoS, Domain Name System (DNS) reflection attack	Retaliation to cyberbunker (an anonymous hosting service) being blacklisted by Spamhaus	Cyberbunker and anti-Spamhaus group "Stophaus" were named as the attackers.	3
Ashley Madison Data Theft (2015)	Possible insider data theft	Hacktivisim	The stolen data were revealed by the criminal. Data analysis dubiously revealed that it was insider job.	1

DDoS, Distributed Denial of Service
IP, Internet Protocol
HTTP, Hypertext Transfer Protocol
USB, Universal Serial Bus

Resistance Anonymity Techniques

Anonymity allows actions to be taken within cyberspace without the perpetrator likely being caught. For example, ISIS attempts to gain recruits and financial support via cyber means. ISIS tailors its messages posted on the Internet to appeal to frustrated youth regardless of their geographical location. By tailoring a message that reinforces anti-US messages and in a manner that appeals to younger audiences, ISIS generated a global stream of foreign support.[37]

Resistance operations planners conduct their own analysis of the cyberspace environment and seek to use the environment to their advantage. Some methodologies used by resistance movements leveraging the cyberspace environment include using encryption, Tor, online criminal websites, cyber mercenaries, technical misattribution, blending into a crowd, and facilitating or encouraging multiple actors to operate against the same target.

Encryption

Encryption methods are a basic staple of resistance movements, and many of the more advanced anonymizing methodologies use encryption. As described in chapter 4, secure browsing is more secure than regular browsing, although browsers can be tagged even when using secure browsing. In a similar manner, encrypted communications on smart phones are more secure than unencrypted communications, but cellular communications can be hacked by the state security services. Encrypted emails are more secure than unencrypted ones, but the very use of encrypted emails may be a red flag for the state security services to presume a user hiding information.

> 8.5 While encryption can help resistance movements in most cases, especially for protecting data at rest, encryption of communications needs to be used judiciously to not attract unwanted attention from state security services.

Smart phone applications, such as Telegram, that claim secure encrypted communications have increased in use. Telegram includes features that encrypt message traffic from sender to receiver known as end-to-end encryption and contains a message self-destruct feature, which deletes saved messages after a specified period. Governments expressed concern over criminal and terrorist usage of end-to-end encryption because it is difficult to crack.[38, 39] ISIS used Telegram to broadcast propaganda and used "invite only" features to pass sensitive information such as bomb construction.[40]

In April 2017, a metro station in St. Petersburg, Russia was subjected to a terrorist bombing, claiming fifteen lives and wounding forty-five. The Russian Federal Security Service reported that the terrorists used Telegram to synchronize attack preparations.[41] In July 2017, Indonesia's Ministry of Communication contacted Telegram administrators over concern that the app's public channels were used to broadcast terrorism-related content and propaganda.[42] Telegram announced that it would establish a team of moderators with a proficient understanding of Indonesian language to identify and remove terrorism-linked content.[43]

High-Strength Anonymization: Tor

An anonymization technique, known as Tor, is a program funded by the US government, among other organizations, that enables users to communicate and share content anonymously. Tor enables both the defeat of network-level observation, path analysis, and traffic analysis. By simply installing Tor software on a computer, a user can have very strong anonymization in accessing and sharing information. Tor adds new layers of encryption as a message passes through the next Tor node and then removes layers of encryption as the message is delivered to the intended recipient (see Figure 8-4).

Tor became the Internet communication of choice for those trying to avoid attribution, whether part of a resistance movement or criminal activity such as child pornography. Tor remains in use around the world due to its regular success at retaining the anonymity of its users.

> 8.6   *Resistance movements can avoid most forms of technical*
> *attribution by using Tor to facilitate communications.*

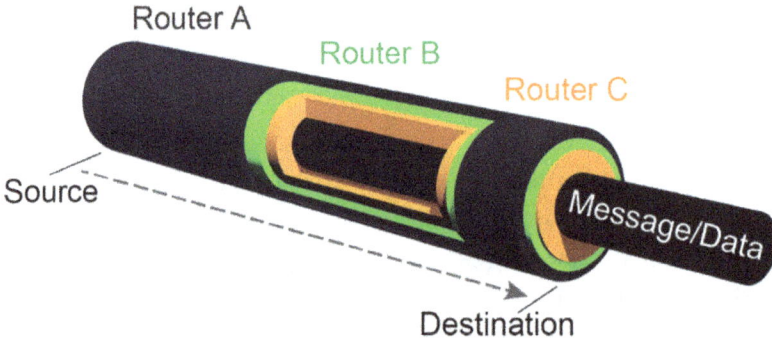

Figure 8-4. Layers of encryption used in Tor.

Leveraging Online Criminal Websites

Tor has been used by a number of illicit sites on the dark web, such as Silk Road, Alpha Bay, and Hansa. Follow-on analysis from the Alpha-Bay and Hansa deactivations indicate that the seizure of one dark web site increases the use of others that have yet to be shut down.[44]

Silk Road was a Tor-hosted website used as an online marketplace to execute anonymous transactions for illegal drugs, forged documents, and illegal services such as hackers.[45] All exchanges on Silk Road were executed via Bitcoin to avoid any traditional currency transaction tracking methods.[46]

The government conducted a two-year investigation that concluded with the arrest of the operator of Silk Road and the disestablishment of the website. Authorities located Tor software associated with servers in Iceland, Romania, and Latvia.[47] Through mutual legal assistance treaties, the government received from law enforcement copies of the servers in the foreign locations. Technical attribution of the foreign servers enabled investigators to map the Silk Road network and its transactions.[48] Federal authorities conducted technical and human attribution, which enabled undercover agents to penetrate the website, locate and dismantle web servers, and identify and arrest of the website operator.[49, 50]

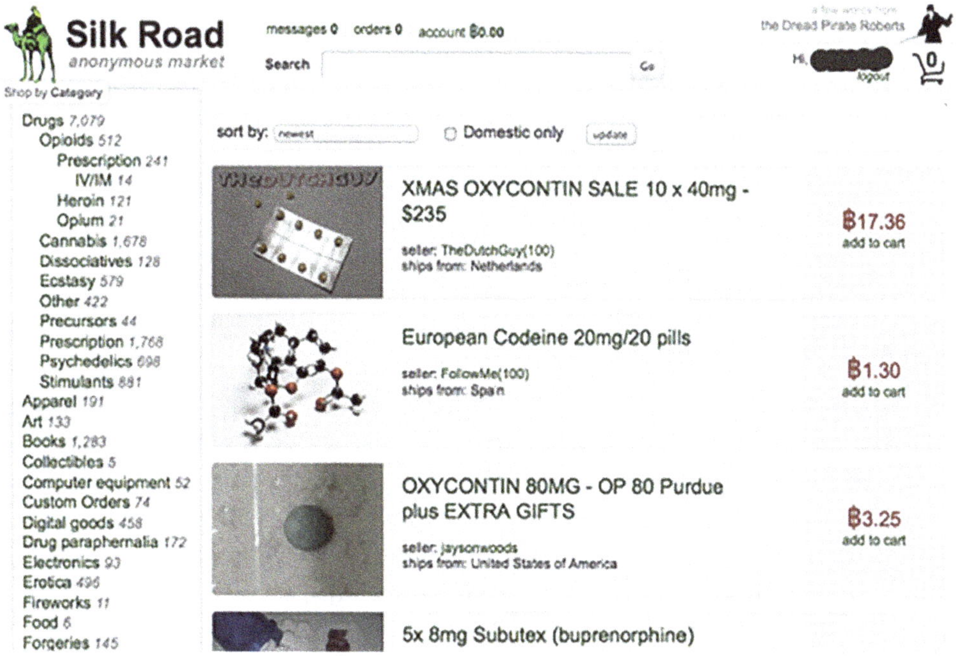

Figure 8-5. Silk Road webpage screenshot.

In July 2017, the black market sites AlphaBay and Hansa were seized and taken offline. Both of these sites operated as Tor hidden services and were accessible only via Tor clients. International cooperation between law enforcement agencies, undercover agents, and human attribution enabled the sites to be located, supporting devices seized, and the sites deactivated.

AlphaBay had over two hundred thousand users and forty thousand vendors before being shut down.[51, 52, 53] The site was used to sell malware, controlled substances, chemicals, weapons, stolen financial information, and counterfeit documents.[54] AlphaBay was upfront about being a black market, advising vendors how to avoid law enforcement detection.[55]

Concurrently, Dutch law enforcement seized and deactivated the black market site Hansa.[56] Hansa was shut down to prevent the sale of drugs, weapons, and malware.[57] Dutch authorities located and arrested site administrators in Germany and seized servers supporting the websites in the Netherlands, Germany, and Lithuania.[58]

Resistance movements can also use an online, criminal, commercial site as the basis to communicate tasks to be undertaken and compensation provided to its distant supporters. At the same time, resistance movements need to disguise their identities if and when such an online criminal network is seized so that the transactions by resistance movements with that site remain unattributable to its members. This leads to Key Takeaway 8.7.

> *8.7   Resistance movements can leverage online criminal sites for hiring hackers, buying weapons, laundering money, or other activities and should hide their identities even when using these sites.*

Employing Cyber Mercenaries

Resistance movements can leverage online, criminal, commercial site for goods and services and directly hire cyber mercenaries. Resistance personnel frequently do not have to be highly skilled in advanced cyber operations. Instead, a resistance movement can hire a cyber mercenary group to execute cyberspace operations on its behalf. The IT magazine *Dataquest* assesses that cyber mercenaries will continue to be sought, grow in number, and increase in usage due to employment simplicity, offering their skills to any third party willing to pay. Kaspersky Laboratories also assesses that cyber mercenaries and other hit-and-run-like groups will continue to grow in number and make themselves available for hire.[59]

Kaspersky Laboratories named and tracked a group it called Icefog since 2011. Icefog is an advanced, persistent threat detected in the networks of the Japanese and South Korean governments. Icefog is assessed to be a small group of highly skilled technical personnel possessing the ability to penetrate a network, pinpoint the targeted data, or precisely generate the effect they want, and then exfiltrate in a hit–and-run-like tactic. Most attacks have been in South Korea and Japan and involved data exfiltration from specific computers.[60] Variants of Icefog code included code that sought and interfaced with Korean- or Japanese-language programs. Icefog attacks appeared to aim against specific targets from which data was exfiltrated, and then the target was disengaged. Kaspersky Laboratories assessed that Icefog knew in advance what was desired from its victims.[61] Since its discovery, additional Icefog activity was detected in the United States, China, Australia, the United Kingdom, Italy, Germany, Austria, Singapore, Belarus, and Malaysia.[62]

Additionally, *Dataquest* assesses that cyber mercenaries will sell digital access to hacked high-profile targets as an access service scheme.[63] Pre-accessed targets potentially decrease the time to implement a cyber attack by having third parties select targets already accessible vice paying cyber mercenaries to access a currently unaccessed target.

> 8.8   *Resistance movements can significantly increase their offensive cyber capabilities by hiring mercenary hackers to perform attacks or provide initial access to desired targets.*

Using Technical Misattribution

Another technique used by malicious actors is to appear to be someone they are not or to disguise the online source and path of their action.[64] As an example of trying to misattribute the identity of the originator, code is available to obfuscate configuration files, such that configuration files written originally in one language (such as English) is published instead in another language (such as Chinese).

Disguising the source IP used by the resistance movement is also essential to misattribution. The Vault7 cyber tool released on WikiLeaks, for example, enables an actor to execute an attack from an IP space but disguises it as an attack from another.[65] Resistance movements can also misattribute the path of their activities by compromising other machines as hop points for their operations. Hop points are compromised cyber assets that are located anywhere in the world. An actor in Russia, for example, could compromise a device in China and launch a cyber attack against the United States from a Chinese-owned IP address. Much like burner phones, resistance movement hop points will likely be used only once or a few times before being abandoned. This ability to quickly shift the apparent path of the communications helps prevent the state security services from tracing back the path from the target through the hop point to the real source of the cyber action by the resistance movement.

For example, a cyberspace attack can be directed from an actor physically residing in Argentina, initiate an attack from Australian IP space, and affect a target in the United States (see Figure 8-6). Separating the attack source and the target enables asymmetric warfare because an actor can put global distance between its location, the target's location, and the connectivity to the target.[66]

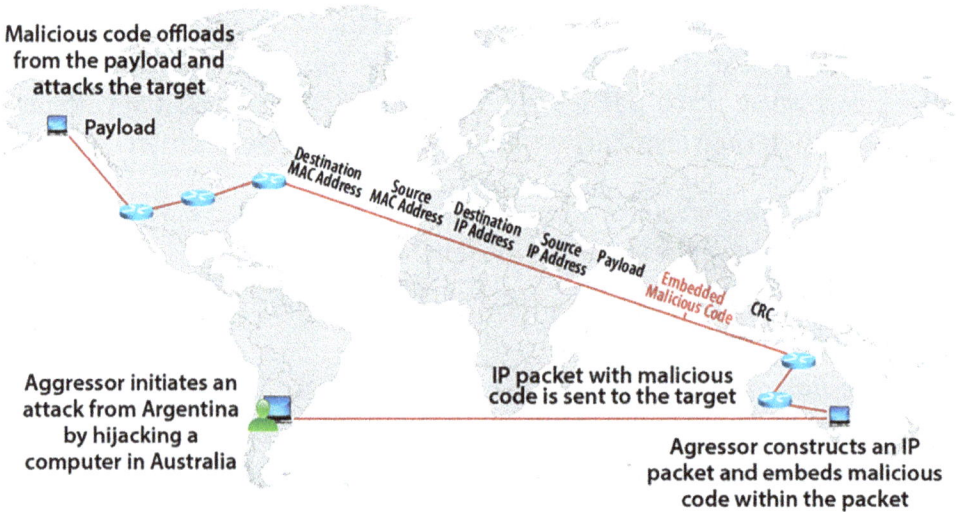

Figure 8-6. Source and path misattribution

> *8.9   Resistance movements need to leverage identity, source, and path data to preclude being attributed by state security services.*

Blending into a Crowd

Another way for a resistance movement to avoid attribution is to use tools available to anyone around the world. This approach enables an actor to blend into a crowd. For example, the Vault7 releases mentioned earlier include tools reportedly used against forty targets across sixteen countries.[67] The Vault7 postings enable anyone to download the attack tools, thereby making any such user a suspect.

In a similar manner, using open-source tools, such as MetalSploit described earlier, can hinder state security services from performing technical attribution to the resistance movement. However, as previously mentioned, simply possessing such tools may be fatal when under a repressive regime.

> *8.10 Using publically available tools can help make it difficult for the security services to perform technical attribution against resistance members, but the very possession of such tools might be sufficient cause for persecution by a repressive regime.*

Facilitating Multiple Actors against the Same Target

Another attribution consideration is that there may be multiple unaffiliated actors attacking the same target. Such a situation can arise when an individual, organization, or device is attacked by multiple hostile parties, but the attacks are not a synchronized effort. For example, during the 2000 Palestinian-Israeli cyber conflict, hackers from around the world joined in attacking either Israeli or sometimes Palestinian-supportive websites.[68]

Conversely, geographically distributed attacks may be synchronized, such as the Russian-organized DDoS attacks on Georgia in 2008. The Russian-provided botnet software was made available for public use from RBN pages and the website stopgeorgia.ru.[69] Russian "patriots," whether physically living in Russia or abroad, could use the RBN website to attack Georgian websites from their personal computers. The RBN provided the means, but Russian supporters located around the world expanded the number of actors involved.[70] The Russian government denied any involvement or control of the cyber attacks, and some scholars argue that it is difficult to hold the Russian government accountable because many of the attackers were civilians recruited from social media sites.[71]

There are multiple, real-time, Internet-based communications capabilities, such as voice-over IP. The book *Cyber Borders No Boundaries* describes cyberspace as a seamless environment unlike traditional geographic borders.[72] This global reach enables seamless, real-time communication between actors that may be geographically separated by thousands of miles. Because of this real-time communication, resistance coordinators can rally and synchronize global support to a common cause.[73]

If a resistance movement can generate sufficient public support from participants abroad, it may be able to generate a DDoS attack against a target of significant political value. If sufficient numbers of people from countries outside the control of the state security services participate in the attack, this magnifies the effect of the attack while protecting the participants from attribution and subsequent retribution.

KEY TAKEAWAYS

While normal communications on the web are fairly easy to attribute the source, path, and sender, there are many techniques available to resistance movements to disguise all of the above, thereby maintaining their anonymity on the web. Operating on the dark web using Tor can significantly increase the probability of maintaining anonymity. However, repressive regimes do not need complete proof of the identity of the sender to identify, arrest, and interrogate likely suspects. Therefore, resistance movement members need to make sure that any encryption, obfuscation, and anonymization software is not readily discoverable as being in their possession. Hiring cyber mercenaries can also provide increased anonymity to resistance movement members.

The following is a summary of the ten principles presented in this chapter:

> 8.1 Resistance movements employ the web, the deep web and the dark web, depending on their purpose and anonymity needs. Most resistance movements rely on the dark web for internal communications due to its strong encryption and counter-attribution capabilities.

8.2 Resistance movements will continue to use botnets for offensive cyber operations, as will cyber criminals and nation-states.

8.3 Resistance movements need to continue to use tools to avoid technical attribution to avoid compromise in the physical world of the individual or the resistance movement.

8.4 Members of resistance movements need to ensure that any attribution, obfuscation, or misattribution capabilities are not readily accessible on their persons, property, or communications devices.

8.5 While encryption can help resistance movements in most cases, especially for protecting data at rest, encryption of communications needs to be used judiciously to not attract unwanted attention from state security services.

8.6 Most resistance movements can avoid most forms of technical attribution by using Tor to facilitate communications.

8.7 Resistance movements can leverage online criminal sites for hiring hackers, buying weapons, laundering money, or other activities and should hide their identities even when using these sites.

8.8 Resistance movements can significantly increase their offensive cyber capabilities by hiring mercenary hackers to perform attacks or provide initial access to desired targets.

8.9 Resistance movements need to leverage identity, source, and path data to preclude being attributed by state security services.

8.10 Using publically available tools can help make it difficult for the security services to perform technical attribution against resistance members, but the very possession of such tools might be sufficient cause for persecution by a repressive regime.

Before the Internet, resistance movements were primarily limited to a geographic region. Through Internet connectivity, the cyberspace environment can link disparate parties to a unified action. Future cyberspace attacks will be difficult to counter as resistance movements maximize resource access through cyberspace and minimize attribution through the use of obfuscation technologies, third parties, and patriotic nationals. Chapter 10 presents a fictional case of "The Red Berets" to describe how a resistance movement can maintain a desired level of attribution, nonattribution, or misattribution.

ENDNOTES

1   In chapter 5, we described that the government-sponsored "50-Cent bloggers" often use multiple online personae to argue with themselves online, thereby leading the online dialogue in the desired direction.

2   John Koetsier, "China Bans Internet Anonymity," *Venturebeat.com*, December 28, 2012, https://venturebeat.com/2012/12/28/china-bans-internet-anonymity/.

3   Other underlying technologies of interest to the Cyber Operator include the Domain Name System (DNS) that enables the programmatic resolution of Domain Names (such as www.example.com) into IP addresses (such as 93.184.216.34). Other Internet technologies, such as electronic mail, rely on TCP/IP but not HTTP.

4   The Onion Router (Tor) is a routing system that encrypts the original message with Tor software at the source, sends the message to a Tor router. The first Tor router peels away the initial encryption layer to reveal the next destination/Tor router, encrypts the message again, and sends the message to the next Tor router where this process is repeated. The decryption peel away-re-encryption pattern repeats until the message reaches the last Tor router prior to its destination where the encryption is removed and the message can be read. Additional details about TOR are provided later in this chapter.

5   Alex Hern, "US Government Increases Funding for Tor, Giving $1.8m in 2013," *Guardian*, July 29, 2014. https://www.theguardian.com/technology/2014/jul/29/us-government-funding-tor-18m-onion-router.

6   Associated Press, "A Look at Estonia's Cyber Attacks in 2007," *NBC News*, July 9, 2009. http://www.nbcnews.com/id/31801246/ns/technology_and_science-security/t/look-estonias-cyber-attack/.

7   Kim Zetter, "Hacker Lexicon: Botnets the Zombie Computer Armies Earn Hackers Millions," *Wired*, December 12, 2012. https://www.wired.com/2015/12/hacker-lexicon-botnets-the-zombie-computer-armies-that-earn-hackers-millions/.

8   Ionut Arghire, "IoT Botnets Fuel DDoS Attacks Growth: Report," *Security Week*, January 24, 2017. https://www.securityweek.com/iot-botnets-fuel-ddos-attacks-growth-report.

9   Wentao Chang, An Wang, Aziz Mohaisen, and Songqing Chen, "Characterizing Botnets-as-a-Service," ACM SIGCOMM *Computer Communication Review* 44, no. 4 (August 2014): 585.

10 "Using Spamhaus BGPF in a Production Environment," (Geneva, Switzerland: The Spamhaus Project, April, 2013), 4.

11 John Markoff and Nicole Perlroth, "Firm Is Accused of Sending Spam, and Fight Jams Internet," *New York Times*, March 26, 2013, https://www.nytimes.com/2013/03/27/technology/internet/online-dispute-becomes-internet-snarling-attack.html.

12 G. Shrikanth, "Cyber Mercenaries to Challenge Enterprise Security in 2016," *Dataquest*, November 23, 2015, https://www.dqindia.com/cyber-mercenaries-to-challenge-enterprise-security-in-2016/.

13 James R. Clapper, *Statement for the Record Worldwide Threat Assessment of the US Intelligence Community* (Washington, DC: House Appropriates Subcommittee on Defense, March 24, 2015).

14 Zack Whittaker, "Researchers Obtain a Command Server Used by North Korean Hacker Group," *TechCrunch*, March 3, 2019, https://techcrunch.com/2019/03/03/north-korea-lazarus-hackers/?renderMode=ie11; Catalin Cimpanu, "How US Authorities Tracked Down the North Korean Hacker

Behind WannaCry," *ZDNet*, September 6, 2018, https://www.zdnet.com/article/how-us-authorities-tracked-down-the-north-korean-hacker-behind-wannacry/.

15 Kim Zetter, "The Sony Hackers Were Causing Mayhem Years Before They Hit the Company," *Wired*, February 24, 2016, https://www.wired.com/2016/02/sony-hackers-causing-mayhem-years-hit-company/.

16 "North Korea Denies 'Righteous' Hack Attack on Sony," *BBC News*, December 7, 2014, https://www.bbc.com/news/world-asia-30366449.

17 Ibid.

18 Michael Edison Hayden, "A Timeline of the WannaCry Cyberattack," *ABC News*, May 15, 2017, https://abcnews.go.com/US/timeline-wannacry-cyberattack/story?id=47416785.

19 Andy Greenberg, "The Untold Story of NotPetya, the Most Devastating Cyber-attack in History," *Wired*, August 22, 2018, https://www.wired.com/story/notpetya-cyberattack-ukraine-russia-code-crashed-the-world/.

20 W. Earl Boebert, "A Survey of Challenges in Attribution," in *Proceedings of a Workshop on Deterring Cyber Attacks: Informing Strategies and Developing Options for US Policy*, ed. National Research Council (Washington, DC: The National Academies Press), 49.

21 "Cyber Banking Fraud Global Partnerships Lead to Major Arrests," FBI, October 1, 2010, https://archives.fbi.gov/archives/news/stories/2010/october/cyber-banking-fraud.

22 Jawad Shamsi, Fareha Sheikh, Sherali Zeadally, and Angelyn Flowers, "Attribution in Cyberspace: Techniques and Legal Implications," *Security and Communication Networks* 9, no. 15 (April 26, 2016): 5.

23 Boebert, "A Survey of Challenges in Attribution," 43.

24 Shamsi et al, "Attribution in Cyberspace: Techniques and Legal Implications."

25 Dmitri Alperovitch, "Bears in the Midst: Intrusion into the Democratic National Committee," *CrowdStrike*, June 15, 2016, https://www.crowdstrike.com/blog/bears-midst-intrusion-democratic-national-committee/.

26 "What is Stuxnet?," *McAfee*, https://www.mcafee.com/enterprise/en-us/security-aware-ness/ransomware/what-is-stuxnet.html.

27 Kevin Townsend, "Microsoft Proposes Independent Body to Attribute Cyber Attacks," *Security Week*, July 6, 2016, https://www.securityweek.com/microsoft-proposes-independent-body-attribute-cyber-attacks.

28 Collin S. Allan, "Attribution in Cyberspace," *Chicago-Kent Journal of International and Comparative Law* 13, no. 2 (November 2013): 59.

29 John Markoff, "Before the Gunfire, Cyberattacks," *New York Times*, August 12, 2008, https://www.nytimes.com/2008/08/13/technology/13cyber.html.

30 Ibid.

31 Scott Depasquale and Michael Daly, "The Growing Threat of Cyber Mercenar-ies," *Politico*, October 12, 2016, https://www.politico.com/agenda/story/2016/10/the-growing-threat-of-cyber-mercenaries-000221.

32 Heather Kelly, "83 Million Facebook Accounts are Fakes and Dupes," *CNN*, August 3, 2012, https://www.cnn.com/2012/08/02/tech/social-media/facebook-fake-accounts/index.html.

33 Jim Finkle, "Iranian Hackers Use Fake Facebook Accounts to Spy on U.S., Others," *Reuters*, May 29, 2014, https://www.reuters.com/article/us-iran-hackers/iranian-hackers-use-fake-facebook-accounts-to-spy-on-u-s-others-idUSKBN0E90A220140529.

34 Ibid.

35 Ibid.

36 Shamsi et al, "Attribution in Cyberspace: Techniques and Legal Implications."

37 Spencer Ackerman, "US Central Command Twitter account hacked to read 'I love you ISIS'," *Guardian*, January 12, 2015, https://www.theguardian.com/us-news/2015/jan/12/us-central-command-twitter-account-hacked-isis-cyber-attack.

38 Nicole Perlroth, "Security Experts Oppose Government Access to Encrypted Communication," *New York Times*, July 7, 2015, https://www.nytimes.com/2015/07/08/technology/code-specialists-oppose-us-and-british-government-access-to-encrypted-communication.html.

39 James Comey, "Encryption, Public Safety, and 'Going Dark'," *Lawfare*, July 6, 2015, https://www.lawfareblog.com/encryption-public-safety-and-going-dark.

40 Sherisse Pham, "Telegram Promises to Act Faster on Terror Content as Indonesia Threatens Ban," *CNN*, July 17, 2017, https://money.cnn.com/2017/07/17/technology/telegram-block-indonesia-terrorism/index.html.

41 "Russia Says Telegram App Used in Saint Petersburg Attack," *MSN.com*, June 26, 2017.

42 Pham, "Telegram Promises to Act Faster on Terror Content as Indonesia Threatens Ban."

43 Ibid.

44 Leo Kelion, "Dark Web Markets Boom after AlphaBay and Hansa Busts," *BBC News*, August 1, 2017.

45 Donna Leinwand Leger, "How FBI Brought Down Cyber-underworld Site Silk Road," *USA Today*, October 21, 2013.

46 Bitcoin is a blockchain-based cryptocurrency. Bitcoin can be purchased with actual currencies (e.g. US dollars) and used for financial transactions. Being a digital non-state currency, Bitcoin is not government-regulated and offer further anonymity as they are exchanged user to buyer and not through a bank or payment service.

47 Leger, "How FBI Brought Down Cyber-underworld Site Silk Road."

48 Ibid.

49 Ibid.

50 Ibid.

51 John Leyden, "Cops Harpoon Two Dark Net Whales in Megabust: AlphaBay and Hansa," *Register*, July 20, 2017.

52 Joseph Cox, "Stolen Uber Customer Accounts Are for Sale on the Dark Web for $1," *Motherboard*, March 27, 2015.

53 David Nield, "Stolen Uber Accounts on Sale for $1 Each," *Digital Trends*, March 28, 2015.

54 UNITED STATES OF AMERICA v. ALEXANDRE CAZES, Verified Complaint for Forfeiture in REM, Case 1:117 at 00557 document 1, filed July 19, 2017.

55 Ibid, 10.

56 Kelion, "Dark Web Markets."

57 Teri Robinson, "International Operation Takes Down AlphaBay, Hansa Dark Web Markets," *SC Magazine*, July 20, 2017.

58 Ibid.

59 Ibid.

60 Kaspersky Lab Global Research and Analysis Team, "The 'Icefog' Apt: A Tale of Cloak and Three Daggers," Version 1.00, 2013, https://media.kaspersky.com/en/icefog-apt-threat.

pdf; Kaspersky, "Kaspersky Unmasks 'Icefog' Cyberattacks," 2019, https://usa.kaspersky.com/resource-center/threats/icefog-cyberattacks.

61 Kaspersky, "The Icefog."

62 Kenneth Rapoza, "Kaspersky Lab Uncovers New Cyber Hit-n-Run Op Called 'Icefog,'" *Forbes*, September 25, 2013.

63 Shrikanth, "Cyber Mercenaries," 2.

64 Ibid.

65 The Latch Key, "It is IMPOSSIBLE to 100% Verify the Origin of a Cyber Attack. Vault 7 Documents Prove You Can Spoof Certain Variables to Obfuscate the Original Location," *Reddit Blog*, March 31, 2017.

66 Ronald Keys and Kendra Simmons, "Cyberspace Security and Attribution," National Security Cyberspace Institute, July 20, 2010, 4.

67 Symantec, "Longhorn: Tools Used by Cyberespionage Group Linked to Vault 7: First Evidence Linking Vault 7 Tools to Known Cyberattacks," *Symantec Official blog*, April 10, 2017.

68 Patrick D. Allen and Chris C. Demchak, "The Palestinian-Israeli Cyberwar," *Military Review* 83, no. 2 (March-April 2003).

69 Markoff, "Before the Gunfire."

70 Allan, "Palestinian-Israeli Cyberwar," 62.

71 Ibid., 76.

72 Timothy R. Sample and Michael Swetnam, *#CyberDoc No Borders-No Boundaries* (Arlington, VA: Potomac Institute Press, 2012), 12.

73 Ibid.

CHAPTER 9.
LEGAL ISSUES IN CYBER RESISTANCE

INTRODUCTION

The growth of information technologies and their increasing use worldwide by civilian populations, US allies, and adversaries continues to have a significant impact on military operations. These technologies are accessible to non-state adversaries who can leverage them for anything from intelligence collection to recruitment, often without legal and policy impediments. For SOF, cyberspace activities, or IRCs[1] are relevant across core SOF missions. However, as the focus of this compendium is on cyber and resistance, the discussion focuses on IRC as related to UW,[2] particularly in the context of advice and support to resistance groups. In addition, MISO, which may rely extensively on IRCs to influence foreign governments, organizations, and groups, is also relevant, especially as it may synchronize with UW.

The objective of this chapter is to highlight key legal and policy issues that arise in the specific context of IRCs used in support of resistance. The goal is to give the reader an understanding of important legal considerations and identify emerging issues around unsettled law or policy. It is important to note that popular media and some resistance groups themselves may refer to activities using the term "cyber." This term may be loosely seen as a colloquial and common word for IRC, meaning the means through which a group acts to affect the information environment. As used here, the term "resistance" refers generally to nonviolent or armed opposition to a standing government and all actors that may be components thereof. It is important to note that more broadly within the US military, IRCs may be part of IO when used in an integrated way during military operations "in concert with other lines of operation to influence, disrupt, corrupt, or usurp the decision making of adversaries."[3] IO operations are conducted in accordance with specific executive orders, the details of which are not discussed here. Rather, the focus remains on the implications for SOF conducting activities as part of IO.

CONTEXTS IN WHICH LEGAL QUESTIONS EMERGE

Groups may use cyber for various purposes, for example, webpages or social media as a venue for recruitment, as a means for group leaders to disseminate information and distribute orders, or for followers to self-organize. It may be used to enable extraterritorial growth of intelligence networks, conduct training via videos of techniques and the posting of manuals, and provide a mechanism for financing through digital currency such as Bitcoin. Cyber may also be a mechanism for launching an attack that may corrupt a website or computer software or, at the more extreme end, impact physical infrastructure, such as interruption of the normal operations of a government building, or temporary power outages due to an attack on a portion of the electric grid. The activities can grow in scale and effect depending on the sophistication of the group. These examples are essentially IRCs used by resistance groups, and they are also some of the capabilities SOF may be advising or supporting a group to develop or use as part of a UW campaign.

Particularly for special forces operators deployed to advise and support resistance movements, it is important for them to be generally aware of the potential legal consequences of actions a resistance group may take. For any IRCs used within the context of UW, key questions include: what is being done, who is conducting the activity, where is it being conducted, will the actions open the group to criminal charges, is it part of an ongong conflict, does international humanitarian law (IHL) apply, and can the actions be traceable to the US military? Another consideration is whether US domestic law imposes restrictions on the actions of US military personnel.

> *9.1 For IRCs used in the context of UW, it is important to understand in advance: what action is being taken, who is conducting it and where, and whether the actions may expose the resistance or SOF personnel to domestic criminal charges or violations of international law.*

Social Media

The use of social media by resistance groups is widely documented, and these platforms are an important tool for disseminating or countering narratives. The analysis of social media also has significant potential for US military operations by providing insight into how large audiences communicate and on which issues. However, there are several issues with the use and analysis of social media. The DoD is prohibited to collect on communications of US persons, both under Title 10 limitations on domestic operations and under Executive Order 12333, which prohibits the acquisition, storage, and dissemination of US person data. These prohibitions do not account for the modern information sharing environment, where the co-mingling of US and foreign actor communications, is the norm, as characterized by social media.

In the context of support to a resistance, if SOF wanted to analyze social media data to assess the demographics a group is targeting and whether the group could improve its messaging, it is quite possible the collection of that information for analysis, whether done by SOF or another DoD element in support of the UW campaign, could result in the accidental collection of communications of US persons. Moreover, aside from analysis of social media, the DoD cannot target the US public with information aimed at foreign audiences.[4] Therefore, a relevant question becomes, if SOF is supporting a group in developing a social media message, because of the reach of social media, could their efforts to disseminate the message inadvertently constitute propaganda toward the US public?

A military information support team (MISO) charged with counter-messaging could run into the same scenario, and indeed a DoD contractor conducting analysis of opportunities to perform MISO suggested that the US military develop a messaging campaign based on comments posted on a website of a Somali-American, who was later investigated by the FBI for potential support to terrorism.[5] Here, the problem with intermingling of domestic and overseas information is underscored. In the case of support to resistance, it is possible the SOF personnel, in an advising capacity, are a step removed such that they would not be implicated if this issue arose, but it is important to understand the vulnerability. As discussed later, existing international legal standards are informative when it comes to analyzing the accountability of US personnel for certain actions overseas.

Figure 9-1 provides a summary explanation of the various categories of resistance and the activities that affect a group's status. It also summarizes the applicable body of law that applies and when it applies, although the application of the law is highly dependent on the context of the situation, especially in cyber. It is included as a helpful reference for scenarios detailed later in this chapter.

International Humanitarian Law and Cyber

The law that applies in a given campaign or operation varies depending on the status of the group and the character of the underlying conflict, if any. Traditionally, there are four categories within which a resistance movement may fit: nonviolent resistance, rebellion, insurgency, or belligerency.[6] IHL, also referred to as the law of armed conflict, applies (albeit differently) in the last two categories, but if the resistance movement does not cross those thresholds, then the domestic criminal and civil law of the country where the actions take place apply. IHL imposes obligations on resistance movements in exchange for protections, and the US government position is that US personnel will abide by IHL at all times.[7] However, in cyber, a group's status is more difficult to determine, as is the determination on the origin of the action. Without the ability to identify the group's status and where the action originated, it is nearly impossible to determine whether a country's domestic law applies. Moreover, in cyber, it is often difficult to accurately attribute activities to a specific group or person.

Increasing level of intensity, duration, and organization

Nonviolent Resistance

Use of Legal Processes for Political Advantage

Characteristics: Individuals or groups use legal processes to resist standing government, e.g., social media messaging, peaceful demonstration, canvasing polls.

Corresponding Legal Status: Individuals are subject to HN law. Indigenous status would be as a citizen/resident. UW status is as a tourist subject to HN jurisdiction. FID status would depend on any applicable SOFA.*

Third-Party Involvement: Foreign government support unlawful unless HN consents; discovery of the presence of personnel could prompt diplomatic problems, charges of espionage.

Illegal Political Acts

Characteristics: Individuals or groups resorting to illegal political acts to resist standing government, e.g., refusal to comply with certain laws (civil disobedience) or other disruptive, nonviolent acts.

Corresponding Legal Status: Individuals are subject to HN civil and criminal law. Indigenous status would be as a citizen/resident. UW status is as a tourist subject to HN jurisdiction. FID status depends on any applicable SOFA.* Diplomatic channels can be used to negotiate jurisdiction or release.

Third-Party Involvement: Foreign support to domestic criminals unlawful unless HN consents; discovery of support could prompt diplomatic tension, accusations of aggression, charges of espionage.

Armed Resistance

Rebellion

Characteristics: Short-term, isolated, violent engagements of low intensity by a group (e.g., riots); law enforcement mechanisms are able to suppress the violence; it remains a domestic matter.

Corresponding Legal Status: Rebels are subject to domestic criminal law. UW status is as a tourist subject to HN jurisdiction. FID stand alongside the HN government under applicable SOFA.* Diplomatic channels may be used to negotiate jurisdiction or release. NOT yet an armed conflict so IHL does not apply.

Third-Party Involvement: Support of rebels violates HN sovereignty and contravenes international norms of noninterference, with some exceptions. Support to HN by invitation or with consent is permissible.

Insurgency

Characteristics: Recognition of an insurgency is based on facts and political factors. In general, the fighting is more sustained and intense and cannot be easily suppressed by the government. Other elements include increased levels of the insurgent group's organization and territorial control.

Corresponding Legal Status: As a NIAC, the IHL protections of Common Article 3, and potentially Additional Protocol II, apply. Parties can agree to apply more protections but not fewer.

Third-Party Involvement: Support of insurgents violates HN sovereignty and contravenes norms of noninterference, with some exceptions. Support to HN via COIN permissible with consent. States may engage with insurgents to protect property, commercial interests, nationals.

Belligerency

Characteristics: (1) A general as opposed to local armed conflict, (2) belligerents administer a substantial portion of territory, (3) belligerents follow laws of war and use a command system, and (4) circumstances require states to define their positions in relation to the conflict.

Corresponding Legal Status: LOAC applies. The resistance group is deemed a de facto state if the host nation recognizes the resistance as a belligerency, and its forces receive combatant/POW status. The resistance group and HN and allies are bound to apply LOAC, the customary law of international armed conflict.

Third-Party Involvement: Recognition of a resistance group as a de facto state imposes a duty of neutrality on third-party states. Third parties may support one side or the other, but doing so constitutes an act of war.

* SOFA or other international agreement.

COIN–Counterinsurgency	IHL–International humanitarian law	POW–Prisoner of war
HN–Host nation	LOAC–Law of armed conflict	SOFA–Status of forces agreement
FID–Foreign internal defense	NIAC–Noninternational armed conflict	UW–Unconventional warfare

Figure 9-1. Categories of resistance and corresponding legal protections.

IHL may become applicable to cyber operations in two ways: (1) if the cyber operations accompany an existing armed conflict or (2) they trigger an armed conflict on their own. UW involves a resistance movement against a state government, and while UW could be part of a larger interstate war, it is often noninternational in nature, meaning the full scope of the Geneva Conventions do not apply.[8] Common Article 3 to the Geneva Conventions defines noninternational armed conflicts only in the negative as those "not of an international character."[9] Fortunately, the International Criminal Tribunal for the Former Yugoslavia laid down a more helpful standard, which requires two criteria be met, organization and intensity.[10] Figure 9-2 depicts this spectrum.

Level of Organization and Impact on Status

IHL's standard for organization requires a command structure and coordination of activities.[11] The fighting by an organized group should have a collective character, as opposed to individuals operating separately.[12] Cyber operations with a collective character and a command structure might feature the allocation of targets, developing and sharing tools, cooperating in identifying the adversary's vulnerabilities, or conducting post-attack assessments.[13] The requirement of organization would not be satisfied, however, by a large group of independent actors all targeting the government in response to a call to action because those actors would not be under the direction of an individual or a command structure.[14]

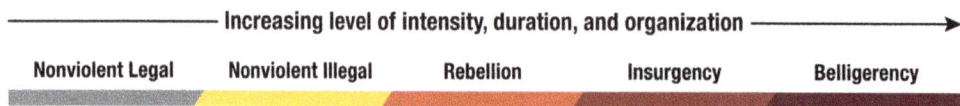

Increasing level of intensity, duration, and organization ⟶

| Nonviolent Legal | Nonviolent Illegal | Rebellion | Insurgency | Belligerency |

Figure 9-2. Spectrum of status and insurgency criteria (intensity, duration, organization).

Consider two hypotheticals. First, a popular opposition leader issues a call on social media for his or her followers to disrupt and interfere with upcoming elections because they are viewed as illegitimate. In response, hundreds of cyber-savvy supporters take it upon themselves to begin conducting DDoS attacks, hacking government websites to deface them or change the information on them and manipulating government messages. This would likely not qualify as organized because even though the supporters have undertaken these actions at the request of the popular opposition leader, their efforts are uncoordinated and independent; they are not directed by any form of leadership. No command structure prescribed or coordinated actions or actors. Indeed, one can see opportunity not only for overlap among the supporters' actions but also interference among their efforts.

Second, the same popular opposition leader issues the same call on social media for followers to disrupt and interfere with upcoming elections. This time, though, two enterprising leaders in the opposition community direct specific individuals to undertake particular actions. They deconflict and assign targets, and they instruct those individuals to undertake certain missions based on their skillsets. Then they ask different supporters to conduct post-attack assessments to determine which tactics demonstrated success. All the individual supporters perform the tasks these two enterprising leaders tell them; they follow orders. This scenario may exhibit a sufficient level of organization to satisfy the legal standard because it includes defined leaders exercising control through a command structure, instead of individuals separately taking actions without any coordination.

These two cases are the two ends of a spectrum with fairly straightforward answers. The difficult cases lie in between these where groups engage in some method of coordination, but it may not be standardized or consistently followed. In those difficult cases, the determination resides with the particular facts. The important takeaway is that whether or not a group is organized, it may be able to achieve intense effects. A traditional interpretation of the law does not consider especially small groups that can achieve outsized impacts. This is a potential gap.

In the cyber domain, it is common for individuals to organize virtually, and it is not necessary to meet in person. It remains unclear whether the organization element requires the group to meet physically and whether virtual organization will be sufficient.[15] The court case establishing these two criteria, however, does not explicitly state a requirement to organize physically in person,[16] so virtual organization as put forth in the hypothetical likely satisfies the requirement. Ultimately, the organization requirement would rule out the relevance of IHL for any cyber attacks performed by non-military, private, individual hackers or groups of hackers operating without coordination and leadership. However, this is perhaps an area for further development, as cyber actions may be capable of causing the same level or more damage as a conventional attack, but without the level of organization required to constitute an insurgency from a legal perspective. Otherwise, only the domestic laws of the country in which they operate and which they attack would apply.

Intensity and Impact on Status

The second criterion, intensity, requires "protracted violence between government authorities and organized armed groups or between such groups within a State."[17] However, "internal disturbances, and tensions, such as riots, isolated and sporadic acts of violence and other acts of a similar nature," do not meet the intensity requirement.[18] Two parts of this standard are difficult to analyze for cyber attacks. First, there must be violence. Scholars make the argument that violence in this context means physical destruction.[19] Accordingly, cyber operations that do not cause physical destruction may not count toward creating a noninternational armed conflict.[20] Second, the violence must be protracted. This presents a particular obstacle for cyber operations because they are often sporadic or occur within short time windows.

> 9.2   The organization and intensity of a group's actions is based on all of their activities, not strictly the cyber activities. Therefore, the requirements of organization and intensity can be satisfied by non-cyber activities, or in conjunction with cyber activities, and IHL applies in that case to the cyber actions.

Consider the hypothetical discussed under the requirement of organization. In that hypothetical, the cyber operations defaced government websites and interfered with their functioning, but they did not destroy physical materials. To satisfy the intensity criterion under the traditional interpretation requires physical destruction. It is unclear under the law whether destruction of data qualifies, but the current argument put forward by scholars holds that the destruction must be analogous to destruction by traditional kinetic weapons.[21] Moreover, for that hypothetical scenario to meet both criteria requires the cyber operations by the opposition not only to physically destroy government objects but to do so over an extended, or protracted, period of time.[22] This means one-off, isolated cyber operations do not satisfy, no matter how intense. It should be noted that the analogous kinetic operations could not be expected to be continuous without any interruption, but must be a sustained series of operations. Likewise, the criterion could be satisfied by a sustained series of destructive cyber operations.

However, the criteria of organization and intensity could be met if the group's activities are analyzed as a whole. It is possible, perhaps even likely, that IRCs are only a part of a group's strategy and, taken with additional actions, pushes the group across the threshold where broader IHL protections apply.

Even though the United States has not signed Additional Protocol II (APII) to the Geneva Conventions, many other countries have, so it is worth noting that APII imposes an additional requirement of controlling enough territory so that the group can carry out sustained and concerted military operations.[23] Cyber operations alone could not meet this requirement. If a special forces operator advises a resistance movement in a country that has signed APII, then cyber operations that meet the requirements of organization and intensity may still not create a noninternational armed conflict that requires applying IHL if the group is not also controlling territory.

The importance of these considerations is that if the cyber operations accompany an existing kinetic armed conflict or create an armed conflict on their own, then IHL applies to cyber operations just like any other operations. If there is no armed conflict, however, only the domestic law of the country in which the resistance takes place applies. Consider the implications of this by walking through three relevant rules of IHL: distinction, proportionality, and direct participation in hostilities.

The rule of distinction means that operations must not target civilian persons and objects.[24] The rule of proportionality deals with the permitted level of loss to civilian life and objects as collateral damage when military personnel and objects are the main target.[25] Without doubt, these principles apply to cyber operations during armed conflicts. How these apply to cyber operations can become difficult. For instance, the rule of distinction means that civilians and civilian objects are not to be targeted, but a question arises as to whether data constitute objects. If they do, then cyber operations must avoid or limit as much as possible any damage to civilian data and civilian data networks. This can be severely limiting because the majority of data networks and other infrastructure across the world are civilian, including those used by militaries. An exception exists, however, for war-sustaining activities and facilities.[26] Traditionally, this permitted targeting factories and warehouses known to produce war materials, such as tanks and munitions, or shipping lanes used to transport personnel and weapons.

Applying this exception in the cyber domain challenges the parameters of this exception. It would seem to permit targeting banks online if those banks are known to provide or channel funding to the adversary.[27] However, consider the vast impact to civilians of a cyber attack by a resistance group on a central banking system. The war-sustaining activity exception to the rule of distinction may appear to permit such an attack, but the rule of proportionality requires the resistance group to limit the impact to civilians.

A common example given about the potential destructive power of cyber operations is interruption to or taking control of infrastructure that relies on networks, such as power grids and dams. Two rules under customary international law IHL prohibit such operations. First, one cannot attack, destroy, remove, or render useless objects indispensable to the survival of the civilian population.[28] This would mean a resistance movement subject to IHL could not use cyber operations to take out a power grid because the grid could be argued to be indispensable to civilian populations. This would not necessarily preclude any and all attacks on power grids. If, for instance, a sophisticated cyber operator could target a limited portion of the grid that supported military installations, that would likely be permissible. This rule would also mean the resistance movement subject to IHL cannot hijack a dam and prevent water from reaching civilians.

Second, one cannot attack structures that contain dangerous forces if the attack releases the dangerous force and causes severe loss to the civilian population.[29] This would mean that a resistance group cannot use cyber operations to hijack a dam and release the water to destroy down river objects if doing so leads to severe loss in the civilian population.

If the resistance movement does not rise to the level of at least an insurgency (see Figure 9-1), then these rules do not apply. Consequently, there would be no rule prohibiting these actions, but the resistance movement would then be subject to domestic criminal and civil law. Thus, if a resistance group uses cyber operations to take down a power grid or hijack and release a dam, those responsible are liable criminally and civilly for the damage, destruction, injuries, and death caused to people and property.

This distinction raises the issue of which law governs nondestructive cyber operations, such as interrupting media, spreading propaganda, or conducting psychological operations. Simply, the IHL does not impose prohibitions or restrictions on these kinds of cyber operations. Indeed, the concept of attacks under IHL does not include embargoes or other non-physical economic warfare either.[30] This does not mean, however, that these kinds of operations are not touched by some source of law. They are subject to the domestic criminal and civil law of the state in which they are undertaken and potentially in the state in which their effect is felt. These laws can include prohibitions on espionage or insurrectionist activities. They can prohibit psychological operations by making certain actions on data networks illegal. Alternatively, some countries maintain laws that make certain speech illegal because it supports terrorism or is considered hate speech. Special forces operators would be well prepared if they know the laws in the country in which they are sent to operate, so that they do not unknowingly place the persons they support in legal jeopardy. They should have that information so they can make an informed decision about whether to take that risk. It is beyond the purview of this chapter to conduct that exhaustive, comparative legal study.

> 9.3 It may not be immediately clear which body of law applies to cyber operations. If certain actions do not invoke IHL, it may be domestic law that applies, which may not be known to special forces advising the resistance.

If only domestic law applies, then cyber operators in resistance movements need to worry about being arrested and prosecuted. If, however, IHL applies, then those in resistance movements conducting cyber operations open themselves to being the target of attack by the national military against which they fight because of the doctrine of direct participation in hostilities; civilians share the protection afforded by the IHL "unless and for such time as they take direct part in hostilities."[31] To be targeted under this rule requires meeting three criteria: (1) "the act must be likely to adversely affect the military operations or military capacity of a party to an armed conflict, or to inflict death, injury or destruction on persons or objects protected against direct attack;" (2) "there must be a direct causal link between the act and the harm likely to result either from that act, or from a coordinated military operation of which that act constitutes an integral part;" and (3) the act must "directly cause the required threshold of harm in support of a party to the conflict and to the detriment of another."[32]

While these criteria are generally accepted, debate arises over how to establish the direct causal link between the act and the harm.[33] For instance, there is disagreement over whether an improvised explosive device maker directly or indirectly causes harm. A similar argument could be made over whether software designers and programmers who create destructive programs and tools cause direct harm. Under this doctrine, cyber operations like military intelligence gathering, disrupting enemy cyber networks, and manipulating military systems qualifies as direct participation, opening the perpetrator to targeting.[34]

Apart from which area of law applies to the cyber operations of a resistance movement, special forces personnel advising these movements abroad should be aware that their connection to the resistance group carries the potential to implicate the United States via the doctrine of state responsibility. That doctrine says essentially that a country can be held responsible and accountable for the actions of irregular forces it supports.[35] It requires a connection between the country and the irregular forces, but the level of connection required has been decreasing over the last few decades. In a case about US support to the contras in Nicaragua, the International Court of Justice (ICJ) said the connection had to be one of effective direction and control.[36] In that case, providing weapons, training, and planning advice did not create enough of a connection to find the United States responsible for the actions of the contras; the United States had to dispatch or direct the contras on operations or be substantially involved in the operations.

> 9.4 US personnel who advise resistance movements, whether the advice pertains to kinetic or cyber operations, can implicate themselves and the United States more broadly, by connection to the resistance movement and the subsequent actions the movement may take based on their advice.

A later case in the International Criminal Tribunal for the Former Yugoslavia lowered the standard for the connection to overall control.[37] In that case, the court held that for the Federal Republic of Yugoslavia to be responsible for the acts of the Bosnian Serb armed groups required more than financing and equipping; it required participation in planning and supervision of military operations. Following September 11, 2001, the United States, supported by North Atlantic Treaty Organization (NATO) and the United Nations (UN) Security Council, invaded Afghanistan in self-defense against al Qaeda without drawing the connection between it and the Taliban government of Afghanistan. The pronouncements of governments and the UN condemned the Taliban government for allowing Afghanistan to be used as a base and sanctuary of al Qaeda, but the self-defense argument was not made on the basis of effective control (the International Court of Justice standard in the Nicaragua case) by the Taliban government over al Qaeda or overall control (the International Criminal Tribunal for the former Yugoslavia standard). Thus, non-state actors who commit an armed attack from within a host nation can trigger the right to self-defense and implicate the nation that hosts them only because it allowed the non-state actors to base there and the host nation did not stop them from committing that armed attack. September 11, 2001 was only one incident to which that standard applied, and it was relatively recent, so the international community may decide in future cases that it was a unique case and requires the overall control test be met. However, when Israel invaded Lebanon in 2006, it did so in response to Hizbollah's terrorism, so the international community appears willing to accept arguments of self-defense in response to non-state group violence.[38] Accordingly, should a non-state group mount cyber operations that meet the threshold of armed attack against a government, and the special forces soldier is identified as being involved by exercising effective control, overall control, or providing the base or safe haven from which to operate, it appears that states are comfortable with holding the United States responsible for the actions of the non-state group because of that connection made by the special forces soldier. This is imperative for a special forces soldier to know because part of the goal of UW is to support irregular armed forces so that the United States does not need to engage directly. Implicating the United States in the resistance group's actions defeats the desire of UW to remain minimally involved in the conflict.

Many of the rules and their consequences mentioned in this chapter depend on the cyber operations being not only detected but also attributed to the perpetrator. The issues of detection and attribution raise serious questions in the law of cyber operations. For instance, an adversary cannot know who to target in response to cyber operations waged against it. Worse, an adversary may not only anonymize its identity and location, but it could also lead forensic analysts toward finding an innocent person or group responsible. This amounts to a false flag operation that has long been forbidden under international law.[39]

Another challenge arises in detection and the difficulty in determining a cyber operation's purpose or when it begins or ends. The doctrine of direct participation in hostilities permits a civilian to be targeted for such time as he or she engages in military operations.

The cyber domain presents two challenges. The first is that some cyber operations may last only seconds or minutes from beginning to end. Others may last for years as the software and tools work through networks to their desired end locations. This leads to the second challenge, determining when a cyber operation begins and ends. Can perpetrators be targeted every time they sit down at a computer? Does the software or other tools have to cause damage before the user is targeted? The right to self-defense suggests otherwise. However, a further problem lies in the ambiguity of cyber weapons. The easy cases are those cyber operations that express clear intentions to impose physical damage and harm on a target or those that clearly express limited purposes of intelligence gathering and no eventual damage or harm to data or physical objects. The difficult cases are those software or tools that hold ambiguous intent or targets. For instance, to effectively attack some targets, reconnaissance is required, so tools are deployed to conduct that reconnaissance. However, it is difficult, if not impossible, to know when those tools are simply in place for the purposes of conducting espionage and when they lay the groundwork for a physically destructive attack. This complicates a response because espionage is a domestic crime, while preparation for a destructive attack may justify self-defense using a military response.

> 9.5 *Cyber strains the traditional legal interpretation of organization and intensity because a group can achieve very high intensity with little organization and few resources.*

KEY TAKEAWAYS

9.1   For IRCs used in the context of UW, it is important to understand in advance: what action is being taken, who is conducting it and where, and whether the actions are likely to expose the resistance or SOF personnel to domestic criminal charges or international law.

9.2   The organization and intensity of a group's actions is based on all of their activities, not strictly the cyber activities. Therefore, the requirements of organization and intensity can be satisfied by non-cyber activities, or in conjunction with cyber activities, and IHL applies in that case to the cyber actions.

9.3   It may not be immediately clear which body of law applies to cyber operations. If certain actions do not invoke IHL, it may be domestic law that applies, which may not be known to special forces advising the resistance.

9.4   US personnel who advise resistance movements, whether the advice pertains to kinetic or cyber operations, can implicate themselves and the United States more broadly, by connection to the resistance movement and the subsequent actions the movement may take based on their advice.

9.5   Cyber strains the traditional legal interpretation of organization and intensity because a group can achieve very high intensity with little organization and few resources.

The status of a resistance group ties to the type of activities it undertakes and whether they are violent or cause destruction. The nature of the activities ties to the character of the conflict (international or noninternational), which determines the body of law that applies to the group and those advising the group. Therefore, it is important to understand the activities a group takes and the legal implications for the resistance, individual special forces, and the US government.

ENDNOTES

1   As described in US Joint Chiefs of Staff, "Information Operations," Joint Publication 3-13 (JP3-13), November 20, 2014, IRCs are "the tools, techniques, or activities that affect any of the three dimensions of the information environment," (physical, informational, and cognitive). See chapter 3 for a detailed description of both cyberspace operations and the information environment.

2   Unconventional warfare "consists of operations and activities that are conducted to enable a resistance movement or insurgency to coerce, disrupt, or overthrow a government or occupying power by operating through or with an underground, auxiliary, and guerrilla force in a denied area" (US Joint Chiefs of Staff, "Special Operations," Joint Publication 3-05 (JP 3-05), July 16, 2014).

3   US Chiefs of Staff, JP 3-13.

4   Ashton B. Carter, "Information Operations" Department of Defense Directive 3600.01, May 4, 2017, 2.

5   Craig Whitlock, "Somali American Caught Up in a Shadowy Pentagon Counterpropaganda Campaign," *Washington Post*, July 7, 2013, https://www.washingtonpost.com/world/national-security/somali-american-caught-up-in-a-shadowy-pentagon-counter-propaganda-campaign/2013/07/07/b3aca190-d2c5-11e2-bc43-c404c3269c73_story.html.

6   Erin N. Hahn and W. Sam Lauber, *Legal Implications of the Status of Persons in Resistance* (Fort Bragg, NC: USASOC, 2015).

7   Stephen Preston, "Department of Defense Law of War Manual" December 2016.

8   Geneva Convention (III) relative to the Treatment of Prisoners of War art. 3, August 12, 1949, 75 U.N.T.S. 135.

9   International Committee of the Red Cross, *Geneva Convention Relative to the Treatment of Prisoners of War (Third Geneva Convention)*, 75 UNTS 135, 12 August 1949, accessed August 14, 2019, https://www.refworld.org/docid/3ae6b36c8.html.

10 Prosecutor v. Tadic, Case No. IT-94-1-A, "Appeals Chamber Decision on the Defence Motion for Interlocutory Appeal on Jurisdiction" (Int'l Crim. Trib. for the Former Yugoslavia: October 2, 1995).

11 Michael N. Schmitt, "Cyber Operations and the *Jus in Bello*: Key Issues," in *International Law and the Changing Character of War*, ed. Raul A. Pedrezo and Daria P. Wollschlaeger, International Law Studies, Volume 87 (Newport, RI: 2011), 98.

12 International Committee of the Red Cross, *Commentary on the Additional Protocols of 8 June 1977 to the Geneva Conventions of 12 August 1949*, eds. Yves Sandoz, Christophe Swinarski and Bruno Zimmerman, (Netherlands: Martinus Nijhoff Publishers, 1987), 511-513.

13 Schmitt, "Cyber Operations and the *Jus in Bello*," 98-99.

14 Ibid.

15 Ibid., 98.

16 Prosecutor v. Tadic.

17 Ibid., para. 70.

18 *Protocol Additional to the Geneva Conventions of 12 August 1949 art. 1, and Relating to the Protection of Victims of Non-International Armed Conflicts*, 1125 U.N.T.S 609, June 8, 1977.

19   Michael Bothe, Karl Josef Partsch, and Waldemar A. Solf, *New Rules for Victims of Armed Conflicts: Commentary on the Two 1977 Protocols Additional to the Geneva Conventions of 1949* (Leiden, Belgium: Martinus Nijhoff Publishers, 1982), 289.

20   Schmitt, "Cyber Operations and the *Jus in Bello*."

21   Michael Schmitt, "'Attack' as a Term of Art in International Law: The Cyber Operations Context," in *4th International Conference on Cyber Conflict*, eds. C. Czosseck, R. Ottis, K. Ziolkowski (Tallinn: NATO CCD COE Publications, 2012), 288.

22   Prosecutor v. Tadic, para. 70.

23   The United States complies with APII despite having not ratified it. See *Protocol Additional to the Geneva Conventions of 12 August 1949, and Relating to the Protection of Victims of Non-International Armed Conflicts*.

24   International Committee of the Red Cross (ICRC), "Chapter 1, Rule 1. The Principle of Distinction between Civilians and Combatants," Customary IHL Database, accessed August 30, 2019, https://ihl-databases.icrc.org/customary-ihl/eng/docs/v1_rul_rule1.

25   ICRC, "Chapter 4, Rule 14, Proportionality in Attack," accessed August 30, 2019, https://ihl-databases.icrc.org/customary-ihl/eng/docs/v1_cha_chapter4_rule14.

26   *Protocol Additional to the Geneva Conventions of 12 August 1949, and Relating to the Protection of Victims of Non-International Armed Conflicts*, art. 52(2).

27   It should be noted that it is the US position, as distinct from the positions other countries, that economic systems can be war-supporting military objectives. Operating in a country with a different interpretation could expose a resistance movement to allegations of war crimes. Schmitt, "Cyber Operations and the *Jus in Bello*," 97.

28   *Protocol Additional to the Geneva Conventions of 12 August 1949, and Relating to the Protection of Victims of International Armed Conflicts*, art. 54.

29   Ibid., art. 56.

30   Bothe, Partsch, and Solf, *New Rules for Victims of Armed Conflicts*, 289.

31   *Protocol Additional to the Geneva Conventions of 12 August 1949, and Relating to the Protection of Victims of International Armed Conflicts*, art. 51.3.

32   Nils Melzer, *Interpretive Guidance on the Notion of Direct Participation in Hostilities Under International Humanitarian Law* (Geneva, Switzerland: International Committee of the Red Cross, 2009), 16-17.

33   Ibid., 52-55.

34   Schmitt, "Cyber Operations and the *Jus in Bello*," 101.

35   International Law Commission, "Draft Articles on Responsibility of States for Internationally Wrongful Acts," Article 8, (Supplement No. 10 (A/56/10), chp.IV.E.1, November 2001), accessed November 30, 2019, https://www.refworld.org/docid/3ddb8f804.html

36   Military and Paramilitary Activities in and Against Nicaragua (Nica. v. U.S.), Merits, Judgment, 1986 I.C.J. 14, para. 115.

37   Prosecutor v. Tadic, para. 120.

38   Michael N. Schmitt, "Cyber Operations in International Law: The Use of Force, Collective Security, Self-Defense, and Armed Conflicts," in *Proceedings of a Workshop on Deterring CyberAttacks: Informing Strategies and Developing Options for U.S. Policy*, ed. National Research Council (Washington, DC: The National Academies Press, 2010), 151-178.

39   ICRC, "Chapter 18, Rule 63, Improper Use of the Flags or Military Emblems, Insignia or Uniforms of Neutral or Other States Not Party to the Conflict," accessed August 30, 2019, https://ihl-databases.icrc.org/customary-ihl/eng/docs/v1_cha_chapter18_rule63.

CHAPTER 10.
APPLICATION IN CASE STUDIES AND
FICTIONAL SCENARIOS

Chapter 10 contains an applied lessons chapter with case studies of the Syrian Civil War, Ukraine, nonviolent movements, and a fictional scenario involving the Redlands Resistance Movement. At the conclusion of each case study, there are a series of questions to prompt readers to integrate key takeaways from the previous chapters in this study. It is important to note that the answers for the historically based cases are based in accurate events. The fictional scenario has a set of hypothetical answers, but the reader should consider alternative solution space. In sum, this chapter enables the ARSOF solider to critically think through the takeaway messages from across the various chapters and think holistically about an operation.

SYRIAN CIVIL WAR

The ongoing civil war in Syria that grew out of nonviolent and later low-level violent resistance to the regime of Bashar al-Assad served to illustrate a range of cyber actors and activities in modern warfare. As examined in detail by Edwin Grohe in his monograph entitled "The Cyber Dimensions of the Syrian Civil War,"[1] individuals, groups, and state-sponsored actors both internal and external to the conflict engaged in cyberspace operations directly related to the conflict.

Supporting the regime, the Syrian Electronic Army (SEA), a loosely state-sponsored organization, initially focused primarily on spreading pro-regime messages.

Overtime, the group increasingly engaged in efforts to reach beyond messaging to more destructive attacks in both cyber and physical space.

Anti-regime forces, while less organized, also mounted cyber campaigns. These primarily focused on delivering their own messages but also included cyber espionage. Interestingly, external actors also engaged in cyberspace operations in the Syrian Civil War.

Cyber was also used by both sides and external entities to gather information then used for attack in physical space.

What medium could the SEA use to spread pro-regime messages?

Related Key Takeaways: 2.6

Spreading pro-regime messages through website defacements attacking a legitimate website and replacing its homepage with one created by the hackers.

What destructive attack in both cyber and physical space would further the reach of their messaging? Cyber to physical?

Related Key Takeaways: 3.4, 4.1, 5.8, 6.5

These methods included DDoS attacks, spreading fake news (one story involved reports of attacks on the White House and harm to President Obama, which resulted in a temporary drop in global stock markets2), and attempts to attack industrial control systems, such as a water facility in Israel.

What cyber espionage actions could reinforce their narrative?

Related Key Takeaways: 3.2, 4.2

Actions include hacking and monitoring al-Assad's personal email account in an attempt to gather derogatory information about the regime.[3]

How could a cyber action enable an attack in the physical space?

Related Key Takeaways: 4.9, 6.2, 6.3

In one widely reported example, an ISIL operative in Syria took a "selfie" in front of a "headquarters" building and posted it online. The facility was subsequently bombed and destroyed shortly thereafter by the US Air Force.[4] This illustrates the interconnected nature of the cyberspace domain and the physical domains, and warfighters must understand that actions in one can lead to effects in the others.

CONFLICT IN UKRAINE

The continuing conflict in Ukraine also offers illustrative examples of the conduct of cyberspace operations in modern conflict. Cyberspace actions included IO using social media as well as website defacements. Most notably, the conflict also included the first publicly acknowledged cyber attacks to result in electric power outages.[5] State-sponsored cyber activity included signals intelligence and cyber espionage targeting the Ukrainian army, to include collection of location data associated with mobile phones and wireless networks.[6] A significant escalation in the cyber conflict occurred in December 2015 when a cyber attack on several Ukrainian electrical power distribution networks, causing power outages lasting several hours that affected approximately 225,000 people.[7]

What cyber threat actors could be at play in the Ukraine conflict?

Related Key Takeaways: 3.3

Cyber threat actors involved could include highly capable, state-sponsored groups and hacktivists, as well as other lower tier, less capable individuals and organizations.

What actions could have taken place in the physical layer? Logical layer?

Related Key Takeaways: 6.2, 6.4

In the physical layer, Russian special forces seized an IXP in Crimea.[8] In the logical layer, the following actions were taken: Website defacements, DDoS attacks,[9] and advanced malware.[10]

What type of cyber action would a pro-Russian hacktivist engage in?

Related Key Takeaways: 4.1, 4.7, 7.9

"CyberBerkut" engaged in a variety of cyber attacks. These attacks included confidentiality attacks enabling subsequent IO. One such case involved publicizing a confidential phone call between a US diplomat and the US ambassador to Ukraine[11] by uploading it to YouTube. This organization also compromised the Ukrainian Central Election Commission during the presidential election. While limited in effect, this attack delayed software intended to provide real-time vote results by approximately twenty hours.[12]

What TTPs would an actor employ in the cyber attack on the Ukrainian electrical power networks?

Related Key Takeaways: 3.5, 8.3

This attack displayed a variety of TTPs, including the use of spearphishing, malware, and the use of VPNs[13] to traverse the target networks.[14] The attackers appeared to gain access to the electrical power networks for at least six months,[15] and the three separate power companies were attacked within thirty minutes of each other,[16] affecting at least twenty-seven separate electrical substations.[17]

CYBER IN NONVIOLENT MOVEMENTS

Unlike the case studies presented earlier, there is no single, compelling, nonviolent movement or event in which cyber operations played a substantial role to date. Rather, there are several anecdotal examples associated with nonviolent movements, as well as the example of groups such as Anonymous that appear to support a range of movements in a manner generally consistent with a collection of principles they seem to be attempting to advance.

One example of a nonviolent movement for which there was some cyber aspect was Occupy Wall Street, a protest against the perceived excesses of the financial industry in September 2011. While this movement primarily focused around a physical occupation of Zuccotti Park in New York City, it also demonstrated a multi-faceted cyber presence.

Another example can be found in the mobilization, coordination, and information sharing aspects visible in the case of the Stop Huntingdon Animal Cruelty campaign, in which organizers hacked the networks of the Huntingdon Life Sciences Corporation, which conducts animal testing for medical purposes.[18]

What are some of the facets of the cyber presence associated with the Occupy Wall Street movement?

Related Key Takeaways: 2.9, 3.5, 7.9

Digital media was used to mobilize support, coordinate efforts, and share information. Armed with cell phones, Occupy participants became media outlets, projecting news, images, and their messages directly from the physical location.[19] Other examinations of this event also noted how digital media enabled the exercise of democratic freedoms of assembly and free speech beyond the physical locality of a single event and further enabled digital age social movements broader reach.[20]

How could stolen information be used to further the Stop Huntingdon Animal Cruelty campaign's agenda?

Related Key Takeaways: 2.3, 2.10

Information stolen in the attack was used to expose the private addresses of employees, investors, clients, and partners. While this information was used primarily for online shaming and harassment, violent acts such as attacking the company's marketing director at his home with pepper spray occurred. Acts incited through digital media resulted in convictions of several activists,[21] but the campaign was also successful in that the company was delisted from the New York Stock Exchange,[22] though whether the cyber component of the campaign was central to this outcome is unclear.

FICTIONAL EXAMPLE: REDLANDS RESISTANCE MOVEMENT

The following fictional example accounts for the aforementioned challenges in attributing an actor to an action and cybersecurity concerns. Within the scenario, the antagonist (the Red Berets) uses the cyberspace environment to support its resistance movement within the state of Redlands. The US geographic combatant command (US Eastern Command) is tasked to provide Redlands with assistance in combating the Red Berets. The Red Berets seek to maximize the advantages that operating within cyberspace enables, and both Redlands and US EASTCOM seek to attribute Red Beret actions and mitigate Red Beret cyberspace threats.

Scenario

Within the western pacific, the nation-state of Redlands has been a pro-US supporter for the past fifty years. Redlands is seen as a reliable, international partner who shares US values. Redlands is a US friendly nation, but recently in the past five years, a resistance movement known as the Redlands National Front or Red Berets for short engaged in political and civil disorder actions against the government. The Red Berets seek a socialist government and state the current democratic republic government is insufficient to govern and meet the needs of the Redlands nation. The government of Redlands asked for US training and assistance in the form of host-nation engagement to counter the Red Berets movement. The United States provides host-nation support to the state of Redlands, which is under the geographic responsibility of US EASTCOM.

US EASTCOM establishes Joint Task Force (JTF) Redlands to provide host-nation assistance to Redlands. Host-nation assistance consists of humanitarian aid, civil affairs, security force training, and combined field training operations with US advisors. The JTF will operate within Redlands as the area of operation is limited to Redlands borders.

How might the Red Berets group assess its cyberspace environment?

Related Key Takeaways: 2.8, 5.2, 7.6

The Red Berets started its movement with multi-domain analysis. Within this analysis is an assessment of the cyberspace environment and how cyberspace can support its intent and operations. The Red Berets assess that it has a maneuver disadvantage in the air, land, and sea domains but has a maneuver advantage within the cyberspace domain. This is due to the cyberspace domain not being bound by physical borders, but the JTF Redlands is bound. Within cyberspace, physical borders are seamless, and the Red Berets intend to link disparate, globally distributed individuals and groups to its cause. The Red Beret planners assess that the anti-US sentiment among actors is enough to entice them to cooperate, which on their own initiative might not have occurred.

What type of C2 structure should the Red Berets employ?

Related Key Takeaways: 7.1, 7.4

The Red Berets employ a command structure that combines the best practices of military doctrine. To this end, the Red Berets have a leader who employs a staff that acts on his guidance to produce a desired end state. To maximize the use of cyberspace, the Red Berets have an IO planner and a cyberspace operations planner. The focus of the IO planner is to use multiple means, to include cyberspace, to project information to influence target audiences—friendly, hostile, and neutral. The focus of the cyberspace planner is to use cyberspace to support IO, which includes but is not limited to defacing adversary websites, disrupting adversary network operations, projecting and broadcasting propaganda and anti-US rhetoric, enabling communications with potential global recruits, and enabling global financial transactions.

What could be components of a global messaging plan by the Red Berets?

Related Key Takeaways: 2.1, 2.8, 5.2, 5.3, 5.9, 5.10, 8.3

The Red Berets IO coordinator developed a messaging plan that addresses three efforts simultaneously. The first effort is a propaganda and anti-US messaging effort. Second within these messages are calls to global audiences to resist the United States and support the Red Berets. Third within certain messages are coordinating instructions for global operations. To avoid targeting by US cyberspace forces, websites used for recruiting and propaganda messaging are often hosted in IP space of other countries. This is deliberate as to create the conditions that if the website is attacked, purportedly by the United States, the chances of creating collateral damage and escalating the conflict by executing an attack outside of the area of operation significantly increases.

How can the Red Berets secure global financing?

Related Key Takeaways: 8.5, 8.6

To secure global financing, coordinating activities are executed via Tor. Prospective sources are provided a link to download the Tor browser, an easy-to-use web browser with a built-in Tor client. To avoid exploitation by US intelligence, the link is provided by a one-time transaction email, and any associated contacted is executed through a VPN. Additionally, financing is executed through financial services similar to Western Union and only by exception via bank accounts that offer anonymity similar to Swiss bank accounts. The globally sourced finances are used to purchase arms to support field operations, transport recruits to Redlands, and pay for third-party cyber expertise.

What global messaging can be used to support Red Berets efforts?

Related Key Takeaways: 5.6, 7.7, 8.6

Global sourcing of personnel also executes through Tor. Prospective recruits are initially contacted via the social media effort. The patriotism of globally dispersed Redlandians, a dislike of the current government, and a dislike of the United States are the primary drivers for recruit support. Red Beret planners depend on this sense of patriotism and dislike of the United States to draw prospective recruits to its cause. The Red Berets source recruits who are not Redlandian by descent, but a dislike of the United States binds many Redlandians and non-Redlandians together. The use of global messaging through the Internet is a key enabler to this effort.

How can the Red Berets use social media for recruiting personnel?

Related Key Takeaways: 4.2, 4.8, 4.9, 7.6, 7.7, 8.6

The IO planner has a staff that methodically combs through social media collaboration sites where anti-US sentiment is common. The IO staff scours posted dialogues to identify potential recruits. Knowing that intelligence agencies are monitoring, the planners often engage other audiences in very broad, non-committal language for two reasons. First is to increase the anti-US feelings within the individual and second to not alarm any open-source or human intelligence analysts monitoring the site. If a planner believes the identified persona could be a potential recruit, then the persona is sent a personal message, and the planner communicates directly with the persona.

The planner uses a false persona. A false persona is often tailored to be very amicable to the target. The planner scours a site and develops a false persona later to deliberately target and dialogue with a potential recruit. At this point, the dialogue becomes increasingly anti-US, but the planner does not hint of any recruiting desires. The planner continues to engage the potential recruit until the persona hints he/she desires to support the Red Berets movement.

At this point, the planner guides the persona to reveal the nature of the desired support for the movement (financing, foreign fighter, blogger, etc.). From this point, the planner lets the recruit know "I will see if I know anyone who can help you." After a few days, another planner contacts the potential recruit and starts another dialogue. The recruit is provided a link to a Tor-based collaboration site within which the potential recruit and the planner continue their dialogue. This Tor-based site is not the primary collaboration site; its purpose is to further screen protective recruits as legitimate, illegitimate (intelligence personnel in disguise), or not worth the effort. The dialogue continues until the planner makes a determination. If the recruit is assessed to be legitimate, it is given a link to one of multiple Tor-based collaboration sites. The Red Berets deliberately run multiple Tor-based sites for redundancy purposes if one or more of the primary sites are compromised.

What covert strategies can be employed to minimize the risk of attribution?

Related Key Takeaways: 8.7, 8.8, 8.9, 9.1

To further disguise its actions, the Red Berets sometimes use a third party. The Red Berets use cyber mercenary groups to execute actions on its behalf. This is deliberate in the belief that the Red Berets assess US intelligence will at least acquire technical attribution. To minimize attribution risk, the Red Berets often employ cyber mercenaries for attacks on US government systems. If technical attribution is achieved, it is the third party that is attributed, but the Red Berets are not.

How can an overt attack be used to distract from a covert operation?

Related Key Takeaways: 3.4, 6.3, 7.10, 8.2, 8.6, 8.7, 8.10

The Red Berets sometimes attack via a combination of overt and covert means. This is deliberate as the overt means are used to distract network defenders, while the covert means are employed. The Red Berets draw upon its global support network to execute the overt attack.

To minimize attribution risk, the Red Berets use publically available tools, which significantly make human attribution more difficult because the public has access to the tools. This and the fact that hundreds of global sympathizers actively support the Red Berets cause make it difficult to attribute overt attacks to the Red Berets.

Concurrent with the overt attacks, the Red Berets use its cyber expertise and third-party expertise to execute covert attacks. Both attacks are designed to penetrate a targeted network, but the overt attack is used to draw resources and attention to that attack while the covert attack takes advantage of the situation and quietly penetrates the same targeted network. The Red Berets demonstrated this combined attack ability when the US EASTCOM's public website was defaced with the Red Berets banner. This was accomplished through a combined overt and covert attack. Overtly, the website was subjected to a DDoS attack, which sent to many website access requests from thousands of bots within a botnet.

Additionally, a spear-phishing attack was conducted against US EASTCOM leadership. Tailored emails with embedded malicious code was sent to the command staff, with the malicious code programmed to search for the website files and exfiltrate the data back to a Red Berets programmer. US EASTCOM network defenders were on the DDoS and how to mitigate the attack without shutting down the website, while missing the spearphishing and resulting exfiltration. Because the malicious code bypassed the security settings, its presence was initially unknown to the network defenders, and it communicated back to the Red Berets cyberspace operators the malicious code location, website configuration, and open entry and exit points. The Red Berets cyberspace operators then redesigned the US EASTCOM website, injected the code back into US EASTCOM's public website, and broadcast the Red Berets banner on the public website. This action forced US EASTCOM network defenders to take the website offline, find and quarantine the code, and search the remaining network topologies for any remaining malicious code.

How could the Red Berets gain access to the networks of deployed forces?

Related Key Takeaways: 4.4, 4.6, 8.8

The Red Berets developed a cyber concept of employment before JTF Redlands deployed. While the US EASTCOM site was defaced, another cyberspace operation to penetrate JTF Redlands was underway. Red Berets made contact with third parties who are sympathetic to the group and have advanced code writing skills at a near US-peer level.

The Red Berets sought to penetrate the networks of deployed forces. To do this, Red Berets cyber planners deliberately developed code unknown to cybersecurity companies because it has not been used before. The Red Berets slowly penetrated US networks stateside, building upon the penetration of US EASTCOM's networks. Because of the network trust between US EASTCOM's networks and JTF Redlands, malicious code transferred undetected into JTF Redlands networks.

Once inside JTF Redlands networks, Red Berets cyber operators navigated to data storage servers, located operations updates, and exfiltrated the data to Red Berets field forces. Additionally, the Red Beret cyber operators located cyberspace key terrain and staged code on it. This is at the request of Red Beret leadership who desires to disrupt US networks at a decisive time.

What US EASTCOM counteractions could be taken?

Related Key Takeaways: 3.6, 4.7, 4.9

Fortunately for JTF Redlands, the J2 assessed that the Red Berets would have advanced cyber support within its global support networks. This prompted the J2, J6, and J3 to identify cyberspace key terrain (CKT) and request cyber protection team (CPT) support to secure and monitor the key terrain. CPT support is a joint, high-demand, low-density asset so support did not arrive until one month after the arrival of JTF Redlands into the area of operation. The CPT built upon the J2/3/6 CKT identification and executed hunt operations with the CKT and the topological connections linking them. After two months of operations, the CPT discovered activity that it could not classify. Although not an immediate indicator of adversary activity, this was a key enabler to locating the adversary penetration. In collaboration with the intelligence community, a forensic analysis was conducted of the activity, and it was determined to be unknown malicious code. The network defense and detection systems were updated, and the code within the network was detected and quarantined.

What actions should the JTF Redlands employ to address the cyberspace asymmetry?

Related Key Takeaways: 4.5, 4.6, 4.7

The Red Berets used the cyberspace domain's seamless borders to link disparate parties separated by thousands of miles. This is a significant asymmetric advantage because cyberspace actors can engage virtually person to person short of physically engaging. This global connectivity enabled the Red Berets to secure global financing, material, and personnel. Fortunately for JTF Redlands, the J2 and J3 assessed that the Red Berets would leverage global support. Because of this, JTF Redlands provided planners to US EASTCOM operational planning teams and developed a cross combatant command support plan. This support plan enabled JTF Redlands to identify an actor with direct impact on its operations but who is outside the area of operation. Upon identification, JTF Redlands coordinated with US EASTCOM, which coordinated with the combatant command within which the actor is located for threat mitigation.

What cyber hygiene measures should the Red Berets employ to protect against offensive cyberspace forces?

Related Key Takeaways: 3.4, 3.5, 4.2, 4.4, 4.5, 8.1, 8.4, 8.5

The Red Berets defaced pro-US rhetoric websites, attacked US networks, recruited foreign fighters, and secured finances through cyberspace means. The Red Beret planners assess that US intelligence will try to identify the communications nodes and personnel, so the Red Berets implemented internal measures to protect its networks and minimize attribution risk. Realizing that the human link is the weakest, it is the primary emphasis for network defense. The Red Berets adopt a thorough cybersecurity hygiene plan to emphasize the role the human link plays in promoting cyber defense and minimizing attribution risk.

To enforce cyber hygiene, the Red Berets implemented the following procedures:

Cell phones: Personnel use burner cell phones. These phones are meant to be used for short periods and then discarded, making it difficult for an opponent to track the phone. Because burner phones can be purchased almost anywhere, it is difficult to compromise the supply chain of burner phones. Only by exception are smart phones used, and the GPS must be disabled as to not geolocate and record the phone's location.

Email: Multiple email accounts are used and changed on a random basis. Personal email is only used by exception, and transactions via email are avoided when face-to-face contact is feasible. All email attachments and hyperlinks are opened in a virtual network to detect any embedded malicious code or beacons. The virtual network uses replicated computer operating systems virtually resident within a physical host computer. By opening hyperlinks in a virtual network, any malicious code will interface with the virtual network computers and not the physical computer a Red Berets operator uses.

Removable media: File copying and data transfer are executed through CD burning. Using a CD minimizes the use of flash software supported by thumb drives. If thumb drives have to be used, then they are opened in a virtual network. Use of a virtual network enables embedded malicious flash software to interface with the virtual software and not harm the actual hosting terminal and network.

IoT devices: The IoT is not used within the Red Berets physical infrastructure. However, from the core members moving outwards, the IoT is used by some supporting and tertiary personnel. Rather than believe all members are 100 percent policy compliant, guidance issued from the Red Berets core is not to use the same username and password for each device.

Cloud: Cloud services are not used for personal smart phones, tablets, laptops, and the storage of data related to daily activities. The Red Berets policy is to store data locally but encrypt the data and restrict who can access the data. If Red Beret personnel have to move, and the risk of devices being captured is moderate or high, then data are uploaded into a cloud service, and the device hard drives are wiped. This measure is key takeaway #4 (separation of cyber capabilities) in action because personnel and devices are at the greatest risk of capture when in transit. For this reason, key takeaway #6 is enacted, and data are uploaded into a cloud before travel, and when the personnel and device reach their destination, the data are downloaded back to the device.

Darknets: The Red Berets host coordinates activities of personnel, material, and financial transactions within a Darknet. The Red Berets divide activities into functional cells (as described in key takeaway #4) because it separates the recruiting activities in the public domain from the resistance movement's coordinating activities. This further protects the location of the Red Berets' Darknet assets if a recruiting activity or site is compromised in the public domain.

ENDNOTES

1   Edwin Grohe, *The Cyber Dimensions of the Syrian Civil War: Implications for Future Conflict* (Laurel, MD: Johns Hopkins University Applied Physics Laboratory, 2015).

2   Ibid.

3   Ibid.

4   Michael Hoffman, "US Air Force Targets and Destroys ISIS HQ Building Using Social Media," *Defensetech*, Military.com. Last modified June 3, 2015. https://www.defensetech. org/2015/06/03/us-air-force-targets-and-destroys-isis-hq-building-using-social-media/.

5   Robert M. Lee, Michael J. Assante, and Tim Conway, "Analysis of the Cyber Attack on the Ukrainian Power Grid Defense Use Case," SANS and Electricity Information Sharing and Analysis Center (E-ISAC), March 18, 2016, https://ics.sans.org/media/E-ISAC_ SANS_Ukraine_DUC_5.pdf, iv.

6   Glib Pakharenko, "Cyber Operations at Maidan: A First-Hand Account," in *Cyber War in Perspective: Russian Aggression Against Ukraine*, ed. Kenneth Geers (Tallinn, Estonia: NATO Cooperative Cyber Defence Center of Excellence, 2015), 63.

7   Lee, Assante, and Conway, "Analysis of the Cyber Attack on the Ukrainian Power Grid Defense Use Case," v.

8   Sven Sakkov, "Foreword" in *Cyber War in Perspective: Russian Aggression Against Ukraine*, ed. Kenneth Geers (Tallinn, Estonia: NATO Cooperative Cyber Defence Center of Excellence, 2015), 11.

9   Ibid.

10 Nikolay Koval, "Revolution Hacking" in *Cyber War in Perspective: Russian Aggression Against Ukraine*, ed. Kenneth Geers (Tallinn, Estonia: NATO Cooperative Cyber Defence Center of Excellence, 2015), 57.

11 Sakkov, "Foreword," 11.

12 Koval, "Revolution Hacking," 56.

13 A VPN uses cryptography to securely emulate a point-to-point link while using a shared or public network.

14 Lee, Assante, and Conway, "Analysis of the Cyber Attack on the Ukrainian Power Grid Defense Use Case," 2.

15 Ibid., 6.

16 Ibid., v.

17 Ibid., 8.

18 P. W. Singer and Allan Friedman, *Cybersecurity and Cyberwar: What Everyone Needs to Know* (New York, NY: Oxford University Press, 2014), 79.

19 Alexander Dirksen, *Occupying the Digital Public Square* (Published online, 2013), http:// static1.squarespace.com/static/534b5427e4b0b51fbb0e5506/t/537e360ee4b0b51f 32c11132/1400780302793/Occupy+Online.pdf.

20 Victoria Carty, "The Indignados and Occupy Wall Street Social Movements: Global Opposition to the Neoliberalization of Society as Enabled by Digital Technology," *Tamara Journal* 13, no. 3 (2015): 21-33.

21 David Kocieniewski, "Six Animal Rights Advocates Are Convicted of Terrorism," *New York Times*, March 3, 2006, https://www.nytimes.com/2006/03/03/nyregion/six-animal-rights-advocates-are-convicted-of-terrorism.html.

22 Singer and Friedman, *Cybersecurity and Cyberwar*, 79-80.

BIBLIOGRAPHY

BIBLIOGRAPHY

"Al-Qaeda Puts Job Ads on Internet – Arab Paper." *Irish Times,* October 6, 2005.

"Army 'Big Brother' Unit Targets Bloggers." *Defense Tech,* October 13, 2006.

"Astroturfing." *Wikipedia,* June 2017. https://en.wikipedia.org/wiki/Astroturfing.

"Cyber Banking Fraud Global Partnerships Lead to Major Arrests." FBI, October 1, 2010. https://archives.fbi.gov/archives/news/stories/2010/october/cyber-banking-fraud.

"Egypt: Military Junta Launches Facebook Page," *Telegraph*, February 17, 2011.

"Massive Networks of Fake Accounts Found on Twitter." *BBC online,* Technology section, January 24, 2017.

"North Korea denies 'righteous' hack attack on Sony," BBC News, December 7, 2014, https://www.bbc.com/news/world-asia-30366449.

"Pizzagate Consipiracy Theory." *Wikipedia.* http://www.wikipedia.org/wiki/pizzagate_conspiracy_theory.

"Russian Web Brigades." *Wikipedia.* https://en.wikipedia.org/wiki/Russian_web_brigades.

"Ten Years of Join Me." *Leader,* March 21, 2011. http://www.join-me.co.uk.

"Toward A Model of Meme Diffusion (M3D) - Spitzberg - 2014 - Communication Theory - Wiley Online Library." Accessed February 28, 2017, http://onlinelibrary.wiley.com/doi/10.1111/comt.12042/full.

"Using Spamhaus BGPF in a Production Environment." Geneva, Switzerland: The Spamhaus Project, April, 2013.

"What is Stuxnet?," McAfee, https://www.mcafee.com/enterprise/en-us/security-awareness/ransomware/what-is-stuxnet.html.

Abdelrahman, Maha. "In Praise of Organization: Egypt between Activism and Revolution." *Development and Change* 44, no. 3 (2013): 569–585.

Abdulrahim, Raja. "Egypt Police Try to Improve Image through Facebook." *Los Angeles Times,* February 19, 2011.

Ablon, Lillian, Martin C. Libicki, and Andrea A. Golay. *Markets for Cybercrime Tools and Stolen Data : Hacker's Bazaar.* Santa Monica, CA: RAND Corporation. 2014. PDF e-book.

Ackerman, Spencer. "US Central Command Twitter Account Hacked to Read 'I love you Isis.'" *Guardian,* January 12, 2015.

Agan, Summer. *Narratives and Competing Messages.* Fort Bragg, NC: US Special Operations Command, 2018.

Alexander, Yonah, and Michael S. Swetnam. *Cyberterrorism and information warfare: Threats and responses.* London: Transnational Publishers, Incorporated, 2001.

Alfano, Mark, J. Adam Carter, and Marc Cheong. "Technological seduction and self-radicalization." *Journal of the American Philosophical Association* 3 (2018):298–322.

Allan, Collin S. "Attribution in Cyberspace." *Chicago-Kent Journal of International and Comparative Law* 13, issue 2 (November 2013).

Allcott, Hunt, and Matthew Gentzkow. "Social Media and Fake News in the 2016 Election." National Bureau of Economic Research, January 2017. https://doi.org/10.3386/w23089.

Allen, Patrick D. *Cloud Computing 101: A Primer for Project Managers,* CreateSpace Independent Publishing Platform, 2015.

Allen, Patrick D., and Chris C. Demchak. "The Palestinian-Israeli Cyberwar." *Military Review* 83, no. 2 (March-April 2003).

Allen, Patrick. *Information Operations Planning.* Boston, MA: Artech House, 2007.

Alperovitch, Dmitri. "Bears in the Midst: Intrusion into the Democratic National Committee." CrowdStrike, June 15, 2016. https://www.crowdstrike.com/blog/bears-midst-intrusion-democratic-national-committee/.

Alvarez, John, Robert Nalepa, Anna-Marie Wyant, and Fred Zimmerman, eds. *Special Operations Forces Reference Manual, Fourth Edition.* MacDill AFB, FL: The JSOU Press, June 2015.

Amato, John. "Has The Beltway Created A Twitter Media Playhouse?" *crooksandliars.com,* October 10, 2012.

Anthony, Sebastian. "Facebook's facial recognition software is now as accurate as the human brain, but what now?" *Extreme Tech,* March 19, 2014.

Arghire, Ionut. "IoT Botnets Fuel DDoS Attacks Growth: Report." Security Week, January 24, 2017. https://www.securityweek.com/iot-botnets-fuel-ddos-attacks-growth-report.

ARIS, *Human Factors Considerations of Undergrounds in Insurgencies*, 2nd Ed. (Alexandria, VA: US Army Publications Directorate, 2013). http://www.soc.mil/ARIS/ARIS.html.

ARIS, *Undergrounds in Insurgent, Revolutionary, and Resistance Warfare*, 2nd Ed. (Alexandria, VA: US Army Publications Directorate, 2013). http://www.soc.mil/ARIS/ARIS.html.

Asch, Solomon E. "Studies of Independence and Conformity: I. A Minority of One against a Unanimous Majority." *Psychological Monographs: General and Applied* 70, no. 9 (1956): 1–70.

Assange, Julian. "The Curious Origins of Political Hacktivism." *Counterpunch*, November 25, 2006. https://www.counterpunch.org/2006/11/25/the-curious-origins-of-political-hacktivism/.

Associated Press. "A Look at Estonia's Cyber Attacks in 2007." NBC News, July 9, 2009. http://www.nbcnews.com/id/31801246/ns/technology_and_science-security/t/look-estonias-cyber-attack/.

Athanasiadis, Iason. "How Hi-Tech Hezbollah Called the Shots." *Asia Times*, September 9, 2006.

Atran, Scott, Hammad Sheikh, and Angel Gomez. "Devoted actors sacrifice for close comrades and sacred cause." *Proceedings of the National Academy of Sciences* 111, no. 50 (2014): 17702–17703.

Ayres, Jeffrey M. "From the Streets to the Internet: The Cyber-Diffusion of Contention." *Annals of the American Academy of Political and Social Science* 566, no. 1 (1999).

Bacon, Donald J. "Second World War Deception – Lessons Learned for Today's Joint Planners." *Air Command and Staff College*, December 1998. https://apps.dtic.mil/dtic/tr/fulltext/u2/a405884.pdf.

Bandura, A., C. Barbaranelli, G.V. Caprara, and C. Pastorelli. "Mechanisms of moral disengagement in the exercise of moral agency." *Journal of Personality and Social Psychology*, 71, 2 (1996): 364–374.

Banks, Emily. "Egyptian President Steps Down Amidst Groundbreaking Digital Revolution." *Mashable, CNN.com*, February 11, 2012.

Barabasi, A., and R. Albert. "Emergence of Scaling in Random Networks." *Science* 286 (1999).

Barberá, Pablo, John T. Jost, Jonathan Nagler, Joshua A. Tucker, and Richard Bonneau. "Tweeting from left to right: Is online political communication more than an echo chamber?" *Psychological Science* 26, no. 10 (2015): 1531–1542.

Bari Atwan, Abdel. *Islamic state: The digital caliphate.* Oakland, CA: University of California Press, 2015.

Barkai, John. "Cultural Dimension Interests, the Dance of Negotiation, and Weather Forecasting: A Perspective on Cross-Cultural Negotiation and Dispute Resolution." *Pepperdine Dispute Resolution Law Journal* 8, issue 3 (April 2008).

Barlett, Christopher P., Douglas A. Gentile, and Chelsea Chew. "Predicting cyberbullying from anonymity." *Psychology of Popular Media Culture* 5, no. 2 (2016): 171–180.

Barnard, Anne. "Children, Caged for Effect, to Mimic Imagery of ISIS." *New York Times*, February 20, 2015. https://www.nytimes.com/2015/02/21/world/middleeast/activists-trying-to-draw-attention-to-killings-in-syria-turn-to-isis-tactic-shock-value.html.

Barnas, Neil B. "Blockchains in National Defense: Trustworthy Systems in a Trustless World." Research Report in Partial Fulfillment of Graduation Requirements, Air University Academic Research Report, June 2016. http://www.dtic.mil/doctrine/education/jpme_papers/barnas_n.pdf

Barry, Ellen. "Sound of Post-Soviet Protest: Claps and Beeps." *New York Times Online*, July 14 2011. http://www.nytimes.com/2011/07/15/world/europe/15belarus.html?pagewanted=all.

Basu, Moni. "Good Morning, Mosul: Pirate radio risks death to fight ISIS on airwaves." *CNN*, October 22, 2016.

Baum, Matthew, David Lazer, and Nicco Mele. "Combating Fake News: An Agenda for Research and Action." Paper for Harvard Kennedy School, Shorenstein Center on Media, Politics, and Public Policy, February 18, 2017. https://shorensteincenter.org/combating-fake-news-agenda-for-research/.

Benigni, M., "Detection and Analysis of Online Extremist Communities." Dissertation for Carnegie Mellon University. Technical Report CMU-ISR-17-108, 2017.

Bennett, W. Lance, and Alexandra Segerberg, "Digital Media and the Personalization of Collective Action." *Information, Communication & Society* 14, no. 6 (September 1, 2011): 770–99. https://doi.org/10.1080/1369118X.2011.579141.

Bennett, W. Lance, and Alexandra Segerberg. "Introduction." In *The Logic of Connective Action: Digital Media and the Personalization of Contentious Politics.* Cambridge: Cambridge University Press, 2013.

Bennett, W. Lance, and Alexandra Segerberg. "Personalized Communication in Protest Networks." In *The Logic of Connective Action: Digital Media and the Personalization of Contentious Politics.* Cambridge: Cambridge University Press, 2013.

Bennett, W. Lance, and Alexandra Segerberg. "The Logic of Connective Action." In *The Logic of Connective Action: Digital Media and the Personalization of Contentious Politics.* Cambridge: Cambridge University Press, 2013.

Bennett, W. Lance, and Alexandra Segerberg. "The Logic of Connective Action: Digital Media and the Personalization of Contentious Politics." *Information, Communication & Society* 15, no. 5 (2012).

Bennett, W. Lance, and Alexandra Segerburg. "Networks, Power, and Political Outcomes." In *The Logic of Connective Action: Digital Media and the Personalization of Contentious Politics.* Cambridge: Cambridge University Press, 2013.

Berger, J. M. "How ISIS Games Twitter: The Militant Group that Conquered Northern Iraq is Deploying Sophisticated Social Media Strategy." *Atlantic*, June 16, 2014. https://www.theatlantic.com/international/archive/2014/06/isis-iraq-twitter-social-media-strategy/372856/.

Berger, J. M., and Jonathon Morgan. "The ISIS Twitter Census: Defining and Describing the Population of ISIS Supporters on Twitter." The Brookings Project on US Relations with the Islamic World Analysis Paper no. 20, March 5, 2015, https://www.brookings.edu/wp-content/uploads/2016/06/isis_twitter_census_berger_morgan.pdf.

Bishop, Jonathan. "The Psychology of Trolling and Lurking: The Role of Defriending and Gamification for Increasing Participation in Online Communities Using Seductive Narratives." In *Virtual Community Participation and Motivation: Cross-Disciplinary Theories.* IGI Global, 2012.

Boebert, W. Earl. "A Survey of Challenges in Attribution." In *Proceedings of a Workshop on Deterring Cyber Attacks: Informing Strategies and Developing Options for US Policy,* edited by National Research Council, 41-52. Washington, DC: The National Academies Press.

Bombardieri, Marcella. "The inside story of MIT and Aaron Swartz." *The Boston Globe,* March 29, 2014. https://www3.bostonglobe.com/metro/2014/03/29/the-inside-story-mit-and-aaron-swartz/YvJZ-5P6VHaPJusReuaN7SI/story.html?arc404=true

Borgatti, S. P., et al. "Network Analysis in the Social Sciences." *Science* 323, no. 5916 (2009).

Borwell, Jildau, Jurjen Jansen, and Wouter Stol. "Human Factors Leading to Online Fraud Victimization: Literature Review and Exploring the Role of Personality Traits." In *Psychological and Behavioral Examinations in Cyber Security.* IGI Global, 2018.

Bos, Nathan D., et al. *Human Factors Considerations of Undergrounds in Insurgencies Second Edition.* Alexandria, VA: US Army Printing Office, 2013.

Bothe, Michael, Karl Josef Partsch, and Waldemar A. Solf. *New Rules for Victims of Armed Conflicts: Commentary on the Two 1977 Protocols Additional to the Geneva Conventions of 1949.* Leiden, Belgium: Martinus Nijhoff Publishers, 1982.

Bowles, Nellie. "How 'Doxxing' Became a Mainstream Tool in the Culture Wars." *New York Times,* August 30· 2017. https://www.nytimes.com/2017/08/30/technology/doxxing-protests.html.

Breitenbach, Daniel L. "Operation Desert Deception: Operational Deception in the Ground Campaign." Naval War College (June 19, 1991).

Brewer, Paul R., and Kimberly Gross. "Values, Framing, and Citizens' Thoughts about Policy Issues: Effects on Content and Quantity." *Political Psychology* 26, no. 6 (December 1, 2005). https://doi.org/10.1111/j.1467-9221.2005.00451.x.

Brownlee, Jason. "The decline of pluralism in Mubarak's Egypt." *Journal of Democracy* 13, no. 4 (2002): 6–14.

Brunetti-Lihach, Nick. "Information Warfare Past, Present and Future." *RealClear Defense,* November 14, 2008. https://www.realcleardefense.com/articles/2018/11/14/information_warfare_past_present_and_future_113955.html.

Buckels, Erin E., Paul D. Trapnell, and Delroy L. Paulhus. "Trolls Just Want to Have Fun." *Personality and Individual Differences, The Dark Triad of Personality.* 67 (September 2014): 97–102. https://doi.org/10.1016/j.paid.2014.01.016.

Calhoun, Craig. "Occupy wall street in perspective." *The British Journal of Sociology* 64, no. 1 (2013): 26–38.

Caliskan, Emin, Tomas Minarik, and Anna-Maria Osula. "Technical and Legal Overview of the Tor Anonymity Network." Tallinn, Estonia: NATO Cooperative Cyber Defence Center of Excellence, 2015.

Carlin, John P. *Dawn of the Code War: America's Battle Against Russia, China, and the Rising Global Cyber Threat.* New York: Public Affairs, 2018.

Carnegie Mellon University, Software Engineering Institute. "Common Sense Guide to Mitigating Insider Threats, 5th Edition." Technical Note CMU/SEI2015-TR-010, December 2016.

Carter, Ashton B. "Information Operations (IO)." Department of Defense Directive 3600.01, Under Secretary of Defense (Policy), May 2, 2013.

Carty, Victoria. "The Indignados and Occupy Wall Street Social Movements: Global Opposition to the Neoliberalization of Society as Enabled by Digital Technology." *Tamara Journal* 13, no. 3. (2015): 21–33.

Castells, Manuel. *The Power of Identity, 2nd edition.* Malden, MA: Blackwell Publishing Ltd, 2004.

Castillo, J. C. "The Mexican Cartels' employment of Inform and Influence Activities (IIA) as tools of asymmetrical warfare." In *Information Security for South Africa* (ISSA), IEEE 2014.

Central Intelligence Agency. "The Office of Strategic Services: Morale Operations Branch." July 29, 2010. https://www.cia.gov/news-information/featured-story-archive/2010-featured-story-archive/oss-morale-operations.html.

Chacos, Brad. "Major DDoS Attack on Dyn DNS Knocks Spotify, Twitter, Github, PayPal, and More Offline." *PCWorld,* October 21, 2016. https://www.pcworld.com/article/3133847/internet/ddos-attack-on-dyn-knocks-spotify-twitter-github-etsy-and-more-offline.html.

Chang, Wentao, An Wang, Aziz Mohaisen, and Songqing Chen. "Characterizing Botnets-as-a-Service." ACM SIGCOMM *Computer Communication Review* 44, no. 4 (August 2014): 585-586.

Chiesa, Raoul, Stefania Ducci, and Silvio Ciappi. *Profiling Hackers: The Science of Criminal Profiling as Applied to the World of Hacking.* Boca Raton, FL: Taylor and Francis, 2012, Kindle.

Choo, Kim-Kwang Raymond. "Cryptocurrency and virtual currency: corruption and money laundering/terrorism financing risks?" In *Handbook of Digital Currency.* Academic Press, 2015.

Cimpanu, Catalin. "How US authorities tracked down the North Korean hacker behind WannaCry." ZDNet, September 6, 2018. https://www.zdnet.com/article/how-us-authorities-tracked-down-the-north-korean-hacker-behind-wannacry/.

Cipriani, Jason. "What you need to know about encryption on your phone." *CNET*, March 10, 2016.

Clapper, James R. *Statement for the Record Worldwide Threat Assessment of the US Intelligence Community.* Washington, DC: House Appropriates Subcommittee on Defense, March 24, 2015.

Cohen, I. Glenn, Sharona Hoffman, and Eli Y. Adashi. "Your Money or Your Patient's Life? Ransomware and Electronic Health Records." *Annals of internal medicine* 167, no. 8 (2017): 587–588.

Coleman, Gabriella, and A. Golub. "The anthropology of hackers." *Atlantic*, September 21, 2010. https://www.theatlantic.com/technology/archive/2010/09/the-anthropology-of-hackers/63308/.

Coleman, Gabriella. "Anonymous: From the Lulz to collective action." New Everyday: a Media Commons Project 6, 2011.

Coleman,Gabriella. *Hacker, Hoaxer, Whistleblower, Spy: The Many Faces of Anonymous.* Brooklyn, NY: Verso Books, 2014.

Colleoni, Elanor, Alessandro Rozza, and Adam Arvidsson. "Echo Chamber or Publc Sphere? Predicting Political Orientation and Measuring Political Homophily in Twitter Using Big Data: Political Homophily on Twitter." *Journal of Communication* 64, no. 2 (April 2014). https://doi.org/10.1111/jcom.12084.

Comey, James. "Encryption, Public Safety, and 'Going Dark'." Lawfare, July 6, 2015. https://www.lawfareblog.com/encryption-public-safety-and-going-dark.

Connor, Michael, and Sarah Vogler. "Russia's Approach to Cyber Warfare." CNA Analysis and Solutions, March 2017, https://www.cna.org/CNA_files/PDF/DOP-2016-U-014231-1Rev.pdf.

Conover, Michael D., Emilio Ferrara, Filippo Menczer, and Alessandro Flammini. "The digital evolution of occupy wall street." *PloS one* 8, no. 5 (2013): e64679.

Cook, Sarah. "China's Growing Army of Paid Internet Commentators." *Freedom House*, October 11, 2011.

Coursera. "Machine Learning." Course offered by Stanford University, https://www.coursera.org/learn/machine-learning (accessed 01 August 2017).

Cox, Joseph. "Stolen Uber Customer Accounts Are for Sale on the Dark Web for $1." *Motherboard*, March 27, 2015.

Cragin, R. Kim, and Ari Weil. "'Virtual Planners' in the Arsenal of Islamic State External Operations." *Orbis* 62, issue 2 (2018): 294–312.

Critical Art Ensemble. *Electronic Civil Disobedience.* Brooklyn, New York: Autonomedia, 1996. http://critical-art.net/electronic-civil-disobedience-1996/.

Crossett, Chuck, and Jason Spitaletta. *Radicalization: Relevant Psychological and Sociological Concepts.* Fort Meade, MD: Asymmetric Warfare Group, 2010.

Cunningham, Vinson. "The Masks in Venezuela and the Pathos of Protest Art." *New Yorker*, May 13, 2017. https://www.newyorker.com/culture/photo-booth/the-masks-in-venezuela-and-the-pathos-of-protest-art?mbid=rss.

Currier, Cora, and Morgan Marquis-Boire. "Secret Manuals Show the Spyware Sold to Despots and Cops Worldwide." *Intercept*, October 30, 2014.

Cyber-Physical Systems Public Working Group. "Framework for Cyber-Physical Systems: Volume 1, Overview." National Institute of Standards and Technology Special Publication 1500-201, Version 1.0, June 2017.

Dardis, Frank E., et al. "Media Framing of Capital Punishment and Its Impact on Individuals' Cognitive Responses." *Mass Communication and Society* 11, no. 2 (April 7, 2008). https://doi.org/10.1080/15205430701580524.

Dasgupta, N., C. Freifeld, C. S. Brownstein, C. M. Menone, H. L. Surratt, L. Poppish, J. L. Green, E. J. Lavonas, and R. C. Dart. "Crowdsourcing black market prices for prescription opioids." *Journal of Medical Internet Research* 15, no. 8 (August 2013): e178, https://openi.nlm.nih.gov/detailedresult.php?img=PMC3758048_jmir_v15i8e178_fig2&req=4.

Dauber, Cori, and Mark Robinson. "ISIL and the Hollywood Visual Style." *Jihadology*, July 6, 2015. http://jihadology.net/2015/07/06/guest-post-ISIL-and-the-hollywood-visual-style/.

Deegan, Arthur, Yasir Khalid, Michelle Kingue, and Aldo Taboada. "Cyber-ia: The Ethical Considerations Behind Syria's Cyber-War." *Small Wars Journal* (2017). https://smallwarsjournal.com/index.php/jrnl/art/cyber-ia-the-ethical-considerations-behind-syria%E2%80%99s-cyber-war.

Delistraty, Cody. "When Wearing a Graphic T-Shirt is a Revolutionary Act." *Garage*, March 30, 2018. https://garage.vice.com/en_us/article/zmgmd4/graphic-t-shirt-revolutions.

DeLuca, Kevin M., Sean Lawson, and Ye Sun. "Occupy Wall Street on the Public Screens of Social Media: The Many Framings of the Birth of a Protest Movement." *Communication, Culture & Critique* 5, no. 4 (December 1, 2012): 483–509. https://doi.org/10.1111/j.1753-9137.2012.01141.x.

Denning, D. E. "Activism, Hacktivism, And Cyberterrorism: The Internet as A Tool for Influencing Foreign Policy" In *Networks and Netwars: The Future of Terror, Crime, and Militancy*, edited by J. Arquilla and D. Ronfeldt. Santa Monica, CA: RAND Corporation, 2001.

Department of Defense, Defense Science Board. "Task Force Report: Resilient Military Systems and the Advanced Cyber Threat." Washington, DC: Office of the Under Secretary of Defense for Acquisitions, Technology and Logistics, January 2013.

Department of Defense. "Summary: Department of Defense Cyber Strategy 2018." September 2018.

Depasquale, Scott and Michael Daly. "The Growing Threat of Cyber Mercenaries." *Politico*, October 12, 2016. https://www.politico.com/agenda/story/2016/10/the-growing-threat-of-cyber-mercenaries-000221.

Derrick, Douglas C., Gina Ligon, Mackenzie Harms, and William R. Mahoney. "Cyber-Sophistication Assessment Methodology for Public-Facing Terrorist Web Sites." *Journal of Information Warfare*16, no. 1 (2017): 13–30.

Derrick, Douglas C., Karyn Sporer, Sam Church, and Gina Scott Ligon. "Ideological rationality and violence: An exploratory study of ISIL's cyber profile." *Dynamics of Asymmetric Conflict* 9, no. 1-3 (2016): 57–81.

Deutsche Telekom AG. "Security Dashboard Shows Cyber Attacks in Real Time." *Press Release,* June 3, 2013.

Dingledine Roger, Nick Mathewson, and Paul Syverson. "Tor: The Second-Generation Onion Router." Paper presented at the 13th USENIX Security Symposium, San Diego, CA, 2004.

Dirksen, Alexander. *Occupying the Digital Public Square.* Published online, 2013. http://static1.squarespace.com/static/534b5427e4b0b51fbb0e5506/t/537e360ee4b0b51f32c11132/1400780302793/Occupy+Online.pdf.

Dobush, Grace. "How Mobile Phones are Changing the Developing World." *Consumer Technology Association Blog,* July 27, 2015.

Dolliver, Diana S., and Kevin Poorman. "Understanding Cybercrime." In P. Reichel and R. Randa, *Transnational Crime and Global Security,* 2018.

Dörrer, Kiyo. "Hello, Big Brother: How China Controls Its Citizens Through Social Media." *Deutsche Welle,* March 31, 2017. http://www.dw.com/en/hello-big-brother-how-china-controls-its-citizens-through-social-media/a-38243388.

Drum, Kevin. "Social Media is Best Used for Distraction, not Argument." *Mother Jones News,* January 17, 2017. https://www.motherjones.com/kevin-drum/2017/01/social-media-best-used-distraction-not-argument/.

Duggan, Patrick M. "Strategic Development of Special Warfare in Cyberspace." *Joint Force Quarterly* 79, no. 4 (2015): 46–53.

Dunbar, Robin. "The Social Brain Hypothesis and Its Implications for Social Evolution." *Annals of Human Biology* 36, no. 5 (August 2009):

Duncan, Kirk A. "Assessing the Use of Social Media in a Revolutionary Environment." Master's thesis, Naval Postgraduate School, 2013.

Eagleton-Pierce, Matthew. "The internet and the Seattle WTO protests." *Peace Review* 13, no. 3 (2001): 331–337.

Earl, Jennifer, and Katrina Kimport. *Digitally Enabled Social Change: Activism in the Internet Age*. Cambridge, MA: The MIT Press, 2011.

Editorial Board. "The New Radicalization of the Internet." *New York Times*, November 25, 2018. https://www.nytimes.com/2018/11/24/opinion/sunday/facebook-twitter-terrorism-extremism.html.

Eidman, C. R., and G. S. Green. *Unconventional Cyber Warfare: Cyber Opportunities in Unconventional Warfare*. Unpublished Master's thesis, Naval Postgraduate School, 2014.

Ellis, Emma Grey. "Whatever your side, doxing is a perilous form of justice." *Wired*, August 17, 2017. https://www.wired.com/story/doxing-charlottesville?mbid=nl_81717_p2&CNDID=13902615.

Ellison, Nicole B., and Danah M. Boyd. "Sociality Through Social Network Sites." *Oxford Handbook of Internet Studies*, January 1, 2013. https://doi.org/10.1093/oxfordhb/9780199589074.013.0008.

Encyclopedia Brittanica, eds. "Radio Free Europe: United States Radio Network." https://www.britannica.com/topic/Radio-Free-Europe.

Erdos, P., and A. Rényi. "On Random Graphs I." *Publicationes Mathematicae* 6 (1959): 290–297.

Everipidis, Romanidis. "Lawful Interception and Countermeasures In the era of Internet Telephony." COS/CCS 2008-20. Master of Science thesis, School of Information and Communication Technology, Royal Institute of Technology, Stockholm, Sweden, 2008.

Facebook. "What is tagging and how does it work?" *Facebook.com*, 2017.

Ferrara, Emilio, et al. "The Rise of Social Bots." *Communications of the ACM* 59, no. 7 (June 24, 2016): 96–104, https://doi.org/10.1145/2818717.

Fink, Clay, et al. "Complex Contagions and the Diffusion of Popular Twitter Hashtags in Nigeria." *Social Network Analysis and Mining* 6, no. 1 (December 1, 2016). https://doi.org/10.1007/s13278-015-0311-z.

Finkle, Jim. "Iranian hackers use fake Facebook accounts to spy on U.S., others." Reuters, May 29, 2014. https://www.reuters.com/article/us-iran-hackers/iranian-hackers-use-fake-facebook-accounts-to-spy-on-u-s-others-idUSKBN0E90A220140529.

Finlayson, Mark A., and Steven R. Corman. "The Military Interest in Narrative." *Sprache Und Datenverarbeitung* 37, no. 1–2 (2013). https://users.cs.fiu.edu/~markaf/doc/j2.finlayson.2013.sdv.37.173.pdf.

Fisher, Walter R. "Narration as a Human Communication Paradigm: The Case of Public Moral Argument." *Communication Monographs* 51, no. 1 (March 1, 1984). https://doi.org/10.1080/03637758409390180.

Florence, Elinor. "D-Day Dummies and Decoys." *Blog,* May 28, 2014. http://elinorflorence.com/blog/d-day-decoys.

Fuchs, C. "The self-organization of cyberprotest." *Advances in Education, Commerce & Governance,* Internet Society II, (2006): 275–295.

Fulton, Bradley. "The Weakest Link: The Human Factor Lessons Learned from the German WWII Enigma Cryptosystem." *Sans Institute Information Security Reading Room* 2 (2001).

Garcia-Navarro, Lulu. "What's An 'Incel'? The Online Community Behind the Toronto Van Attack." *National Public Radio,* April 29, 2018. https://www.npr.org/2018/04/29/606773813/whats-an-incel-the-online-community-behind-the-toronto-van-attack\.

Gerbaudo, Paolo. *Tweets and the Streets: Social Media and Contemporary Activism.* London: Pluto Press, 2018.

Giles, Jim. "Fake Tweets by Socialbot Fool Hundreds of Followers." *New Scientist, Reed Business Information, UK,* March 19, 2011.

Gill, Stephen. "Toward a postmodern prince? The battle in Seattle as a moment in the new politics of globalisation." *Millennium* 29, no. 1 (2000): 131–140.

Gleason, Benjamin. "#Occupy Wall Street: Exploring Informal Learning About a Social Movement on Twitter." *American Behavioral Scientist* 57, no. 7 (2013): 966–982.

Goffman, Erving. *Frame Analysis: An Essay on the Organization of Experience.* Harvard University Press, 1974.

Gold, Steve. "Get Your Head Around Hacker Psychology." *Engineering & Technology* 9, issue 1 (2014): 76–80.

Goldman, Paul, and Alistair Jamieson. "Hamas Used Fake Social Media Accounts to Hack Israeli Soldiers' Phones: IDF." *NBC News,* January 12, 2017.

Goodboy, Alan K., and Matthew M. Martin "The Personality Profile of a Cyberbully: Examining the Dark Triad." *Computers in Human Behavior* 49 (2015): 1–4.

Goode, Luke. "Anonymous and the Political Ethos of Hacktivism." *Popular Communication,* 13, 1 (2015): 74-86.

Goodin, Dan. "Meet 'badBIOS,' the Mysterious Mac and PC Malware That Jumps Airgaps." *Ars Technica,* October 31, 2013.

Goodin, Dan. "New IoT botnet offers DDoSes of once-unimaginable sizes for $20." *Ars Technica*, February 1, 2018. https://arstechnica.com/information-technology/2018/02/for-sale-ddoses-guaranteed-to-take-down-gaming-servers-just-20/.

Gorwa, Robert. "On the Internet, Nobody Knows That You're a Russian Bot." *RealClearDefense*, March 21, 2017, http://www.realcleardefense.com/articles/2017/03/21/on_the_internet_nobody_knows_that_youre_a_russian_bot_111010.html.

Gottschalk, Frederick C. "The Role of Special Forces in Information Operations." Master of Military Art and Science thesis, U.S. Army Command and General Staff College, June 2000.

Graff, Garret M. "Inside the Hunt for Russia's Most Notorious Hacker." *Wired*, March 21, 2017. https://www.wired.com/2017/03/russian-hacker-spy-botnet/.

Graham-Harrison, Emma. "Could ISIL's 'cyber caliphate' unleash a deadly attack on key targets?" *Guardian*, April 12, 2015.

Greenberg, Andy. "How Hackers Hijacked a Bank's Entire Online Operation." *Wired*, April 4, 2017. https://www.wired.com/2017/04/hackers-hijacked-banks-entire-online-operation/.

Greenberg, Andy. "The Untold Story of NotPetya, the Most Devastating Cyberattack in History." *Wired*, August 22, 2018. https://www.wired.com/story/notpetya-cyberattack-ukraine-russia-code-crashed-the-world/.

Greene, Richard. "The Russian Hack Absolutely Affected the Outcome of the 2016 Election." *HuffPost*, December 15, 2016.

Gregory, Paul Roderick. "Inside Putin's Campaign of Social Media Trolling and Faked Ukrainian Crimes." *Forbes*, May 11, 2014.

Gregory, Paul Roderick. "Russian TV Propagandists Caught Red Handed: Same Guy, Three Different People (Spy, Bystander, Heroic Surgeon)." *Forbes*, April 12, 2014.

Griffin, Andrew. "Anonymous 'Trolling Day' Against ISIL Begins, with Group's 'Day of Rage' Mostly Consisting of Posting Mocking Memes.'" *Independent,* December 15, 2011. https://www.independent.co.uk/life-style/gadgets-and-tech/news/anonymous-trolling-day-against-ISIL-begins-with-group-s-day-of-rage-mostly-consisting-of-posting-a6769261.html.

Grohe, Edwin. *The Cyber Dimensions of the Syrian Civil War: Implications for Future Conflict.* Laurel, MD: Johns Hopkins University Applied Physics Laboratory, 2015.

Gunitsky, Seva. "Corrupting the cyber-commons: Social media as a tool of autocratic stability." *Perspectives on Politics* 13, no. 1 (2015): 42–54.

Guynn, Jessica. "Verizon Data from 6 million Users Leaked Online." *USA Today,* July 12, 2017, updated July 13, 2017.

Hadnagy, Christopher. *Unmasking the Social Engineer: The Human Element of Security.* Indianapolis, IN: John Wiley & Sons, 2014.

Hahn, Erin N. and W. Sam Lauber. *Legal Implications of the Status of Persons in Resistance.* Fort Bragg, NC: USASOC, 2015.

Hale, Nate. "Whatever Happened to OPSEC?" *In From the Cold* blog, May 14, 2011.

Hanna, Richard, Andrew Rohm, and Victoria L. Crittenden. "We're All Connected: The Power of the Social Media Ecosystem." *Business Horizons* 54, no. 3 (2011): 265–273.

Harmon, Amy. "'Hacktivists' of All Persuasions Take Their Struggle to the Web." *New York Times,* October 31, 1998. http://www.nytimes.com/1998/10/31/world/hacktivists-of-all-persuasions-take-their-struggle-to-the-web.html.

Hay Newman, Lily. "The Ransomware Meltdown Experts Warned Us About." *Wired,* May 12, 2017. https://www.wired.com/2017/05/ransomware-meltdown-experts-warned/.

Hayden, Michael Edison. "A timeline of the WannaCry cyberattack." ABC News, May 15, 2017. https://abcnews.go.com/US/timeline-wannacry-cyberattack/story?id=47416785.

Headquarters, Department of the Army. "Cyberspace and Electronic Warfare Operations." Field Manual 3-12, April 11, 2017.

Headquarters, Department of the Army. "The Conduct of Information Operations." ATP 3-13.1 (October 2018).

Heath-Kelly, Charlotte, and Lee Jarvis. "Affecting Terrorism: Laughter, Lamentation, and Detestation as Drives to Terrorism Knowledge." *International Political Sociology* 11, no. 3 (2017): 239–256.

Hegghammer, Thomas, ed. *Jihadi Culture.* Cambridge, UK: Cambridge University Press, 2017.

Hern, Alex. "US government increases funding for Tor, giving $1.8m in 2013." *The Guardian,* July 29, 2014. https://www.theguardian.com/technology/2014/jul/29/us-government-funding-tor-18m-onion-router.

Hilton, Scott. "Dyn Analysis Summary of Friday October 21 Attack." *Oracle Vantage Point in the News,* October 21, 2016. https://www.pcworld.com/article/3133847/internet/ddos-attack-on-dyn-knocks-spotify-twitter-github-etsy-and-more-offline.html.

Hoffman, Michael. "US Air Force Targets and Destroys ISIS HQ Building Using Social Media." Defensetech, Military.com. Last modified 3 Jun 2015. https://www.defensetech.org/2015/06/03/us-air-force-targets-and-destroys-isis-hq-building-using-social-media/.

Honan, Mat. "How Apple and Amazon Security Flaws Led to My Epic Hacking." *Wired,* August 06, 2012. https://www.wired.com/2012/08/apple-amazon-mat-honan-hacking/.

Houlka, Max. "What Anonymous Can Tell Us About the Relationship Between Virtual Community Structure and Participatory Form." *Policy Studies* 38, no. 2 (2017): 168–184.

Humud, Carla E., Robert Pirog, and Liana Rosen. "Islamic State Financing and U.S. Policy Approaches." Congressional Research Service Report R43980, April 10, 2015.

Hussain, Murtaza. "The New Information War." *Intercept,* November 25, 2017. https://theintercept.com/2017/11/25/information-warfare-social-media-book-review-gaza/.

Hutchins, Eric M., Michael J. Cloppert, and Rohan M. Amin. "Intelligence-Driven Computer Network Defense Informed by Analysis of Adversary Campaigns and Intrusion Kill Chains." Paper presented at the 6th Annual International Conference on Information Warfare and Security, Washington, DC, 2011.

Institute for Human Rights and Business. "Human Rights Challenges for Telecommunications Vendors: Addressing the Possible Misuse of Telecommunications Systems. Case Study: Ericsson." Case Study No. 2, November 2014.

International Committee of the Red Cross. *Commentary on the Additional Protocols of 8 June 1977 to the Geneva Conventions of 12 August 1949,* edited by Yves Sandoz, Christophe Swinarski and Bruno Zimmerman. Netherlands: Martinus Nijhoff Publishers, 1987.

International Committee of the Red Cross. *Geneva Convention Relative to the Treatment of Prisoners of War (Third Geneva Convention).* 75 UNTS 135. 12 August 1949. Accessed August 14, 2019. https://www.refworld.org/docid/3ae6b36c8.html.

Jackson, Stephen. "NATO Article 5 and Cyber Warfare: NATO's Ambiguous and Outdated Procedure for Determining When Cyber Aggression Qualifies as an Armed Attack." Center for Infrastructure Protection and Homeland Security, George Mason University, last modified August 18, 2016. http://cip.gmu.edu/2016/08/16/nato-article-5-cyber-warfare-natos-ambiguous-outdated-procedure-determining-cyber-aggression-qualifies-armed-attack/.

Jarvis, Jeff. "#OccupyWallStreet and the Failure of Institutions." *Huff-Post,* October 3, 2011, http://www.huffingtonpost.com/jeff-jarvis/occupywallstreet-the-fail_b_991928.html.

John D. Sutter. "Will Twitter war become the new norm?" *CNN online,* November 15, 2012.

Joint Publication 3-05 Special Operations. United States Department of Defense, 2014.

Joint Publication 3-13 Information Operations. United States Department of Defense, 2014.

Joint Task Force Transformation Initiative. "Guide for Applying the Risk Management Framework to Federal Information Systems: A Security Life Cycle Approach." National Institute of Standards and Technology (NIST) Special Publication 800-37 Revision 1, February 2010, updates as of June 2014..

Kahn, Mukhtar A. "The FM Mullahs and the Taliban's Propaganda War in Pakistan." *Terrorism Monitor* 7, issue 14 (May 26, 2009).

Kalpokas, Ignas. "Influence Operations: Challenging the Social Media – Democracy Nexus." *SAIS European Journal of Global Affairs,* 2016.

Kaspersky Lab Global Research and Analysis Team. "The 'Icefog' Apt: A Tale of Cloak and Three Daggers." Version 1.00, 2013. https://media.kaspersky.com/en/icefog-apt-threat.pdf.

Kaspersky. "Kaspersky Unmasks 'Icefog' Cyberattacks." 2019. https://usa.kaspersky.com/resource-center/threats/icefog-cyberattacks.

Kelion, Leo. "Dark Web Markets Boom after AlphaBay and Hansa Busts." *BBC News*, August 1, 2017.

Kelly, Heather. "83 Million Facebook accounts are fakes and dupes." CNN, August 3, 2012. https://www.cnn.com/2012/08/02/tech/social-media/facebook-fake-accounts/index.html.

Keys, Ronald, and Kendra Simmons. "Cyberspace Security and Attribution." National Security Cyberspace Institute, July 20, 2010, 4.

Khazan, Olga. "Gentlemen Reading Each Others' Mail: A Brief History of Diplomatic Spying." *Atlantic*, June 17, 2013.

King, Gary, Jennifer Pan, and Margaret E. Roberts. "How censorship in China allows government criticism but silences collective expression." *American Political Science Review* 107, no. 2 (2013): 326–343.

Kleinberg, Jon. "The Convergence of Social and Technological Networks." *Communications of the ACM* 51, no. 11 (November 2008): 66–72.

Knapp, E.D. "Unconventional Warfare in Cyberspace." Unpublished master's thesis, U.S. Army War College, 2012.

Kocieniewski, David. "Six Animal Rights Advocates Are Convicted of Terrorism." *The New York Times*, March 3, 2006. https://www.nytimes.com/2006/03/03/nyregion/six-animal-rights-advocates-are-convicted-of-terrorism.html.

Koerner, Brendan I. "Why ISIS is Winning the Social Media War." *Wired*, April 2016, https://www.wired.com/2016/03/isis-winning-social-media-war-heres-beat/.

Koetsier, John. "China Bans Internet Anonymity." *Venturebeat.com*, December 28, 2012. https://venturebeat.com/2012/12/28/china-bans-internet-anonymity/.

Koval, Nikolay. "Revolution Hacking." In *Cyber War in Perspective: Russian Aggression Against Ukraine*, edited by Kenneth Geers, 55-58. Tallinn, Estonia: NATO Cooperative Cyber Defence Center of Excellence, 2015.

Kraemer, Romy, Gail Whiteman, and Bobby Banerjee. "Conflict and Astroturfing in Niyamgiri: The Importance of National Advocacy Networks in Anti-Corporate Social Movements." *Organization Studies* 34, issue 5-6 (2013): 823–852.

Kushner, David. "The Autistic Hacker." *IEEE Spectrum* (2011). Available online at: http://spectrum. IEEE. org/telecom/internet/the-autistic-hacker.

Kuznar, Lawrence A., and William H. Moon. "Thematic Analysis of ISIL Messaging." In Multi-Method Assessment of ISIL, edited by H. Cabayan and S. Canna. Washington, DC: Strategic Multilayer Assessment Office, Office of the Secretary of Defense, 2014. http://nsiteam.com/sma-publications/.

Lai, M. Y., C. F. Jin, K. Nie, and J. H. Zhao. „Cyber physical logistics system: The implementation and challenges of next-generation logistics system." *Systems Engineering* 4, 008 (2011).

Lakoff, George, Howard Dean, and Don Hazen. *Don't Think of an Elephant! Know Your Values and Frame the Debate ; the Essential Guide for Progressives.* White River Junction, VT: Chelsea Green Pub. Co, 2004.

Lakomy, Miron. "Cracks in the Online 'Caliphate': How the Islamic State is Losing Ground in the Battle for Cyberspace." *Perspectives on Terrorism* 11, no. 3 (2017).

Lamba, Hemank, Momin M. Malik, and Juergen Pfeffer. "A Tempest in a Teacup? Analyzing Firestorms on Twitter." *IEEE/ACM International Conference*, 2015. https://www.researchgate.net/publication/301444817_A_Tempest_in_a_Teacup_Analyzing_Firestorms_on_Twitter.

Lane, Jill, "Digital Zapatistas." *The Drama Review* 47, no. 2 (2003): 129–144.

Lazer, David M. J., et al. "The Science of Fake News." *Science* 359, no. 6380 (March 9, 2018). https://doi.org/10.1126/science.aao2998.

Lecher, Colin. "Massive Attack: How a weapon of war became a weapon against the web." *Verge*, April 14, 2017. https://www.theverge.com/2017/4/14/15293538/electronic-disturbance-theater-zapatista-tactical-floodnet-sit-in.

Ledingham, R., and R. Mills, "A preliminary study of autism and cybercrime in the context of international law enforcement." *Advances in Autism*, 1, 1 (2015): 2–11.

Lee, Caroline W. "The Roots of Astroturfing." *Contexts* 9, no. 1 (2010): 73–75.

Lee, Robert M., Michael J. Assante, and Tim Conway. "Analysis of the Cyber Attack on the Ukrainian Power Grid Defense Use Case." SANS and Electricity Information Sharing and Analysis Center (E-ISAC), March 18, 2016. https://www.nerc.com/pa/CI/ESISAC/ Documents/E-ISAC_SANS_Ukraine_DUC_18Mar2016.pdf.

Leger, Donna Leinwand. "How FBI brought down cyber-underworld site Silk Road." *USA Today*, October 21, 2013.

Leiner, Barry M., Vinton G. Cerf, David D. Clark, Robert E. Kahn, Leonard Kleinrock, Daniel C. Lynch, Jonathan B. Postel, Lawrence G. Roberts, and Stephen S. Wolff. "Brief History of the Internet." *Internet Society*, 1997.

Lentini, Peter, and Muhmmad Bakashmar. "Jihadist Beheading: A Convergence of Technology, Theology, and Teleology?" *Studies in Conflict & Terrorism* 30, issue 4 (2007): 303–325.

Lerman, Kristina, Xiaoran Yan, and Xin-Zeng Wu. "The 'Majority Illusion' in Social Networks." PLoS ONE 11, no. 2 (2016). https://doi. org/10.1371/journal.pone.0147617.

Leskovec, Jure, Lars Backstrom, and Jon Kleinberg. "Meme-Tracking and the Dynamics of the News Cycle." In *Proceedings of the 15th ACM SIGKDD International Conference on Knowledge Discovery and Data Mining*, KDD '09. New York, NY, USA: ACM, 2009. https://doi. org/10.1145/1557019.1557077.

Levi, Margaret, and David Olson. "The battles in Seattle." *Politics & Society* 28, no. 3 (2000): 309–329.

Levy, **Steven**. *Hackers: Heroes of the computer revolution*. Vol 4. New York: Penguin Books, 2001.

Lewandowsky, Stephan, et al. "Misinformation and Its Correction Continued Influence and Successful Debiasing." Psychological Science in the Public Interest, December 2012.

Lewis, James Andrew. "'Compelling Opponents to Our Will': The Role of Cyber Warfare in Ukraine." In *Cyber War in Perspective: Russian Aggression Against Ukraine*, edited by Kenneth Geers. Tallinn, Estonia: NATO Cooperative Cyber Defence Center of Excellence, 2015.

Leyden, John. "Cops Harpoon Two Dark Net Whales in Megabust: AlphaBay and Hansa." *Register,* July 20, 2017.

Lieberman, Ariel Victoria. "Terrorism, the Internet, and Propaganda: A Deadly Combination." *Journal of National Security Law and Policy*, 9 (2017): 95–124.

Ligon, Gina S., Mackenzie Harms, John Crowe, Leif Lundmark, and Pete Simi. "An Organizational Profile of the Islamic State: Leadership, Cyber Expertise, and Firm Legitimacy." In *Cabayan, Hriar and Canna, Sarah. A Multi-Method Assessment of ISIL*. Office of the Secretary of Defense, Strategic Multilayer Assessment, 2014.

Lim, Merlyna. "Clicks, cabs, and coffee houses: Social media and oppositional movements in Egypt, 2004–2011." *Journal of Communication* 62, no. 2 (2012): 231–248.

Lin, Abe C. "Comparison of the Information Warfare Capabilities of the ROC and PRC." *Infowar*, 2000. https://cryptome.org/cn2-infowar.htm.

Linebarger, Paul M. A. *Psychological Warfare 2nd Edition*. Washington, D.C.: Combat Forces Press, 1954.

Lister, Martin, Jon Dovey, Seth Giddings, Iain Grant, and Kieran Kelly. *New Media: A Critical Introduction, Second Edition*. New York, NY: Routledge, 2009.

Locket, Jon. "'See the World, Join ISIL' Spoof Recruitment Posters for ISIL Appear Across East London." *Sun*, July 29, 2016. https://www.thesun.co.uk/news/1520886/spoof-recruitment-posters-for-ISIL-appear-across-east-london/.

Lokot, Tetyana. "Public Networked Discourses in Ukraine-Russia Conflict: 'Patriotic Hackers' and Digital Populism." *Irish Studies in International Affairs* 28 (2017): 99–116.

Lottridge, Danielle, Frank Bentley, Matt Wheeler, Jason Lee, Janet Cheung, Katherine Ong, and Cristy Rowley. "Third-wave livestreaming: teens' long form selfie." In *Proceedings of the 19th International Conference on Human-Computer Interaction with Mobile Devices and Services*, ACM, 2017.

Lovell, Stanley. *Of Spies & Stratagems: Incredible Secrets of World War II Revealed By a Master Spy*. New York: Prentice-Hall Pocket Books, 1964.

lto, Aki. "A Former Anonymous Hacker's Search for Redemption." *Bloomberg Technology*, March 6, 2018. https://www.bloomberg.com/news/features/2018-03-06/a-former-anonymous-hacker-s-search-for-redemption.

Luo, Xin (Robert), Richard Brody, Alessandro Seazzu, and Stephen Burd. "Social Engineering: The Neglected Human Factor for Information Security Management." *Information Resources Management Journal*, 24, no. 3, (July-September 2011): 1–8.

Lynch Pyon, Andrea. "Disrupting Terrorist Financing: Interagency Collaboration, Data Analysis, and Predictive Tools." *Forensics Journal* 6 (2015): 42–51.

Lynn, III, William J. "Defending a New Domain: The Pentagon's Cyberstrategy." *Foreign Affairs*, September/October 2010.

Maciulis, Tony. "Al-Qaida Recruitment in Shadowy Net World: As Terror Network Becomes Increasingly Tech-Savvy, Is the US Prepared?" *MSNBC*, March 17, 2006.

MacKinnon, Rebecca. "China's 'Networked Authoritarianism.'" *Journal of Democracy* 22, no. 2 (2011): 32–46.

Maher, Shiraz, and Joseph Carter. "Analyzing the ISIS 'Twitter Storm." *War on the Rocks,* June 24, 2014. https://warontherocks.com/2014/06/analyzing-the-isis-twitter-storm/.

Markoff, John and Nicole Perlroth. "Firm Is Accused of Sending Spam, and Fight Jams Internet." *The New York Times*, March 26, 2013. https://www.nytimes.com/2013/03/27/technology/internet/online-dispute-becomes-internet-snarling-attack.html.

Markoff, John. "Before the Gunfire, Cyberattacks." *The New York Times*, August 12, 2008. https://www.nytimes.com/2008/08/13/technology/13cyber.html.

Marks, Paul. "How ISIS Is Winning the Online War for Iraq." New Scientist, June 25, 2014. https://www.newscientist.com/article/dn25788-how-isis-is-winning-the-online-war-for-iraq/.

Marsden, Peter V., and Noah E. Friedkin. "Network Studies of Social Influence." *Sociological Methods & Research* 22, no. 1 (August 1, 1993). https://doi.org/10.1177/0049124193022001006.

Martinez, Michael. "Cyberwar: CyberCaliphate Targets US Military Spouses; Anonymous Hits ISIS." *CNN*, February 11, 2015. https://www.cnn.com/2015/02/10/us/isis-cybercaliphate-attacks-cyber-battles/index.html.

Martins, Ralph. "'Anonymous' Cyberwar Against ISIL and the Asymmetrical Nature of Cyber Conflicts." *Cyber Defense Review* 2, no. 3 (2017): 95–106.

Massanari, Adrienne. "# Gamergate and The Fappening: How Reddit's algorithm, governance, and culture support toxic technocultures." *New Media & Society* (2015): 1461444815608807.

Mattis, Jim. "Summary of the 2018 National Defense Strategy of the United States of America: Sharpening the American Military's Competitive Edge." US Department of Defense, 2018.

Maxwell, David S. "The Cyber Underground-Resistance to Active Measures and Propaganda: 'The Disruptors Motto' – 'Think for Yourself'." *Small Wars Journal* (2017). http://smallwarsjournal.com/jrnl/art/the-cyber-underground-–-resistance-to-active-measures-and-propaganda-"the-disruptors"-mot-0.

McCarthy, John D., and Mayer N. Zald. "Resource Mobilization and Social Movements: A Partial Theory." In *Social Movements in an Organizational Society: Collected Essays*. New Brunswick, NJ: Transaction Publisher, 1987.

McCleave Maharawal, Manissa. "Occupy Wall Street and a radical politics of inclusion." *Sociological Quarterly* 54, no. 2 (2013): 177–181.

McCulloh, Ian. "Social Conformity in Networks." *Connections* 33, no. 1 (2013): 35–42.

McElreath, David H., Daniel Adrian Doss, Leisa McElreath, Ashley Lindsley, Glenna Lusk, Joseph Skinner, and Ashley Wellman. "The Communicating and Marketing of Radicalism: A Case Study of ISIL and Cyber Recruitment." *International Journal of Cyber Warfare and Terrorism (IJCWT)* 8, no. 3 (2018): 26–45.

McGoogan, Cara. "North Korea's Internet Revealed to Have Just 28 Websites." *Telegraph*, September 21, 2016. http://www.telegraph.co.uk/technology/2016/09/21/north-koreas-internet-revealed-to-have-just-28-websites/.

McMullan, John, and Aunshul Rege. "Cyberextortion at online gambling sites: Criminal organization and legal challenges." *Gaming Law Review* 11, no. 6 (2007): 648–665.

Melzer, Nils. *Interpretive Guidance on the Notion of Direct Participation in Hostilities Under International Humanitarian Law*. Geneva, Switzerland: International Committee of the Red Cross, 2009.

Microsoft Corporation. "Mitigating Pass-the-Hash and Other Credential Theft, Version 2." *Microsoft Trustworthy Computing*, July 7, 2014. https://technet.microsoft.com/en-us/dn785092.aspx.

Middleton, Pete. "When Chinese Troops Fired on Two Gloster Meteors at Chongdan, the Australians Made Them Regret It." *Military History*, August 2005.

Miklaszewski, Jim. "US ID's al-Qaida Iraq Boss, Launches Raid." *NBC News*, June 15, 2006.

Milan, Stefania. "When Algorithms Shape Collective Action: Social Media and the Dynamics of Cloud Protesting." *Social Media+ Society* 1, no. 2 (2015): 2056305115622481.

Miller, Dale T., and Leif D. Nelson. "Seeing Approach Motivation in the Avoidance Behavior of Others: Implications for an Understanding of Pluralistic Ignorance." *Journal of Personality and Social Psychology* 83, no. 5 (2002): 1066–1075.

Minchev, Zlatogor, "Hybrid Challenges to Human Factor in Cyberspace." In *Countering Terrorist Activities in Cyberspace* 139 (2018): 32–43.

Mitnick, Kevin. *Ghost in the wires: My adventures as the world's most wanted hacker.* Paris: Hachette UK, 2011.

Mizrach, Steven. "Is there a hacker ethic for 90s hackers." In *The Legitimization of Strategic Information Warfare: Ethical Consideration*, edited by R. Molander and S. Siang. Professional Ethics Report, XI (4), 1997.

Molnar, Andrew R., William A. Lybrand, Lorna Hahn, James L. Kirkman, and Peter B. Riddleberger. *Undergrounds in insurgent, revolutionary, and resistance warfare.* Washington, DC, 1963.

Moore, Jina. "Social media: Did Twitter and Facebook really build a global revolution?" *Christian Science Monitor*, June 30, 2011. https://www.csmonitor.com/World/Global-Issues/2011/0630/Social-media-Did-Twitter-and-Facebook-really-build-a-global-revolution.

Morin, Richard. "Rising Share of Americans See Conflict Between Rich and Poor." *Pew Research Center Social & Demographic Trends*, January 11, 2012. http://www.pewsocialtrends.org/2012/01/11/rising-share-of-americans-see-conflict-between-rich-and-poor/.

Mueller, Jessica, and Ronn Johnson. "11 Emerging trends in technology and forensic psychological roots of radicalization and lone wolf terrorists." *Emerging and Advanced Technologies in Diverse Forensic Sciences* (2018).

Murphy, Jarrett. "Osama's Satellite Phone Switcheroo." *CBS News*, January 21, 2003.

Murphy, Lorraine. "The Curious Case of the Jihadist Who Started Out a Hacktivist." *Hive*, December 15, 2015. https://www.vanityfair.com/news/2015/12/isis-hacker-junaid-hussain.

Murtaza Hussain. "The New Information War." *The Intercept*, November 25, 2017. https://theintercept.com/2017/11/25/information-warfare-social-media-book-review-gaza/.

Nakashima, Ryan. "AP Exclusive: Google Tracks Your Movements, Like It or Not." *Associated Press*, August 13, 2018. https://www.apnews.com/828aefab64d4411bac257a07c1af0ecb.

National Institute of Standards and Technology. "Standards for Security Categorization of Federal Information and Information Systems." Federal Information Processing Standard Publication 199 (FIPS 199), February 2004.

Navy Doctrine Library System. "Cyberspace Operations." Naval Warfare Publication (NWP) 3-12, July 2011.

Naylor, Hugh and Eric Curringham. "Anti-Islamic State activist and his friend beheaded in Turkey." *Washington Post*, October 30, 2015. https://www.washingtonpost.com/world/anti-islamic-state-activist-and-his-friend-beheaded-in-turkey/2015/10/30/c3340038-7f05-11e5-bfb6-65300a5ff562_story.html?utm_term=.5f6082863ea3.

Nicholas, Scott. "DoD Joint Staff Issues Cybersecurity Warning Against Lenovo Computers, Handheld Devices." *ExecutiveGov*, October 25, 2016.

Nield, David. "Stolen Uber accounts on sale for $1 each." *Digital Trends*, March 28, 2015.

Nikiforakis, Nick, and Gnes Acar. "Browser Fingerprinting and the Online-Tracking Arms Race." *IEEE Spectrum*, July 25, 2014.

Nissen, Thomas Elkjer. "Terror. com – IS's social media warfare in Syria and Iraq." *Contemporary Conflicts: Military Studies Magazine* 2, no. 2 (2014).

Norman, Kent L. *Cyberpsychology: An introduction to human-computer interaction.* Cambridge, UK: Cambridge University Press, 2017.

North Atlantic Treaty Organization. "Wales Summit Declaration." Press Release (2014) 120, September 5, 2014, last modified September 26, 2016. http://www.nato.int/cps/en/natohq/official_texts_112964.htm.

Northcutt, Stephen. "Spear Phishing." *SANS Method of Attack Series*, 2017.

Olson, Mancur. *The Logic of Collective Action: Public Goods and the Theory of Groups.* Cambridge, MA: Harvard University Press, 1965.

Olson, Parmy. *We Are Anonymous: Inside the Hacker World of LulzSec, Anonymous, and the Global Cyber Insurgency.* New York: Little, Brown and Company, 2012.

Oriola, Temitope B., and Olabanji Akinola. "Ideational Dimensions of the Boko Haram Phenomenon." *Studies in Conflict & Terrorism* (June 1, 2017): 1–24. https://doi.org/10.1080/1057610X.2017.1338053.

Oriyano, Sean-Philip . *CEH Certified Ethical Hacker Study Guide, Version 9.* Indianapolis, IN: John Wiley & Sons, 2016.

Paganini, Pierluigi. "Crimea – The Russian Cyber Strategy to Hit Ukraine." Infosec Institute, March 11, 2014. http://resources.infosecinstitute.com/crimea-russian-cyber-strategy-hit-ukraine/.

Paget, Francois. "Hacktivism: Cyberspace has become the new medium for political voices." McAfee Labs White Paper, 2012. https://www.mcafee.com/us/resources/white-papers/wp-hacktivism.pdf.

Pakharenko, Glib. "Cyber Operations at Maidan: A First-Hand Account." In *Cyber War in Perspective: Russian Aggression Against Ukraine*, edited by Kenneth Geers, 59-66. Tallinn, Estonia: NATO Cooperative Cyber Defence Center of Excellence, 2015.

Pandya, Ishan, Hitanshu Joshi, Biren Patel, and Harshil Joshi. "Threats that Deep Web Possess to Modern World." JIRST: National Conference on Latest Trends in Networking and Cyber Security, March 2017. http://www.ijirst.org/articles/SALLTNCSP033.pdf.

Paquet-Clouston, Masarah, David Décary-Hétu, and Olivier Bilodeau. "Cybercrime is whose responsibility? A case study of an online behaviour system in crime." *Global Crime* 19, no. 1 (2018): 1–21.

Park, Albert, and Mike Conway. "Harnessing Reddit to Understand the Written-Communication Challenges Experienced by Individuals with Mental Health Disorders: Analysis of Texts from Mental Health Communities." *Journal of Medical Internet Research*, 20, no. 4 (2018): e121.

Pavel, Tal. "Physical Threats in Online Worlds–Technology, Internet and Cyber under Terror Organization Services; a Test Case of 'The Islamic State.'" ICT Information and Communications Technologies 6, issue 1 (2017): 73–82.

Perez, Evan. "How celebrity hacker 'Sabu' helped feds thwart 300 cyber-attacks." *CNN Business*, May 27, 2014.

Perlroth, Nicole, and Clifford Krauss. "A Cyberattack in Saudi Arabia Had a Deadly Goal. Experts Fear Another Try." *New York Times*, March 15, 2018.

Perlroth, Nicole. "Security Experts Oppose Government Access to Encrypted Communication." *The New York Times*, July 7, 2015. https://www.nytimes.com/2015/07/08/technology/code-specialists-oppose-us-and-british-government-access-to-encrypted-communication.html.

Pew Research Center. "Smartphone Ownership and Internet Usage Continues to Climb in Emerging Economies." February 2016.

Pham, Sherisse. "Telegram promises to act faster on terror content as Indonesia threatens ban." CNN, July 17, 2017. https://money.cnn.com/2017/07/17/technology/telegram-block-indonesia-terrorism/index.html.

Phillips, Whitney. "LOLing at Tragedy: Facebook Trolls, Memorial Pages and Resistance to Grief Online." *First Monday* 16, no. 12 (November 28, 2011). http://firstmonday.org/ojs/index.php/fm/article/view/3168.

Prescott, Anna T., James D. Sargent, and Jay G. Hull. "Metaanalysis of the Relationship Between Violent Video Game Play and Physical Aggression Over Time." *Proceedings of the National Academy of Sciences* 115, no. 40 (2018): 9882–9888.

Prosecutor v. Tadic, Case No. IT-94-1-A. "Appeals Chamber Decision on the Defence Motion for Interlocutory Appeal on Jurisdiction." Int'l Crim. Trib. for the Former Yugoslavia: October 2, 1995.

Protocol Additional to the Geneva Conventions of 12 August 1949, and Relating to the Protection of Victims of International Armed Conflicts, 1125 U.N.T.S. 3, June 8, 1977.

Protocol Additional to the Geneva Conventions of 12 August 1949, and Relating to the Protection of Victims of Non-International Armed Conflicts. 1125 U.N.T.S 609, June 8, 1977.

Putnam, Robert D. *Bowling alone: The collapse and revival of American community.* New York, NY: Simon and Schuster, 2001.

Raab, Jorg, and H. Brinton Milward. "Dark Networks as Problems." *Journal of Public Administration Research and Theory* 13, issue 4 (2003): 413–439.

Rand, Emily. "Source: 2.7 terabytes of Data Recovered from bin Laden Compound." *CBS Evening News*, May 6, 2011.

Rapoza, Kenneth. "Kaspersky Lab Uncovers New Cyber Hit-n-Run Op Called 'Icefog.'" *Forbes*, September 25, 2013.

Ratkiewicz, Jacob, et al. "Detecting and Tracking Political Abuse in Social Media." *ICWSM* 11 (2011).

Ray, Bill. "How I Hacked SIM Cards with a Single Text – and the Networks DON'T CARE: US and Euro Telcos Won't Act Until Crims Do, White Hat Sniffs." *Register*, September 23, 2013.

Raymond, Eric S. *The Jargon File* version 4.4.8. http://www.catb.org/~esr/jargon/.

Read, Stephen John, and Lynn Carol Miller. "Stories Are Fundamental to Meaning and Memory: For Social Creatures, Could It Be Otherwise." Knowledge and Memory: The Real Story. Advances in Social Cognition 8, 1995.

Regalado, Daniel, Nart Villeneuve, and John Scott Railton. "Behind the Syrian Conflict's Digital Front Lines." *FireEye Threat Intelligence Special Report*, February 2015.

Reno, William, and Jahara Matisek. "A New Era of Insurgent Recruitment: Have 'New' Civil Wars Changed the Dynamic?" *Civil Wars* (2018): 1–21.

Robinson, Teri. "International Operation Takes Down AlphaBay, Hansa Dark Web Markets." *SC Magazine*, July 20, 2017.

Rogers, M. K. "The psyche of cybercriminals: A Psycho-Social perspective." In *Cybercrimes: A Multidisciplinary Analysis*. Berlin: Springer, 2011.

Rose, Karen, Scott Eldridge, and Lyman Chapin. "The Internet of Things: An Overview. Understanding the Issues and Challenges of a More Connected World." The Internet Society, October 2015.

Rotman, Dana, Sarah Vieweg, Sarita Yardi, Ed Chi, Jenny Preece, Ben Shneiderman, Peter Pirolli, and Tom Glaisyer. "From Slacktivism to Activism: Participatory Culture in the Age of Social Media." In *CHI'11 Extended Abstracts on Human Factors in Computing Systems*. Vancouver, Canada: ACM, 2011.

Runions, K. C., and M. Bak. "Online moral disengagement, cyberbullying, and cyber-aggression." *Cyberpsychology, Behavior, and Social Networking*, 18, 7 (2015): 400–405.

Russell, James A. "A circumplex model of affect." *Journal of personality and social psychology* 39, no. 6 (1980): 1161.

Russell, Jon. "Chinese police are using smart glasses to identify potential suspects." *Tech Crunch*, February 8, 2018.

Ruston, Scott W., and Jeffry R. Halverson. "'Counter' or 'Alternative': Contesting Video Narratives of Violent Islamist Extremism." In *Visual Propaganda and Extremism in the Online Environment*, edited by Carol K. Winkler and Cori E. Dauber, 116. Carlisle Barracks, PA: United States Army War College Press, 2014.

Sakkov, Sven. "Foreword." In *Cyber War in Perspective: Russian Aggression Against Ukraine*, edited by Kenneth Geers. Tallinn, Estonia: NATO Cooperative Cyber Defence Center of Excellence, 2015.

Sample, Timothy R., and Michael Swetnam. *#CyberDoc No Borders-No Boundaries*. Arlington, VA: Potomac Institute Press, 2012.

Samuel, Sigal. "Canada's 'Incel Attack' and Its Gender-Based Violence Problem." *Atlantic*, April 28, 2018. https://www.theatlantic.com/international/archive/2018/04/toronto-incel-van-attack/558977/.

Sanger David E., and Nicole Perlroth. "A New Era of Internet Attacks Powered by Everyday Devices." *New York Times*, October 22, 2016.

Satell, Greg. "How Social Movements Change Minds." *Harvard Business Review*, July 28, 2015. https://hbr.org/2015/07/how-social-movements-change-minds.

Saul, Lawrence H. "Tactical Commander's Handbook Information Operations: Operation Iraqi Freedom (OIF)." Center for Army Lessons Learned (CALL) Combined Arms Center, May 2015, FOR OFFICIAL USE ONLY.

Scahill, Jeremy. *Dirty Wars: The World is a Battlefield*. New York: Nation Books, 2013.

Schmidt, Leonie. "Cyberwarriors and Counterstars: Contesting Religious Radicalism and Violence on Indonesian Social Media." *Asiascape: Digital Asia* 5, no. 1-2 (2018): 32–67.

Schmitt, Michael N. "Cyber Operations and the *Jus in Bello*: Key Issues." In *International Law and the Changing Character of War*, edited by Raul A. "Pete" Pedrezo and Daria P. Wollschlaeger, 89-110. International Law Studies, Volume 87. Newport, RI: 2011.

Schmitt, Michael N. "Cyber Operations in International Law: The Use of Force, Collective Security, Self-Defense, and Armed Conflicts." In *Proceedings of a Workshop on Deterring Cyberattacks: Informing Strategies and Developing Options for U.S. Policy*, edited by National Research Council, 151-178. Washington, DC: The National Academies Press, 2010.

Schmitt, Michael N., ed. *Tallinn Manual 2.0 on the International Law Applicable to Cyber Warfare.* New York, NY: Cambridge University Press, 2017.

Schmitt, Michael N., ed. *Tallinn Manual on the International Law Applicable to Cyber Warfare.* New York, NY: Cambridge University Press, 2013.

Schneider, Nathan. "From Occupy Wall Street to Occupy Everywhere." *Nation*, October 12, 2011. https://www.thenation.com/article/occupy-wall-street-occupy-everywhere/.

Scott-Railton, John, and Seth Hardy. "Malware Attack Targeting Syrian ISIS Critics." Citizen Lab, December 18, 2014, https://citizenlab.ca/2014/12/malware-attack-targeting-syrian-isis-critics/.

Seigfried-Spellar, K. C., C. L. O'Quinn, and K.N. Treadway. "Assessing the relationship between autistic traits and cyber deviancy in a sample of college students." *Behaviour & Information Technology*, 34, 5, (2015): 533–542.

Selk, Avi. "A Twitter campaign is outing people who marched with white nationalists in Charlottesville." *Washington Post*, August 14, 2017. https://www.Washingtonpost.com/news/the-intersect/wp/2017/08/14/a-twitter-campaign-is-outing-people-who-marched-with-white-nationalists-in-charlottesville/?noredirect=on&utm_term=.c5427dd74a72.

Shachtman, Noah. "Under Worm Assault, Military Bans Disks, USB Drives." *Wired*, November 19, 2008.

Shakarian, Paulo, Jana Shakarian, and Andrew Ruef. *Introduction to Cyber-Warfare: A Multidisciplinary Approach*. Syngress, 2013.

Shamsi, Jawad, Fareha Sheikh, Sherali Zeadally, and Angelyn Flowers. "Attribution in Cyberspace: Techniques and Legal Implications." *Security and Communication Networks* 9, issue 15 (April 26, 2016): 2886-2900.

Shao, Chengcheng, Giovanni Luca Ciampaglia, Onur Varol, Kai-Cheng Yang, Alessandro Flammini, and Filippo Menczer. "The Spread of Low-Credibility Content by Social Bots." *Nature Communications* 9, no. 1 (2018): 4787.

Shelton, Martin, Katherine Lo, and Bonnie Nardi. "Online media forums as separate social lives: A qualitative study of disclosure within and beyond Reddit." *iConference 2015 Proceedings*, 2015.

Shreeves, Robin. "Greenpeace and Nestle in a Kat Fight." Forbes, March 19, 2010, https://www.forbes.com/2010/03/18/kitkat-greenpeace-palm-oil-technology-ecotech-nestle.html.

Shrikanth G. "Cyber Mercenaries to challenge enterprise security in 2016." Dataquest, November 23, 2015. https://www.dqindia.com/cyber-mercenaries-to-challenge-enterprise-security-in-2016/.

Sifry, Micah L. "Did Facebook Bring Down Mubarak?" *CNN online*, February 11, 2011.

Silverman, Craig. "This Analysis Shows How Viral Fake Election News Stories Outperformed Real News On Facebook." BuzzFeed. Accessed July 18, 2017, https://www.buzzfeed.com/craigsilverman/viral-fake-election-news-outperformed-real-news-on-facebook.

Singel, Ryan. "Twitter Settles with Feds Over '09 Obama Hack." *Wired*, March 11, 2011.

Singer, P. W., and Allan Friedman. "Cybersecurity and Cyberwar: What Everyone Needs to Know." New York, NY: Oxford University Press, 2014.

Singer, P. W., and Emerson T. Brooking. *LikeWar: The Weaponization of Social Media*. New York: Houghton Mifflin Harcourt Publishing Company, 2018.

Slatella, Michele, and Joshua Quittner. *Masters of deception: The gang that ruled cyberspace*. New York: Harper Collins, 1995.

Slonje, Robert, Peter K. Smith, and Ann Frisén. "The Nature of Cyber-bullying, and Strategies for Prevention." *Computers in Human Behavior* 29, no. 1 (2013): 26–32.

Snow, David A., et al. "Ideology, Frame Resonance, and Participant Mobilization." *International Social Movement Research* 1, no. 1 (1988).

Speckhard, Anne, and Mubin Shaikh. *Undercover Jihadi: Inside the Toronto 18, Al Qaeda Inspired, Homegrown, Terrorism in the West.* Advances Press, 2014.

Speckhard, Anne. "The Hypnotic Power of ISIS Imagery in Recruiting Western Youth." *International Center for the Study of Violent Extremism,* October 20, 2015, http://www.icsve.org/research-reports/the-hypnotic-power-of-isis-imagery-in-recruiting-western-youth/.

Spitaletta, Jason A. "Terror as a Psychological Warfare Objective: ISIL's Use of Ritualistic Decapitation." In *White Paper on Social and Cognitive Neuroscience Underpinnings of ISIL Behavior and Implications for Strategic Communication, Messaging, and Influence,* edited by J. Giordano and D. DiEuliis. Washington, DC: Strategic Multilayer Assessment Office, Office of the Secretary of Defense, 2015. http://nsiteam.com/sma-publications/.

Spitaletta, Jason A. "Use of Cyber to affect neuroS/T based Deterrence and Influence." In *White paper on Leveraging Neuroscientific and Neurotechnological (NeuroS&T) Developments with Focus on Influence and Deterrence in a Networked World,* edited by D. DiEuliis, W. Casebeer, J. Giordano, N. Wright, & H. Cabayan. Washington, DC: Strategic Multilayer Assessment Office, Office of the Secretary of Defense, 2014.

Spitaletta, Jason A., ed., *Bio-Psycho-Social Applications to Cognitive Engagement.* Washington, DC: Strategic Multilayer Assessment Office, Office of the Secretary of Defense, 2016.

Spitaletta, Jason A., Summer D. Newton, Nathan D. Bos, Charles W. Crossett, and Robert R. Leonhard. "Historical Lessons on Intelligence Support to Countering Undergrounds in Insurgencies." *Inteligencia y Seguridad: Revista de AnálISIL y Prospectiva* 2013, no. 13 (2013), 101–128.

Spitaletta, Jason. "Comparative Psychological Profiles: Baghdadi & Zawahiri." In *Multi-Method Assessment of ISIL*, edited by H. Cabayan & S. Canna. Washington, DC: Strategic Multilayer Assessment Office, Office of the Secretary of Defense, 2014. http://nsiteam.com/sma-publications/.

Steele, Claude M., and Joshua Aronson. "Stereotype threat and the intellectual test performance of African Americans." *Journal of Personality and Social Psychology* 69, no. 5 (1995): 797.

Stella, Massimo, Emilio Ferrara, and Manlio De Domenico. "Bots Increase Exposure to Negative and Inflammatory Content in Online Social Systems." *Proceedings of the National Academy of Sciences* (2018): 201803470.

Stouffer, Keith, Joe Falco, and Karen Scarfone. "Guide to Industrial Control Systems (ICS) Security." National Institute of Standards and Technology (NIST) Special Publication 800-82, Revision 2, May 2015.

Stouffer, Keith, Victoria Pilitteri, Suzanne Lightman, Marshall Abrams, and Adam Hahn, "Guide to Industrial Control Systems (ICS) Security," National Institute of Standards and Technology (NIST) Special Publication 800-82, Revision 2, May 2015,

Sunstein, Cass R. *Echo Chambers: Bush v. Gore, Impeachment, and Beyond.* Princeton University Press Princeton, NJ, 2001. https://pdfs.semanticscholar.org/4e7c/434ec3b8eaf26c62642dfbac56be6eef9647.pdf.

Sutter, John D. "Will Twitter War Become the New Norm?" *CNN.com*, November 15, 2012.

Symantec, "Longhorn: Tools Used by Cyberespionage group linked to Vault 7: First Evidence Liking Vault 7 Tools to Known Cyberattacks." *Symantec Official blog*, April 10, 2017.

Tarrow, Sidney. *Power in Movement: Social Movements and Contentious Politics.* Cambridge: Cambridge University Press, 1994.

The Aerodrome.com. "The Aircraft of World War 1." http://www.the-aerodrome.com/ aircraft/statistics.php.

The Latch Key. "It is IMPOSSIBLE to 100% verify the origin of a cyber attack. Vault 7 documents prove you can spoof certain variables to obfuscate the original location." *Reddit Blog*, March 31, 2017.

The Middle East Media Research Institute. *Research by Archives.* memri. org. Accessed August 29, 2017.

Thomas, Timothy L. *Hezballah, Israel, and Cyber Psyop*. Foreign Military Studies Office (Army) Fort Leavenworth KS, 2007.

Thompson, Emily. "RFE During the Occupation of Czechoslovakia." *RadioFreeEurope RadioLiberty Pressroom*, June 28, 2018. https://press-room.rferl.org/a/rfe-in-1968/29325749.html.

Tompkins Jr., Paul J. "Nonviolent Resistance," Chapter 3 in *Legal Implications of the Status of Persons in Resistance*. Assessing Revolution and Insurgent Strategies (ARIS) series, November 2012.

Tompkins Jr., Paul J. *Case Studies in Insurgency and Revolutionary Warfare – Colombia (1964–2009)*, Assessing Revolution and Insurgent Strategies (ARIS) series, 2014.

Tompkins Jr., Paul J. *Undergrounds in Insurgent, Revolutionary, and Resistance Warfare*, Second Edition, Assessing Revolution and Insurgent Strategies (ARIS) series, January 25. 2013, http://www.soc.mil/ARIS/ARIS.html.

Tompkins, Jr., Paul J. *Human Factors Considerations of Undergrounds in Insurgencies*, 2nd edition. edited by Nathan Bos. Alexandria, VA: Alexandria, VA: US Army Publications Directorate, 2013. http://www.soc.mil/ARIS/ARIS.html.

Tompkins, Jr., Paul J. *Legal Implications of the Status of Resistance*, edited by Erin N. Hahn. Fort Bragg: NC: USASOC, 2015.

Townsend, Kevin. "Microsoft Proposes Independent Body to Attribute Cyber Attacks." SecurityWeek, July 6, 2016. https://www.securityweek.com/microsoft-proposes-independent-body-attribute-cyber-attacks.

Travis, Alan. "Battle Against al-Qaida Brand Highlighted in Secret Paper." Guardian, August 26, 2008.

Troianovski, Anton. "A Former Russian Troll Speaks: 'It Was like Being in Orwell's World.'" *Washington Post*, February 17, 2018. https://www.washingtonpost.com/news/worldviews/wp/2018/02/17/a-former-russian-troll-speaks-it-was-like-being-in-orwells-world/.

Trump, Donald J. "National Cyber Strategy of the United States of America." President of the United States, September 2018.

Trump, Donald J. "The National Security Strategy of the United States of America." President of the United States, December 2017.

Tufecki, Zeynep. *Twitter and Tear Gas: The Power and Fragility of Networked Protest*. New Haven, CT: Yale University Press, 2017.

Tzu, Sun. "The Art of War." Translated by Lionel Giles, from http://classics.mit.edu/Tzu/artwar.html.

U.S. Cyber Command. "Home page." https://www.cybercom.mil/.

UN General Assembly. "Universal Declaration of Human Rights." *UN General Assembly*, 1948.

Underwood, Patrick, and Howard T. Welser. "'The Internet Is Here': Emergent Coordination and Innovation of Protest Forms in Digital Culture." *Proceedings of the 2011 iConference* (ACM, 2011): 304–311.

United States Army Special Operations Command, *SOF Support to Political Warfare*. USASOC: White Paper, March 10, 2015. http://maxoki161.blogspot.com/2015/03/sof-support-to-political-warfare-white.html.

United States Marine Corps. "Marine Corps Cyberspace Operations." MCIP 3-40.02, October 6, 2014.

UNITED STATES OF AMERICA v. ALEXANDRE CAZES. Verified Complaint for Forfeiture in REM. Case 1:117 at 00557 document 1, filed July 19, 2017.

United States Senate Commission on Commerce, Science and Transportation. "A 'Kill Chain' Analysis of the 2013 Target Data Breach. Majority Staff Report." March 26, 2014.

US Air Force Doctrine. "Annex 3-12, Cyberspace Operations." To Joint Publication 3-12, "Cyberspace Operations," November 30, 2011. https://doctrine.af.mil/DTM/dtmcyberspaceops.htm.

US Department of Defense. "Dictionary of Military and Associated Terms." September 2018.

US Department of Homeland Security, Cybersecurity and Infrastructure Security Agency (CISA) Cyber + Infrastructure. "Crypto Ransomware." Alert (TA14295A), October 22, 2014, last revised September 30, 2016.

US Department of Homeland Security, Cybersecurity and Infrastructure Security Agency (CISA) Cyber + Infrastructure. "Understanding Denial-of-Service Attacks." Security Tip (ST04-015), November 4, 2009, last revised June 28, 2018.

US Department of Homeland Security, Cybersecurity and Infrastructure Security Agency (CISA) Cyber + Infrastructure. "Heightened DDoS Threat Posed by Mirai and Other Botnets." Alert (TA16-288A), October 14, 2016, last revised October 17, 2017.

US Department of Homeland Security, Cybersecurity and Infrastructure Security Agency (CISA) Cyber + Infrastructure. "Avoiding Social Engineering and Phishing Attacks." Security Tip (ST04-014), October 22, 2009, last revised August 22, 2019.

US Department of Homeland Security, Cybersecurity and Infrastructure Security Agency (CISA) Cyber + Infrastructure. "Heightened DDoS Threat Posed by Mirai and Other Botnets." Alert (TA16-288A), October 14, 2016, last revised October 17, 2017.

US Department of Homeland Security, Cybersecurity and Infrastructure Security Agency (CISA) Cyber + Infrastructure. "Shamoon/DistTrack malware (Update B)." ICS Joint Security Awareness Report (JSAR-12-241-01B), October 16, 2012, last revised April 18, 2017.

US Department of Homeland Security, Cybersecurity and Infrastructure Security Agency (CISA) Cyber + Infrastructure. "HatMan—Safety System Targeted Malware (Update A)." Malware Analysis Report (MAR-17-352-01), April 10, 2018.

US Department of Justice, Office of Public Affairs. "U.S. Charges Five Chinese Military Hackers for Cyber Espionage Against U.S. Corporations and a Labor Organization for Commercial Advantage: First Time Criminal Charges Are Filed Against Known State Actors for Hacking." *Justice News*, May 19, 2014. https://www.justice.gov/opa/pr/us-charges-five-chinese-military-hackers-cyber-espionage-against-us-corporations-and-labor.

US Joint Chiefs of Staff. "Counterinsurgency." Joint Publication 3-24 (JP 3-24), April 25, 2018.

US Joint Chiefs of Staff. "Cyberspace Operations." Joint Publication 3-12 (JP 3-12), 8 June 2018.

US Joint Chiefs of Staff. "Information Operations." Joint Publication 3-13 (JP 3-13), November 20, 2014.

US Joint Chiefs of Staff. "Stability." Joint Publication 3-07 (JP 3-07), August 3, 2016.

Valenzuela, Sebastian. "Unpacking the Use of Social Media for Protest Behavior: The Roles of Information, Opinion Expression, and Activism." *American Behavioral Scientist* 57, no. 7 (2013): 920–942.

Vegh, Sandor. "Classifying forms of online activism: The case of cyber-protests against the World Bank." In *Cyberactivism: Online Activism in Theory and Practice,* edited by Martha McCaughey and Michael D. Ayers. New York: Routledge Press, 2013.

Verizon Enterprise. "2017 Data Breach Investigations Report." 10th Edition, 2017. http://www.verizonenterprise.com/verizon-insights-lab/dbir/2017/#report.

Vertuli, Mark D., and Bradley S. London, eds. *Perceptions are Reality: Historical Case Studies of Information Operations in Large-Scale Combat Operations.* Fort Leavenworth, KS: Army University Press, 2018.

Vidal, John. "Mexico's Zapatista rebels, 24 years on and defiant in mountain strongholds." *Guardian,* February 17, 2018. https://www.theguardian.com/global-development/2018/feb/17/mexico-zapatistas-rebels-24-years-mountain-strongholds.

Vishwanath, Arun. "Spear Phishing: The Tip of the Spear Used by Cyber Terrorists." In *Combating Violent Extremism and Radicalization in the Digital Era.* IGI Global, 2016.

Voas, Jeffrey. "Networks of 'Things.'" National Institute of Standards and Technology (NIST) Special Publication 800-183, July 2016.

Wait, Patience. "Special Report: The New DNA: DOD Lab Excavates Bits, Bytes to Dig Out Information." *Government Computer News,* July 31, 2006.

Walters, Joanna. "Women's March on Washington Set to be One of America's Biggest Protests." *Guardian,* January 17, 2017. https://www.theguardian.com/us-news/2017/jan/14/womens-march-on-washington-protest-size-donald-trump.

Watts, D. J., and S. H. Strogatz. "Collective Dynamics of 'Small-World' Networks." Nature 393, no. 6684 (1998): 440–442.

Weedon, Jen. "Beyond 'Cyber War': Russia's Use of Strategic Cyber Espionage and Information Operations in Ukraine." Chapter 8 in *Cyber War in Perspective: Russian Aggression Against Ukraine,* ed. Kenneth Geers. Tallinn, Estonia: NATO Cooperative Cyber Defence Center of Excellence, 2015.

Weise, Elizabeth. "Republican Party Data on 198M Voters Exposed Online." *USA Today,* June 19, 2017, updated June 20, 2017.

Weiwei, Ai. "China's Paid Trolls: Meet the 50-cent Party." *NewStatesman,* October 17, 2012.

Whitaker, Mark P. "Tamilnet.Com: Some Reflections on Popular Anthropology, Nationalism, and the Internet," *Anthropological Quarterly* 77, no. 3 (Summer2 004): 469–498.

Whitlock, Craig. "Somali American caught up in a shadowy Pentagon counterpropaganda campaign." *The Washington Post*, July 7, 2013. https://www.washingtonpost.com/world/national-security/somali-american-caught-up-in-a-shadowy-pentagon-counterpropaganda-campaign/2013/07/07/b3aca190-d2c5-11e2-bc43-c404c3269c73_story.html.

Whittaker, Zack. "Researchers obtain a command server used by North Korean hacker group." TechCrunch, March 3, 2019. https://techcrunch.com/2019/03/03/north-korea-lazarus-hackers/?renderMode=ie11.

Wikipedia. "Lawful Interception." https://en.wikipedia.org/wiki/Lawful_interception.

Wikipedia. "Tor (anonymity network)." https://en.wikipedia.org/wiki/Tor_(anonymity_network).

Wikiquote.org. "Ferdinand Foch." https://en.wikiquote.org/wiki/Ferdinand_Foch.

Williams, Brett T. "The Joint Force Commander's Guide to Cyberspace Operations." *Joint Forces Quarterly* 73, April 2014.

Winslow, Luke. "'Not Exactly a Model of Good Hygiene': Theorizing an Aesthetic of Disgust in the Occupy Wall Street Movement." *Critical Studies in Media Communication* 34, no. 3 (2017): 278–292.

Winter, Charlie. "Documenting the Virtual 'Caliphate.'" *Quilliam*, October 2015. http://www.quilliaminternational.com/wp-content/uploads/2015/10/FINAL-documenting-the-virtual-caliphate.pdf.

Workman, Michael. "Gaining Access with Social Engineering: An Empirical Study of the Threat." *Information Systems Security* 16, issue 6 (2007): 315–331.

Wray, Stefan. "Electronic civil disobedience and the World Wide Web of hacktivism: A Mapping of Extraparliamentarian Direct Action Net Politics." *Nova Iorque,* 1998.

Xenakis Christos, and Christoforos Ntantogian. "Attacking the Baseband Modem of Mobile Phones to Breach the Users' Privacy and Network Security." 7th International Conference on Cyber Conflict: Architectures in Cyberspace, NATO CCD COE Publications, Tallinn, May 26–29, 2015.

Yahyanejad, Mehdi. "The Effectiveness of Internet for Informing and Mobilizing in the Events After the Iranian Presidential Election." MIT CSAIL, Fall 2010. http://groups.csail.mit.edu/mac/classes/6.805/admin/admin-fall-2010/weeks/week12-Yahyenejad.pdf.

Young, R., L. Zhang, and V.R. Prybutok. "Hacking into the minds of hackers." *Information Systems Management*, 24, 4 (2007): 281–287.

Zafarani, Reza, Mohammad Ali Abbasi, and Huan Liu. *Social Media Mining: An Introduction.* New York: Cambridge University Press, Draft Version April 20, 2014.

Zaman, Khuram. "ISIS has a Twitter Strategy and It is Terrifying." *Medium*, November 20, 2015. https://medium.com/fifth-tribe-stories/isis-has-a-twitter-strategy-and-it-is-terrifying-7cc059ccf51b.

Zetter, Kim. "Hacker Lexicon: Botnets the Zombie Computer Armies Earn Hackers Millions." *Wired*, December 12, 2012. https://www.wired.com/2015/12/hacker-lexicon-botnets-the-zombie-computer-armies-that-earn-hackers-millions/.

Zetter, Kim. "Inside the Cunning, Unprecedented Attack of Ukraine's Power Grid." *Wired*, March 3, 2016. https://www.wired.com/2016/03/inside-cunning-unprecedented-hack-ukraines-power-grid/.

Zetter, Kim. "The Sony Hackers Were Causing Mayhem Years Before They Hit the Company." *Wired*, February 24, 2016. https://www.wired.com/2016/02/sony-hackers-causing-mayhem-years-hit-company/

Zimbardo, P. G. *The Lucifer effect: Understanding how good people turn evil.* New York: Random House Publishing Group, 2008.

Zimmerman, Dwight Jon. "Operation Mincemeat: The Story Behind 'The Man Who Never Was' in Operation Husky." *DefenseMediaNetwork*, September 9, 2013. https://www.defensemedianetwork.com/stories/operation-mincemeat-the-story-behind-the-man-who-never-war-in-operation-husky/.

ILLUSTRATION AND TABLE CREDITS

Figure 1-1. Chinese military philosopher, Sun Tzu. Wikimedia Commons contributors, "File:Enchoen27n3200.jpg," *Wikimedia Commons, the free media repository*, https://commons.wikimedia.org/w/index.php?title=File:Enchoen27n3200.jpg&oldid=231426518 (accessed January 12, 2019). Attribution: 663highland [GFDL (http://www.gnu.org/copyleft/fdl.html), CC-BY-SA-3.0 (http://creativecommons.org/licenses/by-sa/3.0/) or CC BY 2.5 (https://creativecommons.org/licenses/by/2.5)], from Wikimedia Commons.

Figure 1-2. Picture of inflatable tank during World War II. Elinor Florence, "D-Day Dummies and Decoys," *Blog* (May 28, 2014), http://elinorflorence.com/blog/d-day-decoys.

Figure 1-3. Seal of USCYBERCOM. U.S. Cyber Command, "Home page," https://www.cybercom.mil/.

Figure 2-2. Comandanta Ramon is perhaps the most famous female Zapatista actor. Wikimedia Commons contributors, "File:Comandanta Ramona by bastian.jpg," *Wikimedia Commons, the free media repository*, https://commons.wikimedia.org/w/index.php?title=File:Comandanta_Ramona_by_bastian.jpg&oldid=298751639 (accessed December 12, 2018).

Figure 2-3. ISIS made extensive use of Western-style advertising and public relations tools. Wikimedia Commons contributors, "File:İD bayrağı ile bir militan.jpg," *Wikimedia Commons, the free media repository*, https://commons.wikimedia.org/w/index.php?title=File:%C4%B0D_bayra%C4%9F%C4%B1_ile_bir_militan.jpg&oldid=294679387 (accessed December 12, 2018). Attribution: bastian (Heriberto Rodriguez) from Chiapas, Mexico [CC BY 2.0 (https://creativecommons.org/licenses/by/2.0)], from Wikimedia Commons.

Figure 2-4. The Occupy movement offers a good example of a movement characterized by a flat power distribution. Wikimedia Commons contributors, "File:Day 14 Occupy Wall Street September 30 2011 Shankbone 49.JPG," *Wikimedia Commons, the free media repository*, https://commons.wikimedia.org/w/index.php?title=File:Day_14_Occupy_Wall_Street_September_30_2011_Shankbone_49.JPG&oldid=122194165 (accessed December 6, 2018). Attribution: David Shankbone [CC BY 3.0 (https://creativecommons.org/licenses/by/3.0)], from Wikimedia Commons.

Figure 3-1. Three layers of cyberspace. U.S. Joint Chiefs of Staff, "Cyberspace Operations," Joint Publication 3-12 (JP3-12), 8 June 2018, I-3.

Figure 3-2. Phases of the cyber kill chain. United States Senate Commission on Commerce, Science and Transportation, "A 'Kill Chain' Analysis of the 2013 Target Data Breach. Majority Staff Report," March 26, 2014.

Figure 3-3. Tor operations. Emin Caliskan, Tomas Minarik, and Anna-Maria Osula, "Technical and Legal Overview of the Tor Anonymity Network," Tallinn, Estonia: NATO Cooperative Cyber Defence Center of Excellence, 2015, 8.

Figure 3-4. Cyber threat taxonomy. Department of Defense, Defense Science Board, "Task Force Report: Resilient Military Systems and the Advanced Cyber Threat," Washington, DC: Office of the Under Secretary of Defense for Acquisitions, Technology and Logistics, January 2013, 21.

Figure 3-5. Internet structure. "Internet Map 1024," Wikimedia Commons image, posted by opte.org, January 15, 2005, https://commons.wikimedia.org/wiki/File:Internet_map_1024.jpg (accessed August 29, 2019). Attribution: The Opte Project [CC BY 2.5 (https://creativecommons.org/licenses/by/2.5)], from Wikimedia Commons.

Figure 4-2. Sample female avatar posting malware links on Facebook. Daniel Regalado, Nart Villeneuve, and John Scott Railton, "Behind the Syrian Conflict's Digital Front Lines," *FireEye Threat Intelligence Special Report*, February 2015, 13.

Figure 4-3. Sample Hamas female avatar to trick IDF soldiers. Paul Goldman, and Alistair Jamieson, "Hamas Used Fake Social Media Accounts to Hack Israeli Soldiers' Phones: IDF," *NBC News*, January 12, 2017. Image courtesy of North American Spokesperson, Israeli Defense Force (IDF), April 9, 2018.

Figure 4-4. Types of clouds and their key elements. Figure adapted from Patrick D. Allen, *Cloud Computing 101: A Primer for Project Managers*, CreateSpace Independent Publishing Platform, 2015.

Figure 4-5. Sample jihadist website. The Middle East Media Research Institute, *Research by Archives*, memri.org, accessed August 29, 2017.

Figure 5-1. Internet users in 2015 as a percentage of a country's population. Wikimedia Commons contributors, "File:InternetPenetrationWorldMap.svg," *Wikimedia Commons, the free media repository*, https://commons.wikimedia.org/w/index.php?title =File:InternetPenetrationWorldMap.svg&oldid=236300846 (accessed January 3, 2019). Attribution: Jeff Ogden (W163) [CC BY-SA 3.0 (https://creativecommons.org/licenses/by-sa/3.0)], from Wikimedia Commons.

Figure 5-2. First Lady Michelle Obama holding a sign with the hashtag "#bringbackourgirls" in support of the 2014 Chibok kidnapping. Posted to the FLOTUS Twitter account on May 7, 2014. Wikimedia Commons contributors, "File:Michelle-obama-bringbackourgirls.jpg," *Wikimedia Commons, the free media repository*, https://commons.wikimedia.org/w/index.php?title=File:Michelle-obama-bringbackourgirls.jpg&oldid=252238924 (accessed December 12, 2018). Attribution: Michelle Obama, Office of the First Lady [Public domain], from Wikimedia Commons.

Figure 6-1. ICS system layout. Keith Stouffer, Victoria Pilitteri, Suzanne Lightman, Marshall Abrams, and Adam Hahn, "Guide to Industrial Control Systems (ICS) Security," National Institute of Standards and Technology (NIST) Special Publication 800-82, Revision 2, May 2015, 2–6.

Figure 7-1. Covert and overt functions of an underground. Paul J. Tompkins, Jr., *Human Factors Considerations of Undergrounds in Insurgencies*, 2nd edition, edited by Nathan Bos (Alexandria, VA: Alexandria, VA: US Army Publications Directorate, 2013) http://www.soc.mil/ARIS/ARIS.html.

Figure 7-2. Cells in series and parallel. Ibid.

Figure 8-4. Layers of encryption used in Tor. Figure adapted from Harrison Neal (aka HANtwister), "SVG Diagram of the 'Onion Routing' Principle," *Wikimedia Commons*, 12 March 2008. Attribution: English Wikipedia user HANtwister [CC BY-SA 3.0 (http://creativecommons.org/licenses/by-sa/3.0/)], from Wikimedia Commons.

Figure 8-5. Silk Road webpage screenshot. N. Dasgupta, C. Freifeld, C. S. Brownstein, C. M. Menone, H. L. Surratt, L. Poppish, J. L. Green, E. J. Lavonas, and R. C. Dart, "Crowdsourcing black market prices for prescription opioids," *Journal of Medical Internet Research* 15, no. 8 (August 2013): e178, Figure 2, https://openi.nlm.nih.gov/detailedresult.php?img=PMC3758048_jmir_v15i8e178_fig2&req=4.

Figure 9-1. Categories of resistance and corresponding legal protections. Adapted from Paul J. Tompkins, Jr., *Legal Implications of the Status of Resistance*, edited by Erin N. Hahn, (Fort Bragg: NC: USASOC, 2015).

Figure 9-2. Spectrum of status and insurgency criteria (intensity, duration, organization). Adapted from... ARIS study (pending confirmation from Sam).

Table Credits:

Table 6-1. IT / ICS system differences. Table adapted from National Institute of Standards and Technology Special Publication 800-82, Revision 2 Natl. Inst. Stand. Technol. Spec. Publ. 800-82, Rev. 2, 247 pages (May 2015), 2-16.

Table 6-2. Example ICS adversarial incidents. Table adapted from National Institute of Standards and Technology Special Publication 800-82, Revision 2 Natl. Inst. Stand. Technol. Spec. Publ. 800-82, Rev. 2, 247 pages (May 2015), C-10.

Table 8-2. Cyberspace attribution examples. Jawad Shamsi, Fareha Sheikh, Sherali Zeadally, and Angelyn Flowers, Attribution in Cyberspace: Techniques and Legal Implications: SCN-SI-088 (Security and Communication Networks, Oct 2016).

www.ingramcontent.com/pod-product-compliance
Lightning Source LLC
Chambersburg PA
CBHW052108020426
42335CB00021B/2675